Vadapalli Chandrasekhar

Inorganic and Organometallic Polymers

T0241115

Vadapalli Chandrasekhar

Inorganic and Organometallic Polymers

 Springer

Prof. Dr.
Vadapalli Chandrasekhar
Indian Institute of Technology
Department Chemistry
208016 Kanpur
India
e-mail: vc@iitk.ac.in

ISBN 978-3-642-06144-8   e-ISBN 978-3-540-26215-2

Springer is a part of Springer Science+Business Media
springeronline.com

© Springer-Verlag Berlin Heidelberg 2010
Printed in Germany

The use of designations, trademarks, etc. in this publication does not imply, even in the absence of a specific statement, that such names are exempt from the relevant protective laws and regulations and therefore free for general use.

Cover design: KünkelLopka, Heidelberg

Printed on acid-free paper   2/3130XT 5 4 3 2 1 0

To
My Parents

Preface

This book has its origins in courses taught by the author to various undergraduate and graduate students at the Indian Institute of Technology, Kanpur, India. The diversity of inorganic chemistry and its impact on polymer chemistry has been profound. This subject matter has grown considerably in the last decade and the need to present it in a coherent manner to young minds is a pedagogic challenge. The aim of this book is to present to the students an introduction to the developments in Inorganic and Organometallic polymers.

This book is divided into eight chapters. Chapter 1 provides a general overview on the challenges of Inorganic polymer synthesis. This is followed by a survey of organic polymers and also includes some basic features of polymers. Chapters 3-8 deal with prominent families of inorganic and organometallic polymers. Although the target group of this book is the undergraduate and graduate students of chemistry, chemical engineering and materials science it is also hoped that chemists and related scientists in industry would find this book useful.

I am extremely thankful to my wife Sudha who not only encouraged me throughout but also drew all the Figures and Schemes of this book. I also thank my children Adithya and Aarathi for their constant concern on the progress of this book. I express my acknowledgment to the editorial team of Springer-Verlag for their cooperation.

July 2004   Vadapalli Chandrasekhar, Kanpur

Contents

1 Inorganic Polymers: Problems and Prospects 1
1.1 Introduction 1
1.2 Procedures for Synthesizing Organic Polymers 2
1.3 Inorganic Polymers: A Review of Synthetic Strategies............... 5
    1.3.1 Unsaturated Inorganic Compounds 6
    1.3.2 Inorganic Polymers from Acyclic Monomers 7
    1.3.3 Ring-Opening Polymerization of Cyclic Inorganic
          Compounds.. 16
    1.3.4 Polymers Containing Inorganic Rings or Motifs as
          Pendant Groups 21
1.4 Summary of Polymerization Procedures for Inorganic Polymers . 22
References 25

2 Organic Polymers – A Brief Survey.. 27
2.1 Introduction 27
2.2 Natural Polymers 28
2.3 Synthetic Polymers 29
2.4 Free-radical Polymerization... 30
    2.4.1 Initiators.. 31
    2.4.2 Mechanism of Polymerization... 33
    2.4.3 Branching 34
    2.4.4 Chain Transfer 35
    2.4.5 Summary of the Main Features of Free-radical
          Polymerization.. 36
2.5 Polymerization by Ionic Initiators 36
    2.5.1 Anionic Polymerization... 37
    2.5.2 Cationic Polymerization 40
2.6 Ziegler-Natta Catalysis 43
    2.6.1 Mechanism of Polymerization of Olefins Using the
          Ziegler-Natta Catalysts.. 45
    2.6.2 Supported Ziegler-Natta Catalysts.................................... 50
    2.6.3 Recent Developments in Ziegler-Natta Catalysis.............. 50
    2.6.4 Summary of the Ziegler-Natta Catalysis 55

2.7 Olefin Metathesis Polymerization 56
2.8 Atom Transfer Radical Polymerization 58
2.9 Ring-opening Polymerization 60
2.10 Group-Transfer Polymerization 62
2.11 Condensation Polymers 63
    2.11.1 Polyesters 64
    2.11.2 Polycarbonates 67
    2.11.3 Polyamides 67
    2.11.4 Polyimides 69
    2.11.5 Polysulfones 71
    2.11.6 Thermoset Resins 71
2.12 Some Basic Features About Polymers 75
    2.12.1 Molecular Weights of Polymers 75
    2.12.2 Glass Transition Temperature of Polymers 76
    2.12.3 Elastomers, Fibers and Plastics 77
References 80

3 Cyclo- and Polyphosphazenes 82
3.1 Introduction 82
3.2 Cyclophosphazenes 83
    3.2.1 Preparation of Chlorocyclophosphazenes 86
    3.2.2 Preparation of Fluorocyclophosphazenes and other
        Halogeno- and Pseudohalogenocyclophosphazenes 89
    3.2.3 Chlorine Replacement Reactions of
        Chlorocyclophosphazenes 90
    3.2.4 Other Methods of Preparing P-C Containing Compounds ... 94
    3.2.5 Reactions of Chlorocyclophosphazenes with Difunctional
        Reagents 95
    3.2.6 Isomerism in Cyclophosphazenes 99
    3.2.7 Mechanism of the Nucleophilic Substitution Reaction 100
    3.2.8 Cyclophosphazenes as Ligands for Transition Metal
        Complexes 104
    3.2.9 Structural Characterization of Cyclophosphazenes 106
    3.2.10 Nature of Bonding in Cyclophosphazenes 108
3.3 Polyphosphazenes 112
    3.3.1 Historical 112
    3.3.2 Poly(dichlorophosphazene) 113
    3.3.3 Condensation Polymerization of $Cl_3P=N-P(O)Cl_2$ 116
    3.3.4 Polymerization of $Cl_3P=NSiMe_3$ 118
    3.3.5 ROP of Substituted Cyclophosphazenes 123
    3.3.6 Poly(organophosphazene)s Prepared by Macromolecular
        Substitution of $[NPCl_2]_n$ 126

3.3.7 Preparation of Polyphosphazenes Containing P-C Bonds
by Thermal Treatment of Phosphoranimines 131
3.3.8 Modification of Poly(organophosphazene)s........................ 136
3.3.9 Modification of Poly(alkyl/arylphosphazene)s.................... 140
3.3.10 Structure and Properties of Polyphosphazenes................... 142
3.3.11 Potential Applications of Polyphosphazenes..................... 150
References 153

4 Cyclophosphazene-Containing Polymers 155
4.1 Introduction 155
4.2 Polymers Containing Cyclophosphazenes as Pendant Groups 156
4.2.1 Synthesis of Cyclophosphazene Monomers Containing
Vinyl Groups 158
4.2.2 Polymerization of Cyclophosphazene Monomers............... 165
4.2.3 Other Ways of Making Polymers Containing
Cyclophosphazene Pendant Groups 172
4.2.4 Applications of Polymers Containing Cyclophosphazene
Pendant Groups 173
4.3 Cyclolinear and Cyclomatrix Polymers............................ 178
References 181

5 Other Inorganic Polymers that Contain Phosphorus, Boron
and Sulfur 183
5.1 Poly(heterophosphazene)s 183
5.2 Poly(carbophosphazene)s 184
5.3 Poly(thiophosphazene)s.. 189
5.4 Poly(thionylphosphazene)s... 193
5.4.1 Polymerization of [{NPCl$_2$}$_2${NS(O)Cl}] and
[{NPCl$_2$}$_2${NS(O)F}] 194
5.5 Other P-N-S Polymers 199
5.6 Metallacyclophosphazenes 200
5.7 Other Phosphorus-Containing Polymers 203
5.7.1 Polyphosphinoboranes.. 203
5.7.2 Polymers Containing P=C bonds................................ 204
5.8 Poly(alkyl/aryloxothiazenes) 205
5.9 Borazine-based Polymers 206
References 207

6 Polysiloxanes 209
6.1 Introduction 209
6.2 Historical Aspects of Poly(dimethysiloxane) 215
6.3 Ring-Opening Polymerization of Cyclosiloxanes 219

6.3.1 Cyclosiloxanes... 220
6.3.2 Ring-opening Polymerization of Cyclosiloxanes by
       Ionic Initiators 224
6.3.3 Preparation of Copolymers by Ring-Opening
       Polymerization.. 230
6.3.4 Summary of the Ring-Opening Polymerization of
       Cyclosiloxanes.. 231
6.4 Condensation Polymerization....................................... 232
6.5 Crosslinking of Polysiloxanes 234
    6.5.1 Other Ways of Achieving Crosslinking.............................. 234
6.6 Hybrid Polymers.. 239
6.7 Properties of Polysiloxanes.. 241
    6.7.1 Glass-Transition Temperatures and Conformational
           Flexibility 241
6.8 Applications of Polysiloxanes 244
References 246

7 Polysilanes and Other Silicon-Containing Polymers...................... 249
7.1 Introduction 249
7.2 Historical 251
7.3 Synthesis of Polysilanes 254
    7.3.1 Synthesis of Polysilanes by Wurtz-type Coupling
           Reactions 255
    7.3.2 Polymerization by Anionic Initiators 260
    7.3.3 Polysilanes Obtained by Catalytic Dehydrogenation 265
    7.3.4 Summary of the Synthetic Procedures for Polysilanes........ 268
7.4 Modification of Polysilanes... 269
7.5 Physical Properties of Polysilanes 272
    7.5.1 NMR of Polysilanes 273
    7.5.2 Electronic Spectra of Polysilanes 276
7.6 Optically Active Polysilanes.. 278
7.7 Applications of Polysilanes 280
7.8 Polysilynes... 281
7.9 Polycarbosilanes and Polysiloles.................................... 284
    7.9.1 Polycarbosilanes.. 284
    7.9.2 Polysiloles... 288
References 292

8 Organometallic Polymers.. 296
8.1 Introduction 296
8.2 Polygermanes and Polystannanes 297
    8.2.1 Synthesis of Polygermanes.. 298

8.2.2 Synthesis of Polystannanes 300
8.2.3 Electronic Properties of Polygermanes and Polystannanes .. 302
8.3 Ferrocene-Containing Polymers 303
8.3.1 Synthetic Strategies 304
8.3.2 Strained Ferrocenophanes 310
8.3.3 Poly(ferrocenylsilane)s 312
8.3.4 Other Ferrocenyl Polymers Prepared from Strained
       Ferrocenophanes 318
8.3.5 Other Related Polymers 320
8.3.6 Properties and Applications of Poly(ferrocenylsilane)s
       and Related Polymers 321
8.4 Polyyne Rigid-rod Organometallic Polymers 324
8.4.1 Properties and Applications of Rigid-Rod Polyynes 332
References 333

Index 337

Contents   XVII

1 Inorganic Polymers: Problems and Prospects

1.1 Introduction

One of the indicators of the progress of human civilization is the type of materials that are accessible to society. Just as the Metal (Bronze and Iron) Age marked the beginning of a new chapter of human civilization at the end of the Stone Age, the advent of organic polymers in the last century has heralded a new era. Conventional materials such as iron, steel, wood, glass or ceramics are today being either supplemented or replaced by polymeric materials. From commonplace and routine (elastomers, fibers, thermoplastics) to special and high technology applications (electronic, electrical, optical, biological) polymers are being increasingly utilized. This dramatic impact of polymers has become possible because of two important factors. Firstly, the intrinsic versatility of carbon to form bonds to itself and to other hetero atoms such as oxygen, nitrogen or sulfur has enabled the extension of the principles of organic chemistry to polymer synthesis. This methodology has allowed the assembly of a large variety of polymers that possess varying properties which can fulfill wide and diverse needs. Secondly, the accessibility of inexpensive petroleum feed stocks from the beginning of the last century has allowed a large-scale availability of the building blocks of polymer synthesis and this in turn has helped the mass production of polymeric materials. From exotic laboratory materials polymers have become bulk commodities [1-3].

In spite of such wide-ranging applications and ubiquitous presence carbon-based polymeric materials probably do not have the capability to fulfill *all* the demands and needs of new applications [2]. For example, many organic polymers are not suitable for applications at extreme temperatures. Typically, at very low temperatures they become very brittle while at high temperatures they are oxidatively degraded. Also, most organic polymers are a fire hazard because of their excellent flammability properties. Most often it is required to blend fire-retardant additives to organic polymers to make them less hazardous with respect to inflammability. Another important limitation is that the petroleum feed stocks are not going to last for ever. Clearly this stock is going to run out fairly quickly given the rate of its consumption. Even if one considers that coal reserves are much larger and hence may provide the necessary basis for the continuation of the organic polymers there is going to be a need for supplementing these conventional systems with polymers that contain inorganic elements. It is to be noted that silicon is the second most widely present element in the earth's crust (27.2 % by weight), while carbon ranks seventeenth in the order of abundance (180 ppm). If one takes into account the presence of carbon in oceans and atmosphere also, its abundance goes up to the fourteenth position [4]. Thus, it makes chemical sense to look for alternative polymeric systems based on non-carbon backbones. Lastly, another compelling reason to look for newer polymeric systems containing inorganic elements is the line of thinking that if organic polymers themselves are so diverse in terms of their structure and property, it should be possible to find entirely new types of polymeric systems with completely different properties in some combination of inorganic elements. These are a few reasons that have motivated research scientists across the world to find ways of assembling new types of polymers that are based on inorganic elements [5-9]. It must be stated at the outset that most of these efforts have resulted in polymers which not only contain inorganic elements but also have organic groups either as side groups or in the main-chain itself. While there are some examples of *pure inorganic polymers* they still remain *rare*. In order to appreciate the problems of inorganic polymer synthesis it is worthwhile to briefly examine the synthesis of organic polymers. A more detailed survey of organic polymer synthesis is presented in Chap. 2.

1.2 Procedures for Synthesizing Organic Polymers

Most organic polymers can be synthesized by using any of the following three general synthetic methods: (1) Polymerization of unsaturated organic monomers. (2) Condensation of (usually) two difunctional monomers with each other. (3) Ring-opening polymerization of cyclic organic rings to linear chain polymers [1-3].

Vinyl monomers such as $CH_2=CH_2$ (and others such as $CF_2=CF_2$), mono substituted ethylenes $CH_2=CH(R)$ (such as propylene, styrene, vinyl chloride, acrylonitrile, methyl methacrylate, etc.), some disubstituted olefins $CH_2=CRR'$ (such as isobutylene), dienes (such as 1,3-butadiene, isoprene) and also monomers such as acetylene can be polymerized by various polymerization methods to afford linear chain polymers. In all of these polymerization reactions two new C-C single bonds are created in place of one C=C double bond. This process is thermodynamically favorable. Representative examples of polymers that have been synthesized by this method are shown in Fig. 1.1.

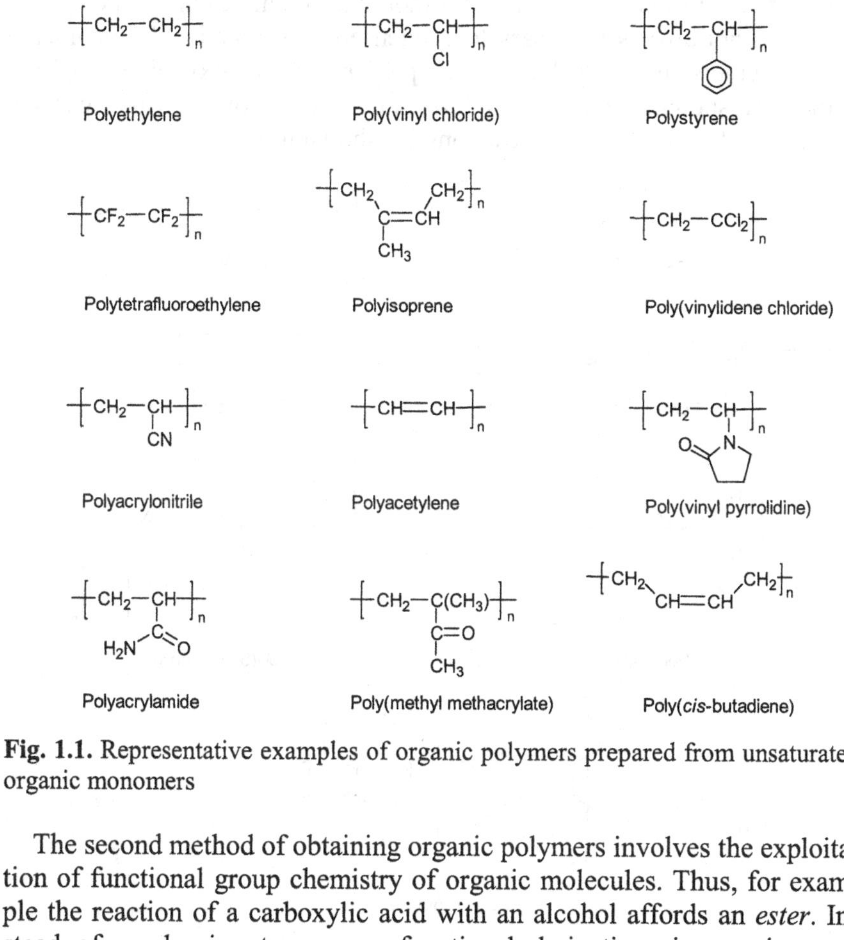

$$\begin{array}{ccc}
\left[CH_2-CH_2\right]_n & \left[CH_2-CH\right]_n & \left[CH_2-CH\right]_n \\
 & Cl & \bigcirc
\end{array}$$

Polyethylene   Poly(vinyl chloride)   Polystyrene

$$\left[CF_2-CF_2\right]_n \qquad \left[CH_2 \quad CH_2\right]_n \qquad \left[CH_2-CCl_2\right]_n$$
$$C=CH$$
$$CH_3$$

Polytetrafluoroethylene   Polyisoprene   Poly(vinylidene chloride)

$$\left[CH_2-CH\right]_n \qquad \left[CH=CH\right]_n \qquad \left[CH_2-CH\right]_n$$
$$CN$$

Polyacrylonitrile   Polyacetylene   Poly(vinyl pyrrolidine)

$$\left[CH_2-CH\right]_n \qquad \left[CH_2-C(CH_3)\right]_n \qquad \left[CH_2 \quad CH_2\right]_n$$
$$H_2N-C=O \qquad C=O \qquad CH=CH$$
$$CH_3$$

Polyacrylamide   Poly(methyl methacrylate)   Poly(cis-butadiene)

Fig. 1.1. Representative examples of organic polymers prepared from unsaturated organic monomers

The second method of obtaining organic polymers involves the exploitation of functional group chemistry of organic molecules. Thus, for example the reaction of a carboxylic acid with an alcohol affords an *ester*. Instead of condensing two mono functional derivatives in reacting two difunctional compounds (dicarboxylic acids and diols) one obtains a *polyester* (see Eq. 1.1).

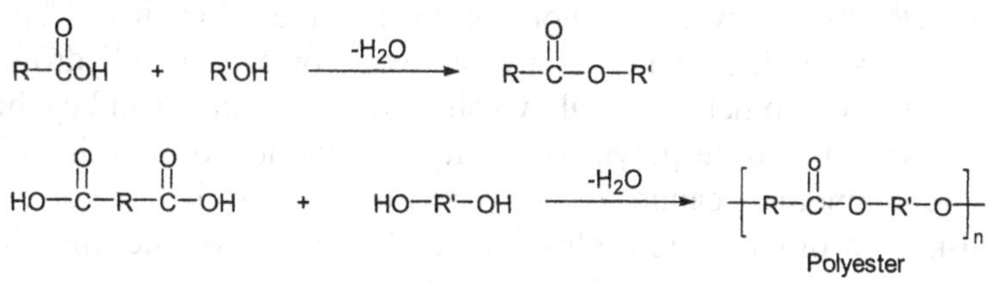

$$R-\overset{O}{\overset{\|}{C}}OH \ + \ R'OH \ \xrightarrow{-H_2O} \ R-\overset{O}{\overset{\|}{C}}-O-R' \qquad (1.1)$$

$$HO-\overset{O}{\overset{\|}{C}}-R-\overset{O}{\overset{\|}{C}}-OH \ + \ HO-R'-OH \ \xrightarrow{-H_2O} \ \left[R-\overset{O}{\overset{\|}{C}}-O-R'-O\right]_n$$
$$\text{Polyester}$$

This methodology is quite general and can be utilized to prepare several types of polymers such as polyamides, polyimides, polyurethanes, polyethers etc. The polymer properties depend on the type of functional groups that link the polymer building blocks. Further modulation is achievable by varying the nature of the difunctional monomer within each class of polymers. It is not always necessary to condense two difunctional monomers. Some polymers such as polyethers are prepared by the oxidative coupling of the corresponding phenols. A few examples of polymers that can be prepared by the condensation reactions are shown in Fig. 1.2.

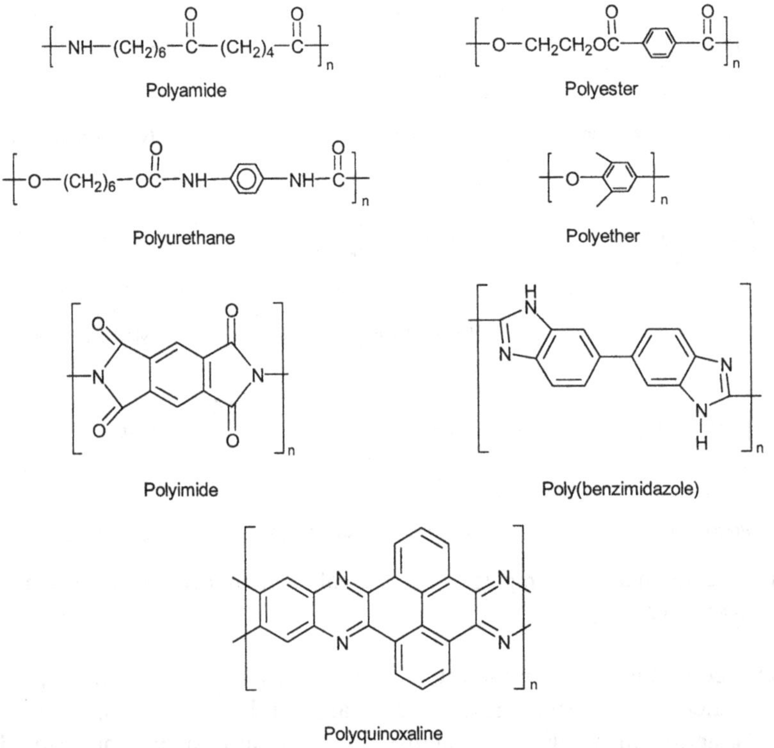

Polyamide

Polyester

Polyurethane

Polyether

Polyimide

Poly(benzimidazole)

Polyquinoxaline

Fig. 1.2. Polymers prepared from the condensation reactions of organic monomers

The third method of polymer preparation involves a ring-opening polymerization (ROP) of cyclic monomers to polymeric chains. Thus, monomers such as ethylene oxide, propylene oxide or even tetrahydrofuran can be used as monomers for ROP. Cyclic amides (lactams) and cyclic esters (lactones) can also be polymerized. It is important to note that all cyclic organic compounds cannot be converted into linear chains. For example, well-known organic molecules such as benzene, cyclohexane, dioxane, tetrahydropyran etc., cannot be polymerized to the corresponding polymers. Some examples of organic polymers that can be obtained by ROP are shown in Fig. 1.3.

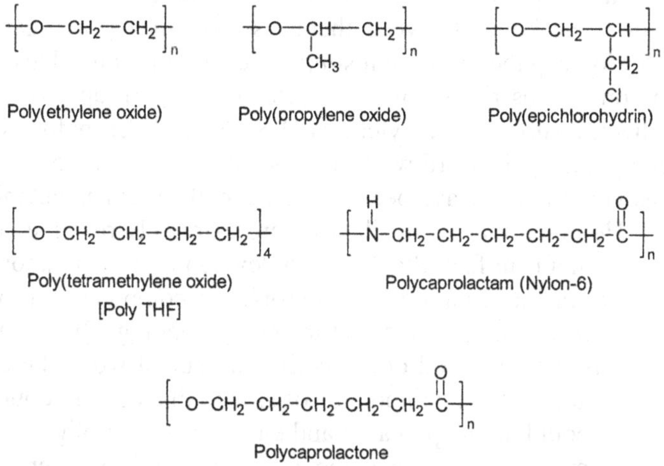

| Poly(ethylene oxide) | Poly(propylene oxide) | Poly(epichlorohydrin) |

Poly(tetramethylene oxide)
[Poly THF]

Polycaprolactam (Nylon-6)

Polycaprolactone

Fig. 1.3. Polymers prepared by the ring-opening polymerization of cyclic organic monomers

1.3 Inorganic Polymers: A Review of Synthetic Strategies

In light of the previous discussion on the synthesis of organic polymers we can now look at the strategies that would be needed for assembling inorganic polymers. It must be borne in mind that many inorganic compounds such as inorganic oxides (for example, SiO_2, B_2O_3, Al_2O_3 etc.,) or inorganic nitrides (for example, Si_3N_4, BN etc.,) are technically polymeric substances which are built from simple structural blocks [4]. However, substances of this type are not considered in this discussion. Only inorganic polymeric substances that are analogous in terms of their solution behavior to organic polymers are considered. Thus, most organic polymers are soluble in some kind of organic solvents (some such as Kevlar are soluble in mineral acids) and more importantly they retain their macromolecular structure in solution. Further, many organic polymers retain their properties upon melting. These properties are possible because of the essentially covalent linkages that are present in these organic polymer chains. Inorganic polymer synthesis needs to address the issue of formation of covalent bonds between similar or dissimilar inorganic elements so that polymeric structures are realized. As will be shown in subsequent sections such element-element linkages can be accomplished in many ways.

1.3.1 Unsaturated Inorganic Compounds

We will first look at the possibility of utilizing inorganic compounds that contain double or triple bonds between them, as monomers for organic synthesis. In order to keep the discussion simple we can limit ourselves by looking at such compounds belonging to the elements of Groups 14 and 15. Thus, the silicon analogue of ethylene is $H_2Si=SiH_2$. Even before we consider this compound it is worthwhile to note that many silicon analogues of alkanes viz., Si_nH_{2n+2} have been prepared by the German chemist Alfred Stock [4]. He noted that these silanes were extremely reactive and burnt spontaneously in air. In fact, Stock had to develop hitherto unknown vacuum-line techniques for handling such pyrophoric compounds. From this reactivity behavior it is anticipated that the corresponding silicon analogues of ethylene in particular and other olefins in general would be extremely reactive. For a long time compounds of silicon that contain double bonds between them could not be prepared and either cyclic- or oligomeric compounds, (or in some instances even some polymeric products) were the result of such investigations [10]. Similarly, it was not possible to isolate compounds of other main-group elements which contain double bonds between them. Several such failures have led to the *promulgation* of a *double-bond role*. This stated that heavier main-group compounds containing double bonds cannot be prepared. The rationale for this rule emanates from two possible reasons. Heavier main-group elements are larger in size and hence their inter-atomic distances are also longer (in comparison to carbon). Consequently, the additional *p*-orbital overlap between heavier main-group elements to generate a double bond would be quite weak. The double-bond rule held its own for a long time until its demise in 1981. In this year the double bond rule was firmly thrown out and a series of main group compounds containing double bonds were isolated and characterized [10-11]. The synthetic technique that contributed to this success was the utilization of sterically hindered groups for kinetically stabilizing these reactive compounds. The idea was to use large-sized substituents to *sterically protect* the double bonds once they were formed. Using this methodology stable compounds containing Si=C, Si=Si, P=C, P=P, P=As, Sb=Sb and even Bi=Bi double bonds were synthesized and characterized [12] (Fig. 1.4). It can be seen that the size of the sterically hindered group increases with the increase in the size of the main-group element. Although, the problem of preparing multiple-bonded compounds containing heavy main-group elements was thus solved, the very method of stabilizing these compounds defeats the possibility of using them as monomers for polymer synthesis. For example, it was found that the *in situ* generated disilene

$i\text{Pr}_2\text{Si}=\text{Si}i\text{Pr}_2$ undergoes spontaneous oligomerization to the cyclic tetramer $[i\text{Pr}_2\text{Si}]_4$ (see Eq. 1.2) [13].

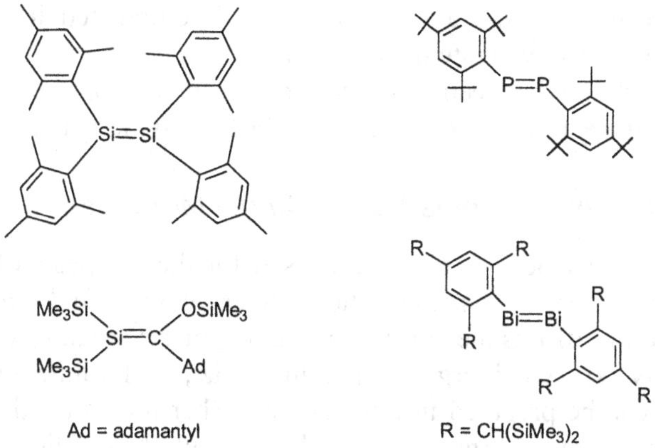

Fig. 1.4. Compounds of main-group elements that contain double bonds between them

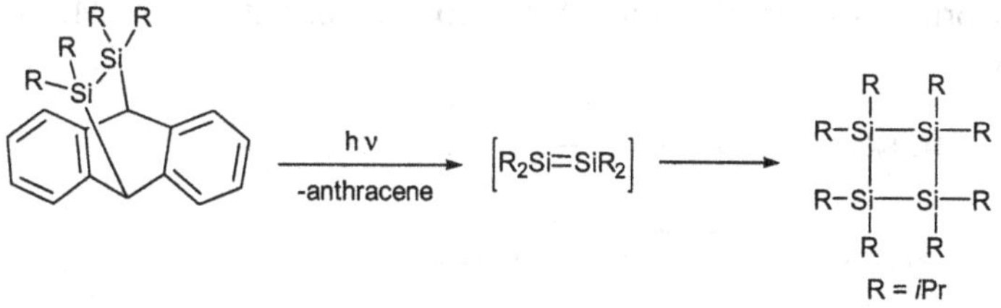

$$\tag{1.2}$$

Unsaturated organic compounds such as olefins also are reactive; however, their reactivity can be suitably directed to afford various kinds of products including polymers. Manipulating the reactivity of the newly discovered *unsaturated* inorganic compounds, to afford polymeric products still remains a synthetic challenge. Recently it was observed that $\text{MesP}=\text{CPh}_2$ could be polymerized by either radical or ionic initiators to afford moderate molecular weight polymers, $[\text{MesPCPh}_2]_n$ (Mes = 2,4,6-$\text{Me}_3\text{-C}_6\text{H}_2\text{-}$) [14].

1.3.2 Inorganic Polymers from Acyclic Monomers

Various types of polymerization reactions involving appropriate *inorganic acyclic monomers* have been quite effectively utilized for the preparation of polymers containing inorganic elements or inorganic groups. Some of these examples are illustrated in this section. These polymerization strategies are of two types.

1. A reaction where a *small molecule* is eliminated from a *single monomer* to generate the polymer.
2. A reaction between *two appropriate difunctional monomers* to eliminate a small molecule and afford a polymeric material.

1.3.2.1 Polymer Synthesis from a Single Monomer

Various types of reactions have been used for the synthesis of polymers starting from a *single monomer*. Thus, for example, *dehydration* of diorganosilane diols affords the corresponding polymeric organosiloxanes. The most widely known inorganic polymer viz., poly(dimethylsiloxane), $[Me_2SiO]_n$, can be prepared in this manner (other routes for the synthesis of this polymer are also known) [15-17]. Although the reaction involved in the synthesis of poly(dimethylsiloxane) is the hydrolysis of dimethyldichlorosilane, the immediate product of the reaction is $Me_2Si(OH)_2$ which is the actual monomer that undergoes self condensation by the elimination of water. This process of dehydration leads to the formation of $[Me_2SiO]_n$ (see Eq. 1.3).

$$Me_2SiCl_2 \ + \ 2H_2O \ \xrightarrow{-2HCl} \ Me_2Si(OH)_2 \qquad (1.3)$$

$$nMe_2Si(OH)_2 \ \xrightarrow{-nH_2O} \ [Me_2SiO]_n$$

The dehydration reaction has also been used to prepare ladder-type polymers. Thus, the hydrolysis of $PhSiCl_3$ affords $PhSi(OH)_3$ which undergoes a self condensation reaction to generate a ladder-like polymer which is known as polysilsesquioxane (see Eq. 1.4) [15, 16].

$$4n \ RSi(OH)_3 \ \xrightarrow{-6H_2O} \qquad\qquad (1.4)$$

Polysilsesquioxane

Metal-assisted *dehalogenation* reaction, analogous to the well-known Wurtz reaction has been very effectively employed for the preparation of polysilanes which contain catenated organosilicon units. Thus, the reaction of a variety of diorganodichlorosilanes with finely divided sodium in a high boiling solvent such as toluene affords the corresponding polysilanes. The reaction involves elimination of sodium chloride. The yields of polysilanes vary from being low to moderate. For example, the reaction of phenylmethyldichlorosilane with sodium affords poly(methylphenylsilane) (see Eq. 1.5) [17-19]. This approach has been used for preparing many poly(dialkylsilane)s as well as poly(diarylsilane)s. Polysilanes containing mixed substituents have also been successfully assembled from this approach [19]. However, the Wurtz-type coupling reaction is not tolerant of reactive groups (such as halogens or the hydroxyl groups).

$$
\begin{array}{c}
\underset{\underset{\text{Me}}{\overset{\text{Ph}}{\mid}}{\overset{\text{Cl}}{\underset{\text{Cl}}{\text{Si}}}}
\end{array}
\quad
\xrightarrow[\substack{\text{110 °C} \\ \text{-NaCl}}]{\substack{\text{Na} \\ \text{Toluene}}}
\quad
\left[\begin{array}{c}\overset{\text{Ph}}{\underset{\text{Me}}{\mid}}\text{Si}\end{array}\right]_n
\tag{1.5}
$$

Poly(methylphenylsilane)

Wurtz-coupling has also been successfully employed for the preparation of a number of copolymers of polysilanes. For example, reacting a 1:1 mixture of phenylmethyldichlorosilane and dimethyldichlorosilane with molten sodium in toluene affords a high-molecular-weight copolymer containing a random arrangement of [PhMeSi] and [Me$_2$Si] units (see Eq. 1.6) [19]. A remarkable feature of the polysilanes is the presence of σ-conjugation in their backbone. Many of these polymers, consequently, are being increasingly investigated for novel electronic applications. Another interesting feature of polysilanes is their sensitivity to light. Thus, many polysilanes undergo photoscission upon exposure to light. This radiation sensitivity is far greater in solution than in the solid-state. Although this might at first sight sound detrimental for any applications, this property of photobleaching allows these polymers to be used as photoresists for microlithography [19].

$$
\underset{\underset{\text{Me}}{\overset{\text{Ph}}{\mid}}{\overset{\text{Cl}}{\underset{\text{Cl}}{\text{Si}}}}
\quad + \quad
\underset{\underset{\text{Me}}{\overset{\text{Me}}{\mid}}{\overset{\text{Cl}}{\underset{\text{Cl}}{\text{Si}}}}
\quad
\xrightarrow[\substack{\text{110 °C} \\ \text{-NaCl}}]{\substack{\text{Na} \\ \text{Toluene}}}
\quad
\left[\left(\overset{\text{Ph}}{\underset{\text{Me}}{\mid}}\text{Si}\right)_x\left(\overset{\text{Me}}{\underset{\text{Me}}{\mid}}\text{Si}\right)_y\right]_n
\tag{1.6}
$$

Copolymer

The dehalogenation reaction has also been used for preparing unusual polymers that contain the Si-Si σ-bonds in conjugation with an organic π-system. Thus, reaction of dichlorosiloles with lithium metal results in the formation of 1,1'-polysiloles (see Eq. 1.7) [20, 21]. Every silicon in these polymers is also a part of a five-membered unsaturated heterocyclic ring.

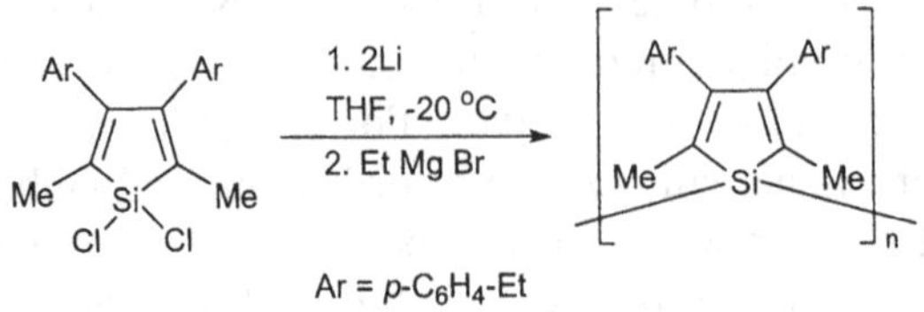

$$(1.7)$$

Ar = p-C$_6$H$_4$-Et

A high-molecular-weight *polystannane* was also prepared by using the dehalogenation reaction [22]. Di-n-butyltindichloride undergoes a dechlorination upon reaction with sodium metal in toluene. 15-Crown-5-ether was used as a structure-specific sequestering agent for sodium ions. It was found that the optimum reaction temperature for obtaining a high-molecular-weight polystannane, [R$_2$Sn]$_n$ was about 60 °C (see Eq. 1.8). Polystannanes are extremely sensitive to light and photo-bleach rapidly to form low-molecular-weight compounds. This can cause difficulties in synthesis unless precaution is taken to avoid light during the reaction and the subsequent work-up.

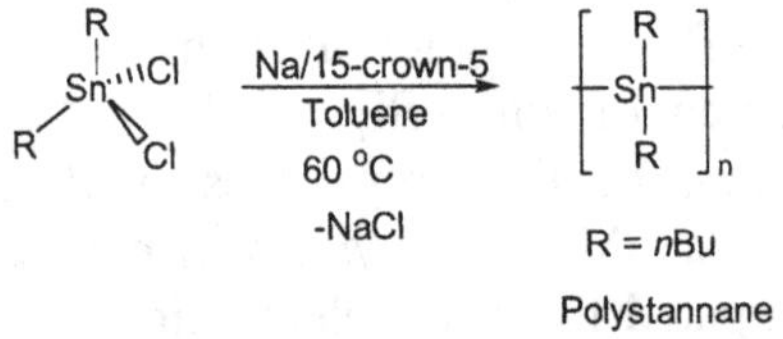

$$(1.8)$$

R = nBu

Polystannane

Similar to polystannanes a number of poly(dialkylgermane)s have also been prepared by the Wurtz-type coupling reaction (see Eq. 1.9) [23, 24].

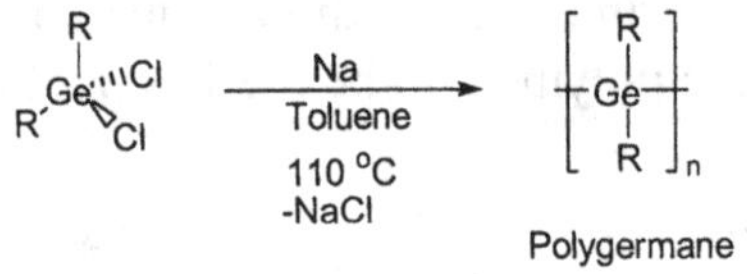

$$(1.9)$$

Polygermane

Catalytic dehydrogenation is an alternative method for the preparation of polysilanes. This reaction involves the dehydrogenation of RSiH$_3$ to afford polysilanes of the type [RSiH]$_n$ (see Eq. 1.10).

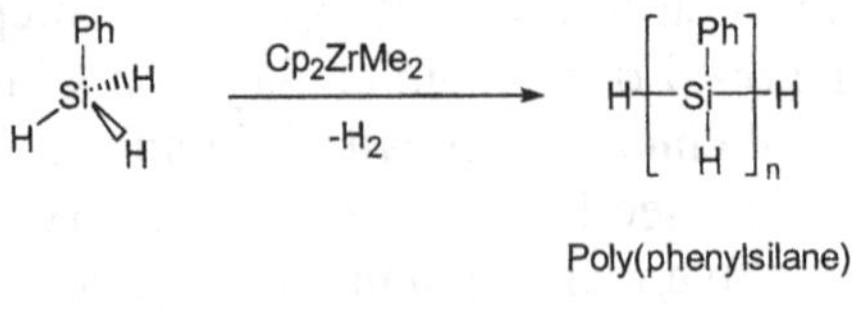

$$(1.10)$$

Poly(phenylsilane)

The dehydrogenation reaction is catalyzed by many types of organometallic complexes of titanium or zirconium [25]. The molecular weights of the polysilanes obtained by dehydrogenation reactions are lower than those obtained by the Wurtz-type dehalogenation method. Another limitation of this methodology is that it seems to be applicable to only arylsilanes; many alkyl silanes with the exception of CH_3SiH_3, are quite unresponsive towards this reaction. Secondly, this method is not very effective for preparing polysilanes of the type $[RR'Si]_n$. However, polysilanes prepared by the dehydrogenation reaction contain a reactive Si-H group that can be used for further elaboration involving hydrosilylation reactions.

Catalytic dehydrogenation reactions have also been used for the preparation of polystannanes. Unlike polysilanes, polystannanes of the type $[R_2Sn]_n$ can be prepared by this method. Thus, the dehydrogenation of dialkyltindihydrides, R_2SnH_2, by zirconocene catalysts provides a good route for the preparation of high-molecular-weight polystannanes, $H(SnR_2)H$ (see Eq. 1.11) [26]. These polymers contain reactive terminal Sn-H groups.

$$(1.11)$$

Heating the inorganic heterocyclic ring $B_3N_3H_6$ leads to the loss of hydrogen to afford a polymer where the borazine rings are interconnected with each other [27] (see Eq. 1.12). These types of polymers have been found to be useful as precursors for the preparation of the ceramic boron nitride.

$$(1.12)$$

Polyphosphinoboranes containing phosphorus and boron linked to each other in polymeric chains have remained elusive for a long time. Recently, catalytic dehydrogenation of the phosphine-borane adducts has been found to be effective to prepare the linear polymer. Thus, thermal treatment of $PhPH_2.BH_3$ in the presence of catalytic amounts of $[(1,5\text{-}COD)Rh(\mu\text{-}Cl)]$ affords the linear polymer poly(phenylphosphinoborane), $[PhHPBH_2]_n$ (see Eq. 1.13) [28-30]. High polymers with M_w of about 33,000 have been isolated by this procedure.

$$(1.13)$$

L = 1,5-Cyclooctadiene   Poly(phenylphosphinoborane)

Polyphosphazenes containing alternate phosphorus and nitrogen atoms in the polymer backbone can be prepared by a variety of synthetic methods. These include condensation polymerization reactions which involve appropriate monophosphazene precursors such as $Me_3SiNPRR'OR''$. These compounds also known as N-silylphosphoranimines eliminate the silylether Me_3SiOR'', upon heating, with the formation of polymeric products. Many, poly(alkyl/arylphosphazene)s can be prepared by this procedure. For example, the preparation of poly(dimethylphosphazene) is accomplished by heating the N-silylphosphoranimine $Me_3SiN=PMe_2$-(OCH_2CF_3) (see Eq. 1.14) [31, 32].

$$(1.14)$$

Poly(dimethylphosphazene)

This method is fairly general and has been used for the preparation of a wide variety of high-molecular-weight poly(alkyl/arylphosphazene)s, $[NPRR']_n$ (R,R$'$ = alkyl; R=alkyl, R$'$= aryl) [31].

Using a similar strategy as above, poly(oxothiazene)s containing alternate sulfur and nitrogen atoms in the backbone have been prepared [33]. Elimination of silylethers from $Me_3SiN=S(O)R(OR')$ or even phenols from $HN=S(O)R(OPh)$ leads to the formation the high polymers (see Eq. 1.15).

$$(1.15)$$

Acyclic monomers have also been found suitable for the preparation of poly(dichlorophosphazene). For example, heating the acyclic phosphazene derivative $Cl_3P=N-P(O)Cl_2$ leads to the elimination of $P(O)Cl_3$ as the by-product and affords the linear polymer, polydichlorophosphazene, $[NPCl_2]_n$ (see Eq. 1.16) [34].

$$PCl_3=NP(O)Cl_2 \xrightarrow[-POCl_3]{250-280\,°C} \left[N=P \begin{array}{c} Cl \\ | \\ | \\ Cl \end{array} \right]_n \quad (1.16)$$

Poly(dichlorophosphazene)

Poly(dichlorophosphazene) has also been prepared by an ambient temperature process involving the reaction of $Cl_3P=NSiMe_3$ with PCl_5. This process affords a high-molecular-weight *living polymer* (see Eq.1.17) [32].

$$nCl_3P=NSiMe_3 \xrightarrow[\substack{CH_2Cl_2 \\ 25\,°C}]{2n\ PCl_5} \left[N=P \begin{array}{c} Cl \\ | \\ | \\ Cl \end{array} \right]_n \quad (1.17)$$

Anionic polymerization also can be used for preparing polyphosphazenes from acyclic monomers. Thus, treatment of $Me_3SiNP(OCH_2CF_3)_3$ with Bu_4NF leads to the formation of the polymer $[NP(OCH_2CF_3)_2]_n$ (see Eq. 1.18) [32].

$$Me_3Si-N=P\begin{array}{c} OR \\ | \\ | \\ OR \end{array}-OR \xrightarrow[\substack{-Me_3SiOR \\ R=CH_2CF_3}]{\substack{Bu_4NF \\ 100\,°C}} \left[N=P \begin{array}{c} OR \\ | \\ | \\ OR \end{array} \right]_n \quad (1.18)$$

Poly(bistrifluoroethoxy-phosphazene)

1.3.2.2 Polymer Synthesis from two Monomers

It is possible to design difunctional inorganic monomers that can react with each other to afford linear polymers. Although, in some cases these inorganic difunctional monomers are different from those encountered in organic polymer synthesis the polymerization principles are very similar. Thus, we have seen that the difunctional organic compounds involved in polymerization processes were diols, dicarboxylic acids, diamines etc. Reaction between them leads to the elimination of a small molecule (such as water, alcohol etc.,) to afford long-chain macromolecules. Exact inorganic analogues containing these reactive functional groups are not always available. In many instances even if it is possible to prepare inorganic compounds containing such functional groups, they are too reactive to be used in a controlled polymerization process.

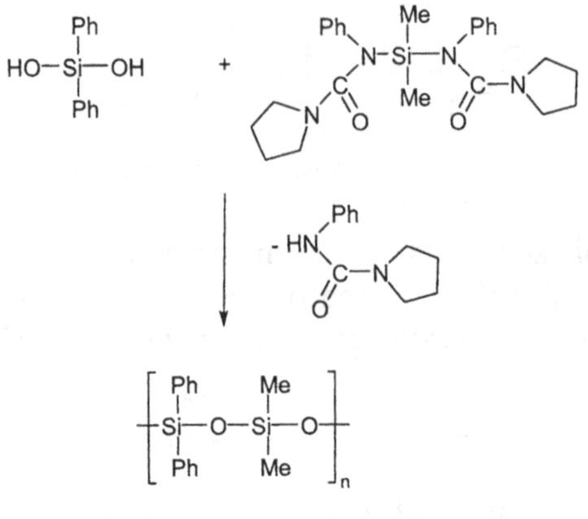

(1.19)

Poly(diphenyl-co-dimethylsiloxane)

Compounds containing Si-OH and SiNR$_2$ are, however, known and have been employed for condensation reactions (compounds containing Si-NH$_2$ groups are still quite rare). However, it will be noticed that during condensation reactions between a compound containing a Si-NR$_2$ unit and a Si-OH unit, elimination of the amine R$_2$NH occurs to form a new Si-O-Si bond. For example, exactly alternating copolymers of the polysiloxane family, containing alternately Me$_2$SiO and Ph$_2$SiO units in the polymer backbone, have been prepared by the condensation of diphenylsilanediol with dimethylbis(ureido)silane. The cleavage of the weak Si-N bonds, formation of new thermodynamically favorable Si-O bonds and the elimination of urea as the insoluble by-product act as the driving force for the polymerization reaction (see Eq. 1.19) [35].

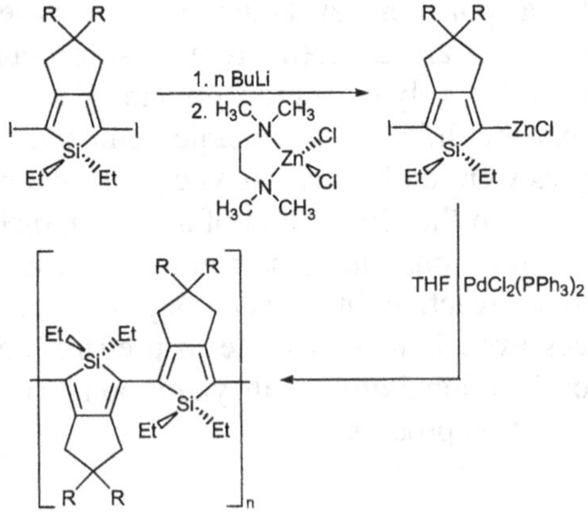

(1.20)

Poly(2,5-silole)

In contrast to poly(1,1'-silole)s which are prepared by the dehalogenation of the silole dichlorides, the preparation of poly(2,5-silole)s has to be carried out by a multi-step procedure (see Eq. 1.20) [36]. Thus, 2,5-diiodosilole can be selectively converted in situ to the monozinc derivative. Palladium-mediated cross-coupling reaction of the monozinc derivative affords poly(2,5-silole) which has a moderate degree of polymerization of about thirteen.

Recently, condensation polymerization has been used for preparing poly(p-phenylene phosphaalkene) which contains P=C double bonds in the polymeric backbone. Thus, the reaction between the silylated phosphane, $(Me_3Si)_2P$-C_6H_4-p-$P(SiMe_3)_2$ and a diacid chloride affords an E/Z mixture of poly(p-phenylene phosphaalkene) [37].

Condensation reactions have also been used to prepare poly(1,1'-ferrocenylene)s containing ferrocene units linked in a chain. Thus, the condensation of 1,1'-dilithioferrocene with 1,1'-diiodoferrocene affords a medium-molecular-weight polymer (see Eq. 1.21) [8, 38]. Condensation of alkylferrocenyldialdehydes such as $[\{\eta^5$-$C_5H_3RCHO\}_2Fe]$ is mediated by Zn/TiCl$_4$ catalysts to afford poly(ferrocenylvinylene)s. These polymers are interesting organometallic analogues of the well-known linear conjugated organic polymers such as poly(1,4-phenylenevinylene).

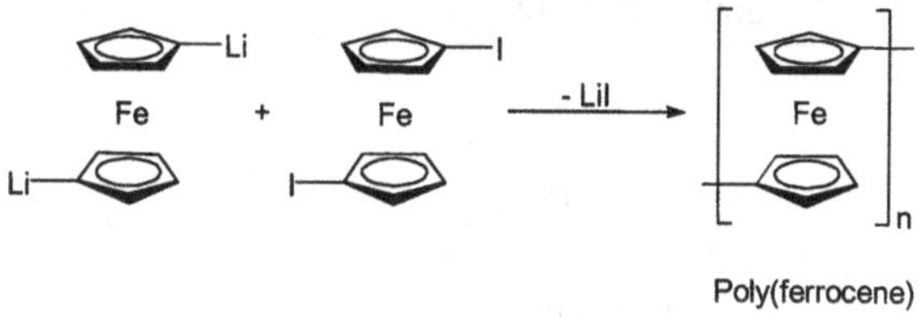

(1.21)

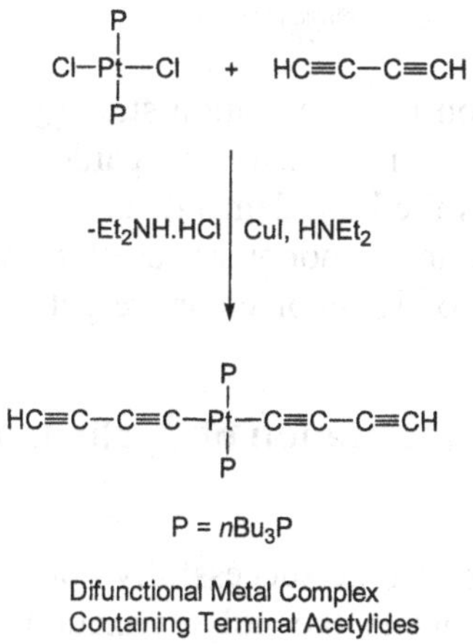

Poly(ferrocene)

(1.22)

$P = nBu_3P$

Difunctional Metal Complex
Containing Terminal Acetylides

Organometallic polymers with a rigid-rod architecture have been prepared by using the condensation strategy involving the condensation of *trans*-Pt(PnBu₃)₂(CCCCH)₂ with *trans*-Pt(PnBu₃)₂Cl₂. This reaction is carried out in amine solvents and is catalyzed by Cu(I) salts (see Eqs. 1.22, 1.23) [39]. The monomer synthesis involves the reaction of *trans*-Pt(II)Cl₂(nBu₃P)₂ with 1,3-butadiyne to afford the alkynylated metal derivative *trans*-Pt(PnBu₃)₂(CCCCH)₂ containing a M-C σ bond (see Eq. 1.22). The difunctional derivative *trans*-Pt(PnBu₃)₂(CCCCH)₂ can be coupled with Pt(II)Cl₂(nBu₃P)₂ to afford linear high-molecular-weight polymers (see Eq. 1.23) [39]. These polymers have rod-like structures because of the rigidity imposed by the alkynyl groups. This type of synthetic strategy is in fact quite general and has been adapted to prepare other rigid-rod type of polymers containing transition metals such as ruthenium, nickel, cobalt etc., [8]. These types of rigid-rod polymers are often characterized by metal-to-alkyne charge transfer transitions. These polymers are also of interest from the point of view of their nonlinear optical properties.

$$HC{\equiv}C-C{\equiv}C-\overset{\overset{\displaystyle P}{|}}{\underset{\underset{\displaystyle P}{|}}{Pt}}-C{\equiv}C-C{\equiv}CH \quad + \quad Cl-\overset{\overset{\displaystyle P}{|}}{\underset{\underset{\displaystyle P}{|}}{Pt}}-Cl \qquad (1.23)$$

$$\downarrow \text{-Et}_2\text{NH.HCl} \quad \text{CuI, HNEt}_2$$

$$\left[\overset{\overset{\displaystyle P}{|}}{\underset{\underset{\displaystyle P}{|}}{Pt}}-C{\equiv}C-C{\equiv}C\right]_n$$

P = nBu₃P

Rigid-rod organometallic polymer

Although the condensation polymerization strategy has been employed quite successfully to prepare a number of inorganic polymers, one of the limitations of this method is the high demands on the purity of the monomer. Unless both the difunctional monomers are rigorously pure, condensation reactions do not lead to high-molecular-weight polymers.

1.3.3 Ring-Opening Polymerization of Cyclic Inorganic Compounds

A number of inorganic rings can be successfully polymerized by the ROP method. The earliest example is the polymerization of rhombic sulfur. Thus, heating S₈ at about 160 °C causes its ROP to lead to the formation of a linear polymer $(S)_n$ (see Eq. 1.24) [4,5]. This is one of the few *true* inorganic polymers. Unfortunately, polymeric sulfur, although elastomeric is not of much use because it undergoes depolymerization upon cooling to room temperature.

$$ \xrightarrow{\Delta} \quad (S)_n \tag{1.24} $$

Polymeric sulfur

Sulfur combines with nitrogen to form several types of rings and cages that are either neutral or ionic [4]. One of the most well-known S-N cages is S_4N_4. This compound, when heated, is converted into the four-membered ring S_2N_2. The latter undergoes a polymerization in the solid-state to afford polythiazyl, $(SN)_n$ (see Eq. 1.25) [4, 5].

$$ \xrightarrow[\text{Vacuum}]{200\text{-}300\ ^\circ\text{C}} \tag{1.25} $$

25 °C

Poly thiazyl

Polythiazyl, also a true inorganic polymer is not soluble in common organic solvents. However, it has attracted considerable interest because of its unusual electrical properties. Thus, $(SN)_n$ shows metal-like conductivity at room temperature and becomes a superconductor at 0.3 K [4, 5]. Although this material itself has not found any applications it has aroused considerable interest in the area of electrically conducting polymeric materials.

The inorganic siloxane rings can be polymerized by both cation and anion initiators to lead to polymeric siloxanes. Thus, for example, the octamethylcyclotetrasiloxane, $[Me_2SiO]_4$ can be polymerized by the use of KOH as an initiator to afford poly(dimethylsiloxane). The polymer $[Me_2SiO]_n$ prepared in this way has a very high molecular weight ($2\text{-}5 \times 10^6$) (see Eq. 1.26) [15-17]. Polysiloxanes are the most important family of inorganic polymers from the commercial point of view.

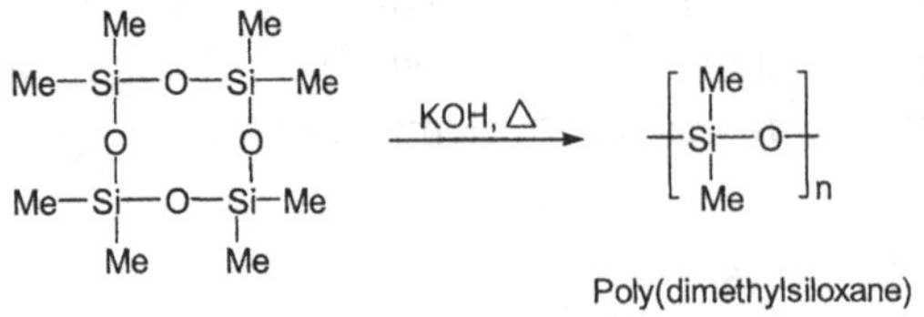

(1.26)

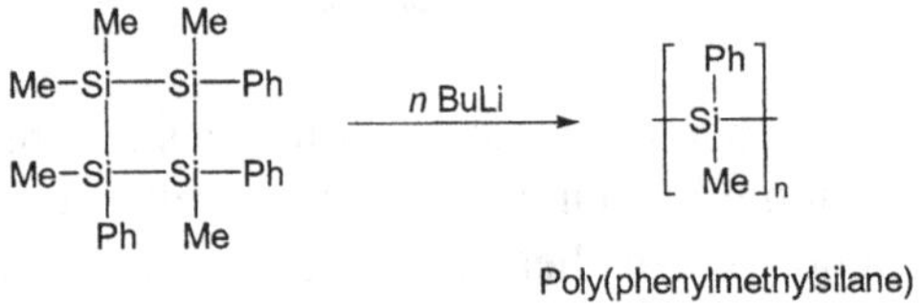

Poly(dimethylsiloxane)

(1.27)

Poly(phenylmethylsilane)

We have seen in the previous section that polysilanes, $[RR'Si]_n$ can be prepared by dehalogenation of $RR'SiCl_2$. These polymers can also be prepared by ROP of two different types of cyclic monomers. The tetrasilane $[PhMeSi]_4$ is polymerized to $[PhMeSi]_n$ by using nBuLi as the initiator (see Eq. 1.27) [40].

n-Butyllithium has also been used as the initiator for polymerizing *masked disilenes*. The latter are essentially disilane compounds which can be viewed as *trapped or masked* disilenes. If the disilene is liberated from this trap it has many choices. It can form a disilene. It can cyclize or polymerize. By a careful choice of substituents on silicon it is possible to use masked disilenes as monomers for polymerization. This method of polymerization has been shown to be quite effective for the preparation of a variety of polysilanes (see Eq. 1.28) [41].

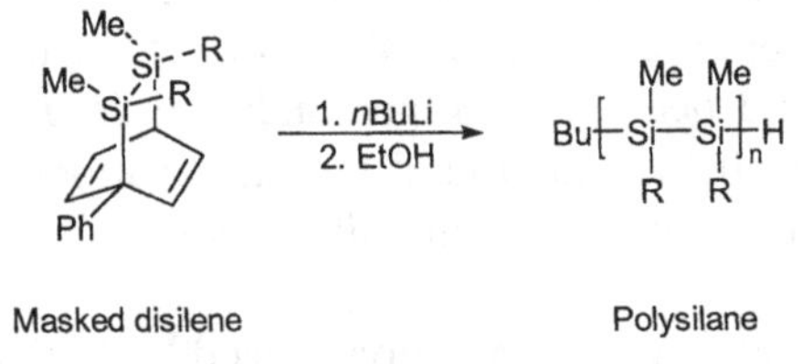

(1.28)

Masked disilene   Polysilane

Polymers containing alternate silicon and carbon centers known as carbosilanes or silylene ethylenes have been originally prepared by the thermal rearrangement of polydimethylsilane, $[Me_2Si]_n$ [19]. More recently, these polymers have been prepared by a more rational route involving the ROP of four-membered disilacyclobutanes. Thus, the ROP of *cyclo*-$[Cl_2SiCH_2]_2$ can be carried out by using H_2PtCl_6 as the catalyst to afford the polymer poly(dichlorosilaethylene),$[Cl_2SiCH_2]_n$. It is possible to reduce the latter with $LiAlH_4$ to afford poly(silaethylene) $[H_2SiCH_2]_n$ (see Eq.

1.29) [42]. The latter can be considered as an analogue of polyethylene where every alternate CH_2 group has been replaced by a SiH_2 group.

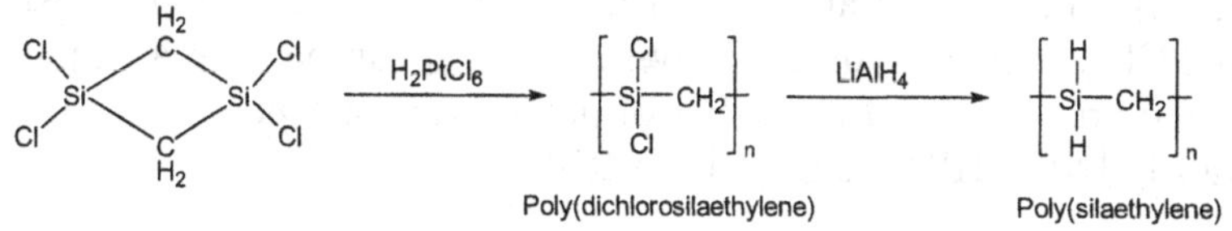

$$(1.29)$$

Poly(dichlorosilaethylene)   Poly(silaethylene)

Polysilaethylenes that contain alkyl groups on the silicon can also be prepared by the H_2PtCl_6 catalyzed ring-opening polymerization of the corresponding silacyclobutanes, $[RR'SiCH_2]_2$ (see Eq. 1.30) [42].

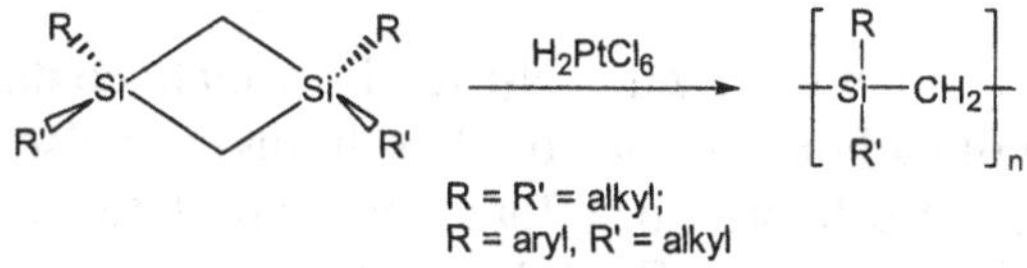

$$(1.30)$$

R = R' = alkyl;
R = aryl, R' = alkyl

Cyclic, silicon-bridged [1]ferrocenophanes have been found to undergo thermal ring-opening polymerization to afford high-molecular-weight poly(ferrocenylsilane)s (see Eq. 1.31) [8, 43]. Many other members of this family of polymers are now known including those where the silicon center is replaced by Sn(IV), P(III), Ge(IV), B(III) etc., [8].

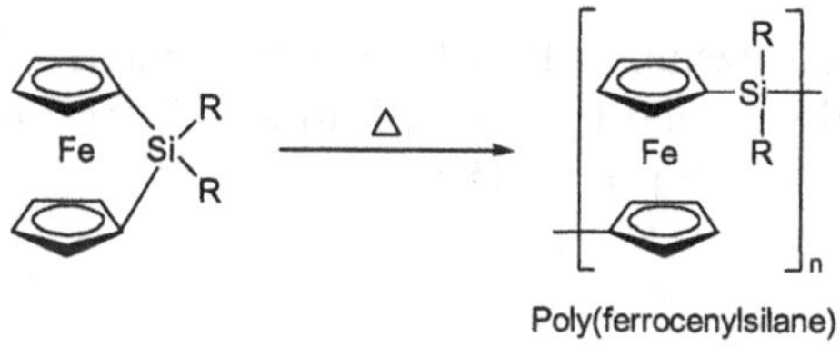

$$(1.31)$$

Poly(ferrocenylsilane)

Another way of preparing ferrocenyl polymers consists of a sulfur-abstraction reaction from a ferrocene monomer where the two cyclopentadienyl units are linked by a trisulfide unit. Sulfur abstraction is brought about by the use of a tertiary phosphine which forms the corresponding phosphine sulfide and leads to a ROP (see Eq. 1.32) [44].

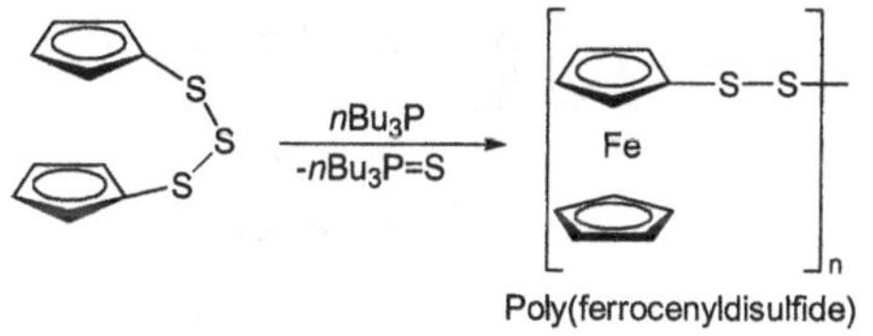

$$(1.32)$$

Poly(ferrocenyldisulfide)

Cyclophosphazenes are a group of inorganic heterocyclic rings containing an alternate arrangement of phosphorus and nitrogen atoms [45-48].

Hexachlorocyclophosphazene, $N_3P_3Cl_6$, undergoes a ring-opening polymerization at 250 °C to afford poly(dichlorophosphazene), $[NPCl_2]_n$ (see Eq. 1.33) [32, 48]. The latter is also one of the pure inorganic polymers. Although, $[NPCl_2]_n$ itself is hydrolytically sensitive, it can be used as a precursor for the preparation of a number of poly(organophosphazene)s such as $[NP(OR)_2]_n$, $[NP(NHR)_2]_n$ and $[NP(NRR')]_n$.

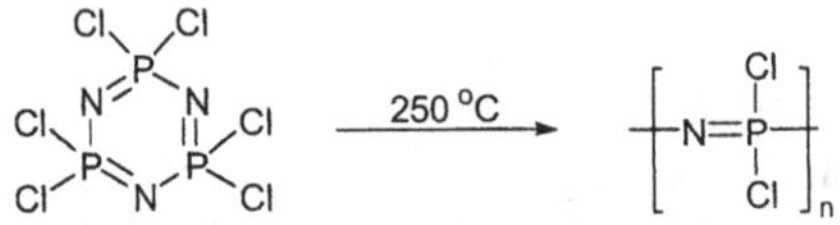

(1.33)

Poly(dichlorophosphazene)

Pentachlorocarbocyclophosphazene, $N_3P_2CCl_5$, which contains a heteroatom in the form of carbon (in place of phosphorus) can also be polymerized by a thermal treatment to afford the hydrolytically sensitive poly(carbocyclophosphazene) (see Eq. 1.34) [49].

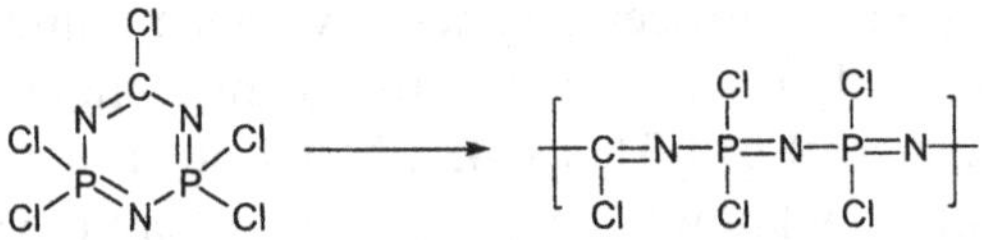

(1.34)

Poly(carbophosphazene)

Pentachlorothiophosphazene, $N_3P_2SCl_5$, which contains a S(IV) as the heteroatom can be polymerized by the ROP to afford the linear polymer, poly(thiophosphazene) (see Eq. 1.35) [50].

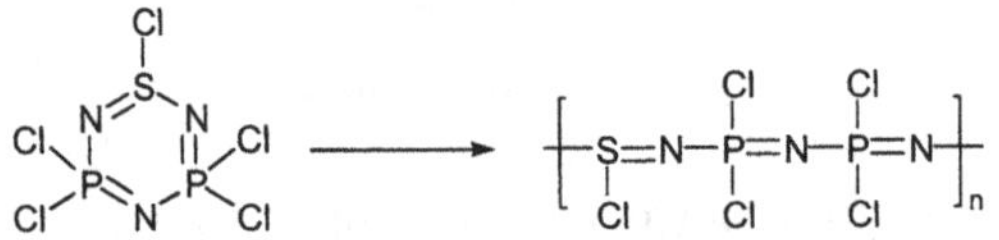

(1.35)

Poly(thiophosphazene)

Pentachlorothionylphosphazene, $N_3P_2S(O)Cl_5$, also can be polymerized by the thermal ROP (see Eq. 1.36) [51].

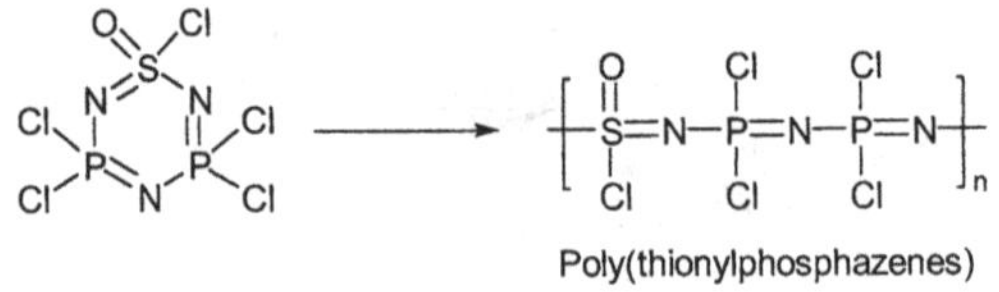

(1.36)

Poly(thionylphosphazenes)

1.3.4 Polymers Containing Inorganic Rings or Motifs as Pendant Groups

Other types of polymers that contain inorganic elements or groups are those that have these units as pendant groups. These polymers contain a carbon backbone. The inorganic units are attached to the backbone as a side chain. In principle almost any inorganic compound can be designed so that it can be a side chain on an organic backbone. The most important of this type of polymer is polystyrene that contains phosphino groups suitable for coordination (see Eq. 1.37) [2, 5].

$$\tag{1.37}$$

Such types of polymers, usually in their crosslinked forms have been used for interaction with transition metals. These heterogeneous metal-carrying polymers are quite valuable as catalysts for organic reactions.

The above strategy has been applied to anchor other inorganic motifs as side-chains. For example, the cyclophosphazene, $N_3P_3Cl_5(OCH=CH_2)$ can be readily polymerized by free-radical polymerization to afford a high-molecular-weight polymer which contains the inorganic heterocyclic ring as regular pendant groups (see Eq. 1.38) [52].

$$\tag{1.38}$$

This methodology can also be used for preparing polymers containing organometallic side groups. Polymers containing ferrocene pendant groups have been prepared by the polymerization of ferrocenylethylene [8, 45]. Similarly polymers containing piano-stool type organometallic complexes as pendant groups can be readily prepared by adopting standard organic polymerization methods (see Eq. 1.39) [53].

$$\text{(1.39)}$$

1.4 Summary of Polymerization Procedures for Inorganic Polymers

In the foregoing we have seen a survey of various procedures that can be used for the preparation of inorganic polymers. Some of these methods are common to those found for the preparation of organic polymers. Some others are different. Many ring-opening polymerization methods are known that involve inorganic hetero- or homocyclic rings. Some of these ROP's are induced by thermal treatment, while some others are catalyzed by transition metal complexes. A number of polymerization methods depend on the loss of a simple molecule such as hydrogen, water, silyl ether, silyl chloride, sodium chloride, lithium iodide etc. Table 1.1 summarizes many of the major types of inorganic polymers and some of the methods utilized to prepare them. The most revealing aspect of the data presented in Table 1.1 is the large variety of inorganic polymers that are already known. The important question that needs to be addressed is as follows. Are these polymers different from organic polymers? Do they have any features that are distinct from organic polymers that make the endeavor of preparing new inorganic polymers worthwhile?

Even a preliminary investigation of the properties of some inorganic polymers reveals that some of them have unexpected properties. Polythiazyl is an anisotropic electrical conductor and shows conductivity that is comparable to metals. At 0.26 K this polymer becomes superconducting [4, 5]. Polysilanes which contain catenated silicon atoms in a polymeric chain have several unusual properties. These polymers have a σ-electron delocalization. They are radiation sensitive and many of them are thermochromic. Many members of this family also show nonlinear optical behavior [17, 19].

Polysiloxanes and polyphosphazenes are among the most flexible polymers known. Because of a combination of factors such as long skeletal bond lengths, wide angles at oxygen and poor intermolecular interaction, poly(dimethylsiloxane) shows many properties not found in organic polymers. These include hydrophobicity, high flexibility, low viscosity, good thermal stability etc., [15]. Such properties have enabled these polymers to be used in various types of applications such as high-temperature insulation, anti-foam applications, bio-transplants, drug-delivery systems, flexible elastomers, personal products etc. Similarly many poly(organophosphazene)s show low temperature flexibility. Besides, many of these polymers are fire-retardant in contrast to many organic polymers that are flammable [32].

Organometallic polymers such as poly(ferrocenylsilane)s have been shown to be precursors for new types of magnetic ceramics [9, 42]. Similarly poly(silyleneethylene)s [41] and some polysilanes are polymeric precursors for silicon carbide ceramics [19].

Thus, inorganic polymers have a unique and distinct place in the family of polymers and they have the potential to function as unique materials. In this book we will examine some of the prominent members of inorganic polymers. There are many polymer types where sufficient examples do not exist or whose synthesis has not yet been shown to be general. Some of these polymers have been alluded to in this Chapter. However, they will not be covered elsewhere in this book. Dendrimeric materials, which are not linear polymers, are also not considered here.

Table 1.1. Various kinds of inorganic polymers – summary of the methods of their preparation

S.No	Polymer	Structural unit	Method of preparation
1	Polyphosphazene	$\left[\begin{array}{c} Cl \\ \mid \\ -P=N- \\ \mid \\ Cl \end{array}\right]_n$	1. ROP of $N_3P_3Cl_6$ 2. Cationic polymerization of $Me_3SiN=PCl_3$ 3. Thermal treatment of $Cl_3P=N-P(O)Cl_2$
2	Polyphosphazenes	$\left[\begin{array}{c} R \\ \mid \\ -P=N- \\ \mid \\ R' \end{array}\right]_n$ R = R' = alkyl R = alkyl; R' = aryl	Thermal polymerization of $Me_3SiN=PR(R')OCH_2CF_3$

Table 1.1. (contd.)

3	Poly(hetero-phosphazene)s		ROP of $N_3P_2MCl_5$

$$\left[-M=N-\underset{\underset{Cl}{|}}{\overset{\overset{Cl}{|}}{P}}=N-\underset{\underset{Cl}{|}}{\overset{\overset{Cl}{|}}{P}}=N-\right]_n$$

M = C(Cl); S(Cl) S(O)Cl

4	Polysiloxanes	$\left[\underset{\underset{R'}{|}}{\overset{\overset{R}{|}}{Si}}-O\right]_n$	1. ROP of $[RR'SiO]_{3,4}$
			2. Hydrolysis of $RR'SiCl_2$ followed by $RR'Si(OH)_2$

5	Polysilanes	$\left[\underset{\underset{R'}{|}}{\overset{\overset{R}{|}}{Si}}\right]_n$	1. Dehalogenation of $RR'SiCl_2$
			2. Anionic polymerization of cyclic poly silanes
			3. Anionic polymerization of masked disilenes

6	Polysilanes	$\left[\underset{\underset{H}{|}}{\overset{\overset{R}{|}}{Si}}\right]_n$	Catalytic dehydrogenation of primary silanes $RSiH_3$

7	Poly(silylene-ethylen)s	$\left[\underset{\underset{R}{|}}{\overset{\overset{R}{|}}{Si}}-CH_2\right]_n$	ROP of $[R_2SiCH_2]_2$

8	Polystannanes	$\left[\underset{\underset{R}{|}}{\overset{\overset{R}{|}}{Sn}}\right]_n$	1. Dehalogenation of Bu_2SnCl_2
			2. Dehydrogenation of R_2SnH_2

9	Polygermanes	$\left[\underset{\underset{R}{|}}{\overset{\overset{R}{|}}{Ge}}\right]_n$	Dehalogenation of R_2GeCl_2

10	Poly(sulfurnitride)	$\left[SN\right]_n$	ROP of S_2N_2

11	Poly(oxo-thiazene)s	$\left[\underset{\underset{R}{|}}{\overset{\overset{O}{\parallel}}{S}}=N\right]_n$	Thermal treatment of $NH=S(O)R(OR')$ or $Me_3SiN=S(O)R(OR')$

12	Poly(phosphino-borane)s	$\left[\underset{\underset{H}{|}}{\overset{\overset{Ph}{|}}{P}}-BH_2\right]_n$	Catalytic dehydrogenation of $PhPH_2.BH_3$

13	Polyferrocene		Condensation reaction of 1,1'-dilithioferrocene and 1,1'-diiodoferrocene

Table 1.1. (contd.)

14	Poly(ferrocenyl si-lanes		ROP of strained ferrocenyl silanes
15	Rigid-rod organometallic polymers		Condensation reaction between *trans* $-(P)_2PtCl_2$ and $L_2Pt(P=Bu_3P; L = -C\equiv C-C\equiv CH$
16	Pendant polymers		Polymerization of the vinyl group attached to the inorganic or organometallic group

References

1. Odian G (2004) Principles of polymerization, 4th Ed. Wiley, New York
2. Allcock HR, Lampe F, Mark J (2004) Contemporary polymer chemistry, 3[rd]Ed.Prentice-Hall, Englewood Cliffs
3. Elias, H.-G (1997) An introduction to polymer science. VCH, Weinheim
4. Greenwood NN, Earnshaw A (1984) Chemistry of the elements. Pergamon, Oxford
5. Mark JE, West R, Allcock HR (1992) Inorganic polymers. Prentice-Hall, Englewood Cliffs
6. Archer RD (2001) Inorganic and organometallic polymers. Wiley-VCH, Weinheim
7. Manners I (1996) Angew Chem Int Ed Engl 35:1602
8. Nguyen P, Gomez-Elipe P, Manners I (1999) Chem Rev 99:1515
9. Manners I (2004) Synthetic metal containing polymers. Wiley-VCH, Weinham
10. West R (1987) Angew Chem Int Ed Engl 26:1201
11. Okazaki R, West R (1996) Adv Organomet Chem 39:231
12. Tokitoh N (2000) J Organomet Chem 611:217
13. Matsumoto H, Arai T, Watanabe H, Nagai Y (1984) Chem Commun 724
14. Tsang CW, Yam M, Gates DP (2003) J Am Chem Soc 125:1480
15. Clarson SJ, Semlyen JA (eds) (1993) Siloxane polymers. Prentice Hall, Englewood Cliffs
16. Rochow EG (1987) Silicon and silicones. Springer-Verlag, Berlin, Heidelberg,New York
17. Jones RG, Ando W, Chojnowski J (eds) (2000) Silicon containing polymers. Kluwer , Dordrecht

18. West R (1986) J Organomet Chem 300:327
19. Miller RD, Michl J (1989) Chem Rev 89:1359
20. Yamaguchi S, Zhi-Jan R, Tamao K (1999) J Am Chem Soc 121:2937
21. Tamao K, Yamaguchi S (2000) J Organomet Chem 611:5
22. Devylder N, Hill M, Molloy KC, Price GJ (1996) Chem Commun 711
23. Miller RD, Sooriyakumaran R (1987) J Polym Sci Polym Chem 25:711
24. Trefonas P, West R (1985) J Polym Sci Polym Chem 23:2099
25. Imori T, Don Tilley T (1994) Polyhedron 13: 2231
26. Imori T, Lu V, Cai H, Don Tilley T (1995) J Am Chem Soc 117:9931
27. Fajan PJ, Beck JS, Lynch AT, Remsen EE, Sneddon, LG (1990) Chem Mater 2:96
28. Dorn H, Singh RA, Massey JA, Nelson JM, Jaska CA, Lough AJ, Manners I (2000) J Am Chem Soc 122:6669
29. McWilliams AR, Dorn H, Manners I (2002) Top Curr Chem 220:141
30. Dorn H, Rodezno JM, Brunnhöfer B, Rivard E, Massey JA, Manners I (2003) Macromolecules 36:291
31. Neilson RH, Wisian-Neilson P (1988) Chem Rev 88:541
32. Allcock HR (2003) Chemistry and applications of polyphosphazenes. Wiley, Hobolen, New Jersey
33. Roy AK, Burns GT, Lie GC, Grigoras S (1993) J Am Chem Soc 115:2604
34. DeJaeger R, Gleria M (1998) Prog Polym Sci 23:179
35. Babu GN, Christopher SS, Newmark RA (1987) Macromolecules 20:2654
36. Yamaguchi S, Jin RJ, Itami Y, Goto T, Tamao K (1999) J Am Chem Soc 121: 10420
37. Wright VA, Gates DP (2002) Angew Chem Int Ed 41:2389
38. Neuse EW, Bednarik L (1979) Macromolecules 12:187
39. Takahashi S, Kariya M, Yatake T, Sonogashira K, Hagihara N (1978) Macromolecules 11:1063
40. Cypryk M, Gupta Y, Matyjaszewski K (1991) J Am Chem Soc 113:1046
41. Sakamoto K, Obata K, Hirata H, Nakajima M, Sakurai H (1989) J Am Chem Soc 111:7641
42. Interrante LV, Liu Q, Rushkin I, Shen Q (1996) J Organomet Chem 521:1
43. Manners I (1999) Chem Commun 857
44. Brandt PF, Rauchfuss TB (1992) J Am Chem Soc 114:1926
45. Chandrasekhar V, Krishnan V (2002) Adv Inorg Chem 53:159
46. Chandrasekhar V, Thomas KRJ (1993) Structure Bonding 81:41
47. Krishnamurthy SS, Sau AC, Woods M (1978) Adv Inorg Rad Chem 21:41
48. Allcock HR (1972) Chem Rev 72:315
49. Manners I, Allcock HR, Renner G, Nuyken O (1989) J Am Chem Soc 111:5478
50. Dodge JA, Manners I, Allcock HR, Renner G, Nuyken (1990) J Am Chem Soc 112: 1268
51. Liang M, Manners I (1991) J Am Chem Soc 113:4044
52. Brown DE, Ramachandran K, Carter KR, Allen CW (2001) Macromolecules 34:2870
53. Pittman CU, Lin C, Rounsefell TD (1978) Macromolecules 11:1022

2 Organic Polymers – A Brief Survey

2.1 Introduction

Polymers are long-chain molecules that are built from simple building blocks, which are known as monomers [1-5]. Polymers are made from either a single monomer or a combination of two or more monomers.

Table 2.1. Examples of monomers and polymers

S.No.	Monomer(s)	Polymer repeat Unit	Polymer
1	$H_2C=CH_2$	$\vphantom{}-[H_2C-CH_2]_n-$	Polyethylene
2	$H_2C=\overset{\displaystyle H}{\underset{\displaystyle Ph}{C}}$	$\left[H_2C-\overset{\displaystyle H}{\underset{\displaystyle Ph}{C}}\right]_n$	Polystyrene
3	$H_2\overset{\displaystyle}{C}\underset{\displaystyle O}{-}CH_2$	$-[H_2C-CH_2-O]_n-$	Poly(ethylene oxide)
4	$HOCH_2CH_2OH$ and $HOOC-\!\!\bigcirc\!\!-COOH$	$\left[O-H_2C-CH_2-O-\overset{O}{\overset{\|}{C}}-\!\!\bigcirc\!\!-\overset{O}{\overset{\|}{C}}\right]_n$	Poly(ethylene terephthalate)
5	$NH_2(CH_2)_6NH_2$ and $HOOC(CH_2)_4COOH$	$\left[NH(CH_2)_6NH-\overset{O}{\overset{\|}{C}}-(CH_2)_4-\overset{O}{\overset{\|}{C}}\right]_n$	Nylon-6,6

Each polymer has a basic structural unit that repeats itself several times. This is also known as the polymer repeat unit. Table 2.1 gives a few examples of monomers and the polymers derived from them.

Many of the polymers are prepared from a single monomer such as ethylene, styrene etc. Other polymers like poly(ethylene terephthalate) or nylon-6,6 are prepared from the reaction of two monomers. There are also polymers which involve the reaction of more than two monomers. Thus, poly(acrylonitrile-butadiene-styrene) is a copolymer and is prepared from polymerizing the monomers acrylonitrile, styrene and butadiene together.

2.2 Natural Polymers

Polymers shown in Table 2.1 are synthetic polymers. However, there are several polymers that also occur naturally. The many known natural polymers include a few famous examples such as *proteins, deoxyribonucleic acid* (DNA), *hemoglobin, insulin, starch, natural rubber* etc. In order to understand the polymeric nature of these natural substances let us look at proteins and see how they are built from simple monomers.

1. Proteins are important naturally occurring polymers that are involved in several biological processes. For example, *enzymes* catalyze most of the chemical reactions in the biological system. Most of the enzymes are proteins.
2. The basic building block (monomer) of all the proteins is an α-amino acid.
3. Every protein has a precisely defined α-amino acid sequence.

An α-amino acid contains an amino group; a carboxyl group, a hydrogen atom and a distinctive R group - all of these are bound to an α-carbon atom. Amino acids in solution at neutral pH are in their zwitterionic form i.e., the -NH$_2$ is present in the form of -NH$_3^+$ and the -COOH is present in the form of a deprotonated COO$^-$. This may be represented in the wedge and Fisher projection forms (Fig. 2.1).

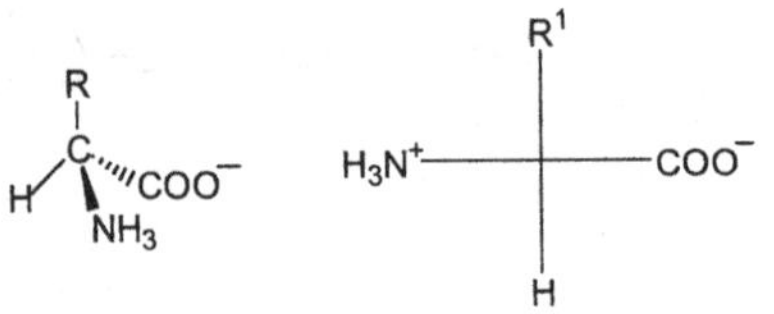

Fig. 2.1. Representations of amino acids

There are twenty amino acids with varying side chains (R groups). All proteins of all species from bacteria to human beings are constructed from the same set of 20 amino acids. These fundamental molecules and the polymers derived from them are at least 2 billion years old!

How are the amino acids linked to each other to form proteins? In proteins the α-carboxyl group of one amino acid is joined to the α-amino group of another amino acid by a peptide bond (amide bond). If two amino acids condense together a dipeptide is formed (Fig. 2.2).

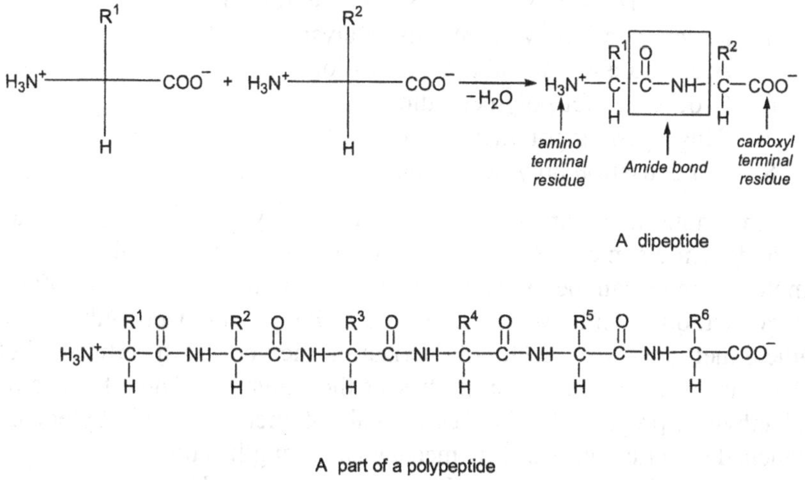

A dipeptide

A part of a polypeptide

Fig. 2.2. Formation of a polypeptide from condensation of amino acids

Many amino acids are condensed together in a sequential manner to form a protein or a polypeptide chain. It may be noted that proteins are not simply linear chains as written above. Such a linear representation showing the sequence of amino acid units is known as the *primary structure* of a protein. A protein possesses complicated *secondary, tertiary and quaternary structures*. The secondary and tertiary structures involve intra- and inter molecular hydrogen bonding interactions. The quaternary structure includes ionic interactions as well. The overall result is that proteins have complex three-dimensional structures.

2.3 Synthetic Polymers

The focus of this book is on *non-traditional polymers* built from inorganic or organometallic monomers [6-17]. Before this subject is dealt with in the subsequent chapters, a brief review on the various preparative methods available for the assembly of organic polymers is presented. This is followed by a summary of some important general characteristics of polymers, which make them different from simple molecules. This background would serve as the platform for the study of some important individual families of inorganic and organometallic polymers. In general, organic polymers are prepared by the following methods of polymerization [1-5].

1. Free-radical polymerization
2. Ionic polymerization
3. Polymerization by Ziegler-Natta type of catalysts
4. Polymerization by metathesis catalysts
5. Atom-transfer radical polymerization
6. Group-transfer polymerization
7. Ring-opening polymerization
8. Condensation polymerization

Some monomers can only be polymerized by specific polymerization methods while some others can be polymerized by many methods. For example, styrene can be polymerized by free-radical, anionic or Ziegler-Natta methods, while vinyl chloride can be polymerized only by free-radical methods. Also, each polymerization method has specific attributes, which are reflected in the properties of the polymers. Thus, for example, polyethylene prepared by the free-radical polymerization of ethylene has a branched structure and has low mechanical strength. This is known as low-density polyethylene (**LDPE**). In contrast, polyethylene prepared by the Ziegler-Natta catalysis of ethylene has a linear structure and has higher mechanical strength. This is known as high-density polyethylene (**HDPE**). Thus, the choice of the polymerization method is dictated not only by the nature of the monomer but also by the properties that are required in the polymer.

2.4 Free-radical Polymerization

Free-radical polymerization is one of the most common ways of preparing organic polymers. This method involves the use of a free-radical initiator and is applicable for the polymerization of a number of vinyl monomers. A number of small molecules containing carbon-carbon double bonds such as ethylene ($CH_2=CH_2$) and related derivatives such as $CH_2=CHR$ can be readily polymerized by free-radical polymerization. Table 2.2 summarizes some of the monomers that can be polymerized by the application of this method. Note that this method is *very tolerant* of the type of substituent R on the olefinic unit and a wide range of derivatives containing substituents of various electronegativities can be polymerized. Even disubstituted olefins such as $CH_2=CCl_2$, methyl methacrylate and fluoro derivatives such as $CF_2=CF_2$ and dienes such as butadiene or isoprene can be polymerized by the use of the free-radical polymerization.

Free-radical polymerization can be carried out in bulk phase (in the absence of a solvent) or in solution. Solution polymerization can be performed in organic solvents or in aqueous media. In some instances an emulsion polymerization is carried out in which a detergent is added to a suspension of the monomer in an aqueous medium. A water-soluble initiator is used which penetrates the emulsion particles to initiate polymerization.

Free-radical polymerization implies that the polymerization proceeds by the generation of a free radical, which rapidly adds other monomer molecules to afford a polymer. Free radicals are generated by heat, radiation or by chemical initiators.

Table 2.2. Some common organic monomers that undergo free-radical polymerization

Compound	Formula	Compound	Formula
Ethylene	$H_2C{=}CH_2$	1,1-Dichloroethylene	$H_2C{=}CCl_2$
Styrene	$H_2C{=}\underset{Ph}{\overset{H}{C}}$	Tetrafluoroethylene	$F_2C{=}CF_2$
Vinyl chloride	$H_2C{=}\underset{Cl}{\overset{H}{C}}$	Butadiene	$H_2C{=}CH{-}CH{=}CH_2$
Acrylonitrile	$H_2C{=}\underset{CN}{\overset{H}{C}}$	Methyl methacrylate	$H_2C{=}C(Me)C(O)OMe$
Acrylamide	$H_2C{=}\underset{CONH_2}{\overset{H}{C}}$	Vinyl esters	$H_2C{=}CHOC(O)R$

2.4.1 Initiators

Typical chemical initiators are (a) benzoyl peroxide (b) 2,2′-azo-bis-isobutyronitrile (AIBN) (c) redox initiators.

Benzoyl peroxide or AIBN decompose when heated to moderate temperatures of 60-80 °C (see Eqs. 2.1, 2.2).

$$C_6H_5{-}\overset{O}{\overset{\|}{C}}{-}O{-}O{-}\overset{O}{\overset{\|}{C}}{-}C_6H_5 \longrightarrow 2\ C_6H_5{-}\overset{O}{\overset{\|}{C}}{-}O\bullet \qquad (2.1)$$

$$C_6H_5-\overset{\overset{\displaystyle O}{\|}}{C}-O\bullet \longrightarrow C_6H_5\bullet + CO_2 \qquad (2.2)$$

The decomposition of benzoyl peroxide occurs by a *homolytic* scission of the peroxide (O-O) bond to generate the benzoyl [C_6H_5-C(O)O•] radical which can further decompose to the phenyl (C_6H_5•) radical by elimination of CO_2. These radicals are quite effective in initiating the polymerization of many vinyl monomers.

AIBN decomposes by the elimination of dinitrogen to generate the iso-butyronitrile radical (see Eq. 2.3).

$$H_3C-\underset{CN}{\overset{\overset{\displaystyle CH_3}{|}}{C}}-N{=}N-\underset{CN}{\overset{\overset{\displaystyle CH_3}{|}}{C}}-CH_3 \xrightarrow[-N_2]{} 2\,H_3C-\underset{CN}{\overset{\overset{\displaystyle CH_3}{|}}{C}}\bullet \qquad (2.3)$$

Both benzoyl peroxide and AIBN are commonly used as initiators when the polymerization is carried out in organic solvents. In aqueous media redox initiators are used. For example, the persulfate ion is reduced by the bisulfite ion to generate [SO_4•]$^-$ and [HSO_3•]$^-$ radical anions. The latter reacts with water to generate [HSO_4]$^-$ and [OH]$^-$ (see Eq. 2.4).

$$\left[O-\overset{\overset{\displaystyle O}{\|}}{\underset{\underset{\displaystyle O}{\|}}{S}}-O-O-\overset{\overset{\displaystyle O}{\|}}{\underset{\underset{\displaystyle O}{\|}}{S}}-O \right]^{2-} + [HSO_3]^- \longrightarrow [SO_4]^{2-} + [SO_4\bullet]^- + [HSO_3\bullet] \qquad (2.4)$$

persulfate   bisulfite

Similarly, thiosulfate can also be used to reduce the persulfate ion (see Eq. 2.5).

$$[S_2O_8]^{2-} + [S_2O_3]^{2-} \longrightarrow [SO_4]^{2-} + [SO_4\bullet]^- + [S_2O_3\bullet]^- \qquad (2.5)$$

persulfate thiosulfate   sulfate   sulfate   thiosulfate
                                  anion   radical   radical
                                                   anion   anion

Free radicals can also be generated by transition metal ion interaction. Thus, ferric ions can oxidize the bisulfite ion to the bisulfite radical (see Eq. 2.6).

$$[HSO_3]^- + Fe^{3+} \longrightarrow [HSO_3\bullet] + Fe^{2+} \qquad (2.6)$$

2.4.2 Mechanism of Polymerization

The polymerization of a vinyl monomer is initiated by the reaction of a free radical R· . This results in the formation of another radical (see Eq. 2.7).

$$R\bullet + H_2C=\overset{\overset{\displaystyle H}{|}}{\underset{\underset{\displaystyle G}{|}}{C}} \longrightarrow R-CH_2-\overset{\overset{\displaystyle H}{|}}{\underset{\underset{\displaystyle G}{|}}{C}}\bullet \qquad (2.7)$$

The propagation of the reaction occurs when the radical that is generated adds onto other monomer molecules (see Eq. 2.8).

$$R-CH_2-\overset{\overset{\displaystyle H}{|}}{\underset{\underset{\displaystyle G}{|}}{C}}\bullet + H_2C=\overset{\overset{\displaystyle H}{|}}{\underset{\underset{\displaystyle G}{|}}{C}} \longrightarrow R-CH_2-\overset{\overset{\displaystyle H}{|}}{\underset{\underset{\displaystyle G}{|}}{CH}}-CH_2-\overset{\overset{\displaystyle H}{|}}{\underset{\underset{\displaystyle G}{|}}{C}}\bullet \qquad (2.8)$$

This mode of propagation is known as the *head-to-tail* propagation. This is the predominant pathway. Other modes of propagation such as head-head or tail-tail are less favored.

How do the growing chains terminate? This occurs in two ways. By (a) coupling reactions or by (b) disproportionation.

In the coupling mechanism two free-radical polymer chains can combine to afford a larger polymer chain (see Eq. 2.9).

$$(2.9)$$

$$R-(CH_2-\overset{H}{\underset{G}{CH}})_n-CH_2-\overset{H}{\underset{G}{C}}\bullet \quad + \quad \bullet\overset{H}{\underset{G}{C}}-CH_2-(\overset{}{\underset{G}{CH}}-CH_2)_m-R$$

$$\downarrow$$

$$R-(CH_2-\overset{H}{\underset{G}{CH}})_n-CH_2-\overset{H}{\underset{G}{C}}-\overset{H}{\underset{G}{C}}-CH_2-(\overset{}{\underset{G}{CH}}-CH_2)_m-R$$

$$(2.10)$$

$$R-(CH_2-\overset{}{\underset{G}{CH}})_n-\overset{H}{\underset{G}{CH}}-\overset{H}{\underset{G}{C}}\bullet \quad + \quad \bullet\overset{H}{\underset{G}{C}}-H_2C-(HC-H_2C)_m-R$$

$$\downarrow$$

$$R-(CH_2-\overset{}{\underset{G}{CH}})_n-CH=\overset{H}{\underset{G}{C}} \quad + \quad H-\overset{H}{\underset{G}{C}}-CH_2-(\overset{}{\underset{G}{CH}}-CH_2)_m-R$$

polymer with unsaturated   polymer with a saturated
         end group   end group

In the disproportionation mode of termination, a transfer of (usually) a hydrogen atom occurs from one polymer radical to the other (see Eq. 2.10). This leads to the formation of one polymer with a saturated end group and another polymer with an unsaturated end group.

2.4.3 Branching

If the reactions involved in free-radical polymerization proceeded precisely in the manner outlined above (initiation, propagation, termination), in all instances, one would obtain long and linear polymer chains. However, in free radical polymerization, linear chains are, in fact, not formed. Instead, extensive branching is observed particularly with monomers such as ethylene. In this context a slight clarification of 'branching' is required. Thus, for a polymer obtained from a substituted vinyl monomer regular side groups are present (Fig. 2.3).

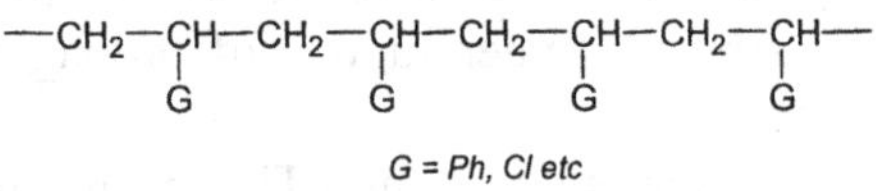

$$-CH_2-CH-CH_2-CH-CH_2-CH-CH_2-CH-$$
$$\quad\ \ \ \underset{G}{|}\qquad\quad \underset{G}{|}\qquad\quad \underset{G}{|}\qquad\quad \underset{G}{|}$$

G = Ph, Cl etc

Fig. 2.3. Polymer chains obtained from $CH_2=CHG$ type of monomers

(2.11)

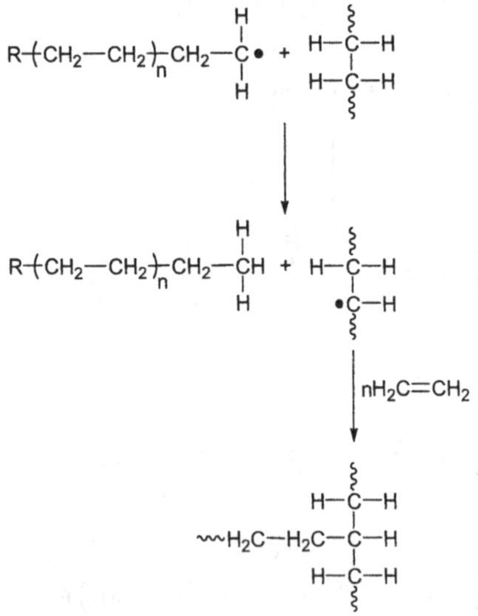

In the context of what we are discussing here these side groups are not considered branches. The concept of branching is clarified using polyethylene as an example. If a growing polyethylene free radical abstracts hydrogen *not from the* terminus of another polymer molecule *but from the*

middle, the latter is left with reactive points along the polymer chain. These can initiate new polymerization reactions (see Eq. 2.11). Thus, the overall effect of such processes is to generate a branched polymer instead of a linear polymer (Fig. 2.4). The consequence of branching is lower crystallinity and inferior mechanical strength. Accordingly, polyethylene prepared by the free-radical polymerization is highly branched and is known as low-density polyethylene (LDPE).

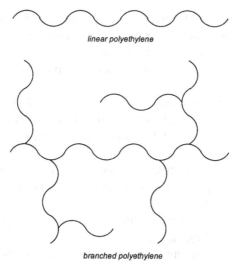

linear polyethylene

branched polyethylene

Fig. 2.4. Linear and branched polyethylene chains

2.4.4 Chain Transfer

Branching is a special case of a process called chain transfer that is operational in free-radical polymerization. Chain transfer simply means the *transfer of the radical* from the growing polymer chain to another species. Effectively chain transfer curtails polymer growth. For example, chlorinated solvents are efficient chain transfer agents (see Eq. 2.12).

$$R\text{---}(CH_2\text{---}CH_2)_n\text{---}CH_2\text{---}\underset{\underset{H}{|}}{\overset{\overset{H}{|}}{C}}\bullet \; + \; CCl_4 \longrightarrow R\text{---}(CH_2\text{---}CH_2)_n\text{---}CH_2\text{---}CH_2\text{---}Cl \; + \; \bullet CCl_3 \qquad (2.12)$$

In many instances molecules like dihydrogen are deliberately introduced in the reaction medium to obtain polymers of desired molecular weights (see Eq. 2.13).

$$R\text{-}(CH_2\text{—}CH_2)_n\text{-}CH_2\text{—}\overset{\overset{\displaystyle H}{|}}{\underset{\underset{\displaystyle H}{|}}{C}}\bullet \ + \ H_2 \longrightarrow R\text{-}(CH_2\text{—}CH_2)_n\text{-}CH_2\text{—}CH_3 \ + \ \bullet H \tag{2.13}$$

2.4.5 Summary of the Main Features of Free-radical Polymerization

1. A large variety of vinyl monomers can be polymerized.
2. The tolerance of the type of vinyl monomers for this type of polymerization is quite remarkable.
3. The polymerization can be done in bulk, solution or emulsion phases.
4. A large variety of initiators are available. Two of the most common chemical initiators for applications involving the use of organic solvents are 2,2'-azo-bis-isobutyronitrile (AIBN) and benzoyl peroxide.
5. Chain-transfer reactions to solvents (such as CCl$_4$), monomer, growing polymer chains or deliberately added reagents (such as H$_2$) are a predominant feature.
6. Polymerization by free-radical methods affords polymers that have considerable branching.
7. At all points of time in the reaction mixture both monomer and polymer are present. Oligomers are not formed. This is because of the chain-growth nature of the polymerization.

2.5 Polymerization by Ionic Initiators

Free-radical polymerization has several advantages, including its applicability to a large variety of monomers, tolerance to impurities etc. However, one of its chief drawbacks is that because of chain-transfer reactions, polymers with branching are obtained. Further, the molecular weight distributions obtained for polymers in this method are generally somewhat broad (in the later part of this chapter we will have a brief look at polymer molecular weights and molecular weight distributions). Ionic polymerization overcomes the problem of branching (to a large extent). In contrast to free-radical initiators, ionic initiators and the polymerization reactions initiated by them are subject to a lot of stringent reaction conditions. The tolerance of monomer is also limited. This means that only certain monomers can be polymerized by the use of these initiators. There are two types of ionic initiators (a) anionic initiators and (b) cationic initiators.

2.5.1 Anionic Polymerization

In contrast to free-radical polymerization, in anionic polymerization, the initiator (R^-) is a carbanion (see Eq. 2.14).

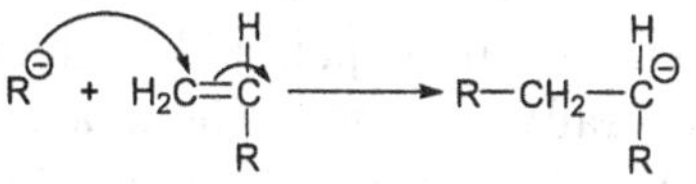

$$R^{\ominus} + H_2C{=}\overset{\overset{\textstyle H}{|}}{\underset{\underset{\textstyle R}{|}}{C}} \longrightarrow R{-}CH_2{-}\overset{\overset{\textstyle H}{|}}{\underset{\underset{\textstyle R}{|}}{C}}{}^{\ominus} \tag{2.14}$$

Because the carbanion is generated on the growing polymer chain the type of monomers that can undergo anionic polymerization are those that have electron-withdrawing substituents. The latter can stabilize the negative charge on the polymer. Hence, the monomers that are most readily polymerized by this method are those that contain the electron-withdrawing groups as well as those that do not have functional groups or reactive groups, which can react with the initiator. For example, the chlorine substituent in vinyl chloride can react with the initiator and hence this monomer is not suitable for being polymerized with anionic initiators. Monomers such as styrene, acrylonitrile or acrylic esters are suitable for polymerization by anionic initiators (Fig. 2.5).

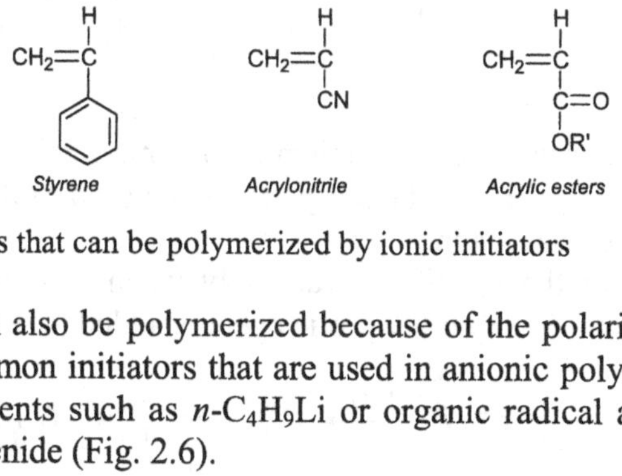

Fig. 2.5. Monomers that can be polymerized by ionic initiators

Aldehydes can also be polymerized because of the polarity of the C=O bond. Some common initiators that are used in anionic polymerization are alkyllithium reagents such as $n\text{-}C_4H_9Li$ or organic radical anions such as sodium naphthalenide (Fig. 2.6).

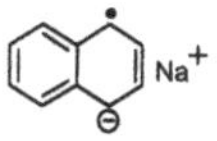

Fig. 2.6. Sodium naphthalenide

Because of the sensitivity of the initiators towards impurities, particularly water and CO_2, rigorous precautions have to be taken to exclude moisture from the reaction medium. It is also logical that chlorinated solvents are not used for these reactions, as they themselves will react with initiators.

2.5.1.2 Mechanism of Anionic Polymerization

Alkyllithiums such as *n*-butyllithium do not exist as such but are present in the form of molecular aggregates. In fact, the X-ray crystal structure of *n*-butyllithium has shown that it is present as a hexamer [18]. It is believed that reagents such as this react as an ion pair, although the covalent character in the Li-C bond is generally accepted. The attack of the nucleophile occurs in such a manner that the negative charge is located on a carbon, which contains the electron withdrawing group (see Eq. 2.15).

$$\tag{2.15}$$

Propagation involves the attack of the growing carbanion on successive monomeric units such that the growing polymer chain resembles a continuous insertion of the monomer molecules between the C^-M^+ moiety (see Eq. 2.16).

$$\tag{2.16}$$

Presumably because the mechanism of polymerization involves the insertion of the monomer into the ion-pair, the addition of the monomers occurs in a stereoregular manner.

Termination of the reaction occurs only when a *terminator* molecule such as CO_2, water or any other protic agent is added (see Eq. 2.17).

$$\tag{2.17}$$

2.5.1.3 Special Features of Anionic Polymerization

1. In anionic polymerization, unlike in free-radical polymerization, there is virtually no chain-transfer. This is especially true for reac-

tions conducted at very low temperatures. As a result of the absence of the chain transfer there is no branching.

2. The polymers obtained by anionic polymerization have a polydispersity index (PDI) close to 1. This is because of the rapidity of the initiation step. Virtually *all* the chains are initiated at the same instant and all of them grow at approximately the same rate until all the monomer is consumed.

3. *Living polymers*. It has been observed that in anionic polymerization if proper precautions are taken the *termination* reaction is virtually absent. In other words termination is only brought out when desired by the addition of protic reagents. This has some interesting consequences. The polymer chains are virtually *living* and if after the consumption of all the monomer more of it is added it participates in the propagation reaction (see Eq. 2.18).

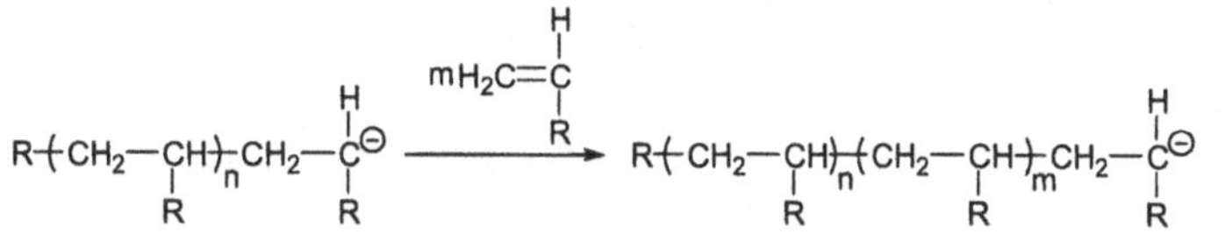

$$(2.18)$$

Such a *living* polymerization has another interesting consequence. If instead of the same monomer a different monomer is added one obtains block copolymers. For example, styrene-butadiene-styrene (SBS) block copolymer can be prepared in this manner (Fig. 2.7).

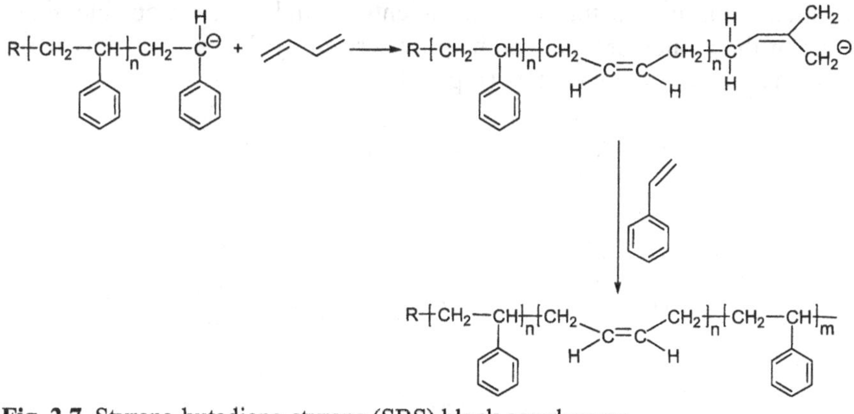

Fig. 2.7. Styrene-butadiene-styrene (SBS) block copolymers

SBS is a hard rubber, which is used in applications such as the soles of shoes, fine treads etc., where durability is important. Polystyrene is a tough hard plastic while polybutadiene is a rubbery material. The latter gives the SBS polymer its rubber-like properties while the former endow it with durability. SBS is also known as a thermoplastic elastomer. These are materials which behave like elastomeric rubbers at room temperature, but when heated can be processed like plastics. SBS and other thermoplastic elastomers manage to be rubbery without being crosslinked.

2.5.2 Cationic Polymerization

In cationic polymerization the end of the growing chain bears a positive charge. Previously, we have seen situations with free-radical chain ends or anionic chain ends (Fig. 2.8).

Fig. 2.8. Polymers with various reactive end-groups

Similar to the situation found in anionic polymerization, all types of vinyl monomers cannot be polymerized by the cationic polymerization. Specifically the monomer should have substituents that can stabilize the carbocation. This means that the substituents should be electron releasing. Some of the monomers that can be polymerized by the cationic polymerization method are shown below (Fig. 2.9).

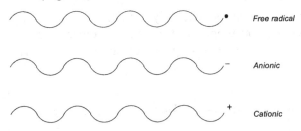

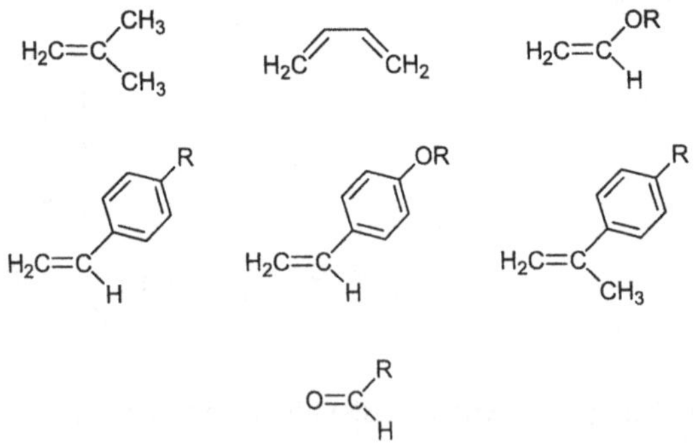

Fig. 2.9. Monomers that undergo polymerization by cationic initiators

Industrially the polymerization of isobutylene by cationic initiators is of importance.

The initiators for cationic polymerization can be protic acids (sulfuric acid, perchloric acid, hydrochloric acid) or Lewis acids (BF_3, BCl_3, $TiCl_4$, $AlCl_3$ etc.,). Protons are the cationic initiators that are generated from protic acids (see Eq. 2.19).

$$H_2SO_4 \rightleftharpoons H^+ + HSO_4^- \tag{2.19}$$

$$HClO_4 \rightleftharpoons H^+ + ClO_4^-$$

With Lewis acids it is noticed that a small amount of protic solvents (such as water or methanol) are required which will result in a Lewis acid-Lewis base adduct (see Eq. 2.20).

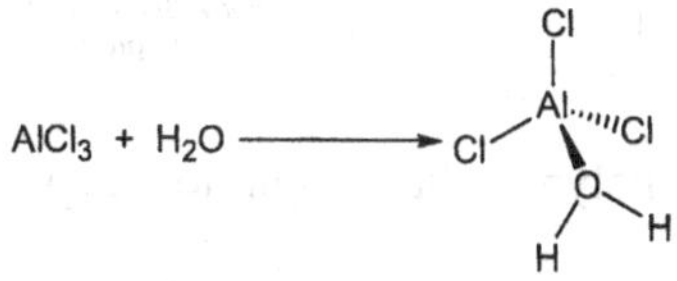

(2.20)

The protons in the $AlCl_3.H_2O$ complex are quite acidic and serve as proton sources (see Eq. 2.21).

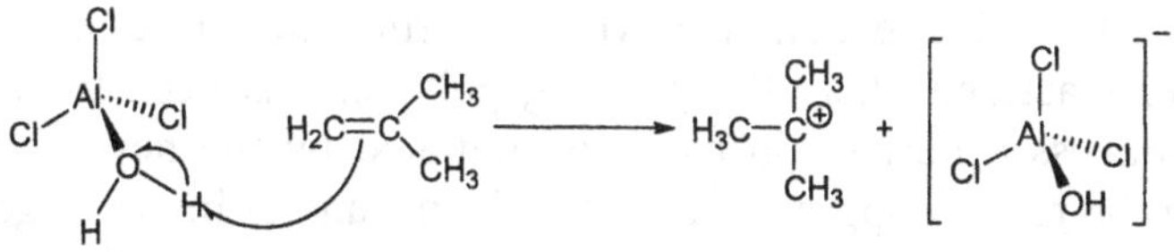

(2.21)

The anion generated will be simply represented as A⁻ and is closely associated with the cation (Fig. 2.10).

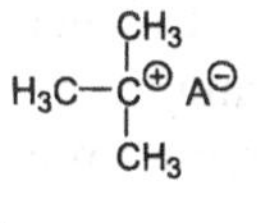

Fig. 2.10. Cation with its counter-anion

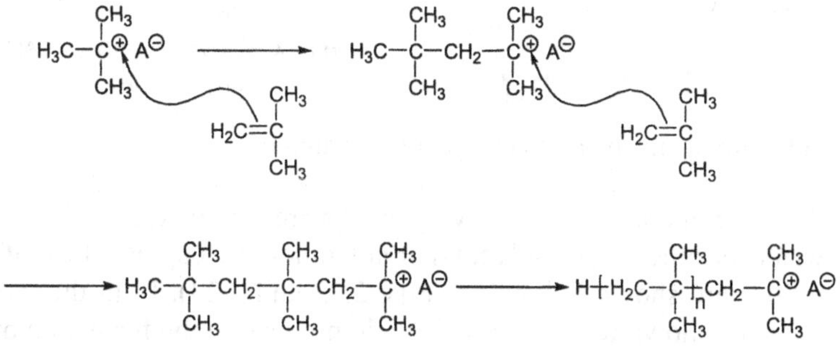

Fig. 2.11. Propagation reaction

Notice also that the addition of the proton occurs in such a manner so as to lead to the formation of the most stable (tertiary) carbocation. The propagation reaction occurs by a successive insertion of the monomer molecules within the C^+A^- ion-pair motif (Fig. 2.11).

The growing polymer chains are terminated by two routes. The first route consists of a proton transfer to another monomer (Fig. 2.12).

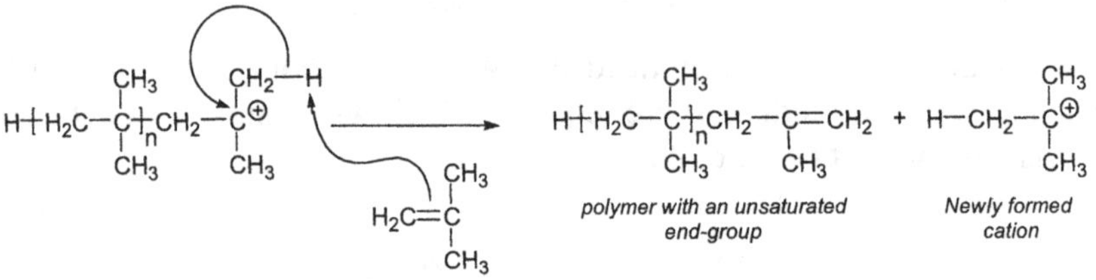

Fig. 2.12. Termination of the polymer chain by proton transfer

From the growing polymer chain a proton is abstracted by the nucleophilic (electron-rich) monomer. This leads to the termination of the polymer chain which is neutral and which now contains an unsaturated end group (Fig. 2.12). Notice that this type of termination is essentially a chain-transfer reaction, where the growing polymer has been terminated and in the process a new chain has been initiated by the newly formed cation. It can be readily appreciated that a chain transfer involves generation of new chains. If more chains are initiated in this way the average chain length of the polymer and hence the molecular weight of the polymer is reduced. Fortunately, the chain-transfer reaction in cationic polymerization is of importance only at higher temperatures and can be considerably reduced at lower temperatures.

The second method of polymer termination can occur by a proton transfer to the anion (Fig. 2.13).

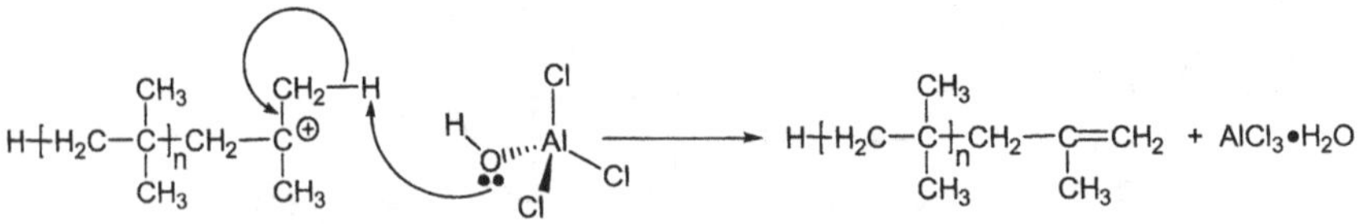

Fig. 2.13. Termination by proton transfer to the anion

At lower temperatures the growing chains are terminated by the reaction with water (or other protic solvents) or any other nucleophile. Two of these possibilities are shown below (see Eqs. 2.22 and 2.23). Both the termination reactions shown in these equations do not lead to the formation of new chains.

$$H\text{---}\!\!\left[\!H_2C\text{---}\underset{\underset{CH_3}{|}}{\overset{\overset{CH_3}{|}}{C}}\!\right]_{\!n}\!\!CH_2\text{---}\underset{\underset{CH_3}{|}}{\overset{\overset{CH_3}{|}}{C}}{}^{\oplus}\,A^{\ominus}\ +\ H_2O\ \longrightarrow\ H\text{---}\!\!\left[\!H_2C\text{---}\underset{\underset{CH_3}{|}}{\overset{\overset{CH_3}{|}}{C}}\!\right]_{\!n}\!\!CH_2\text{---}\underset{\underset{CH_3}{|}}{\overset{\overset{CH_3}{|}}{C}}\text{---}OH\ +\ A\text{-}H \tag{2.22}$$

$$(2.23)$$

$$H\text{---}\!\!\left[\!H_2C\text{---}\underset{\underset{CH_3}{|}}{\overset{\overset{CH_3}{|}}{C}}\!\right]_{\!n}\!\!CH_2\text{---}\underset{\underset{CH_3}{|}}{\overset{\overset{CH_3}{|}}{C}}{}^{\oplus}\ \overset{Cl}{\underset{Cl}{\overset{|}{Al}{}^{\!-}{}_{\text{''''}OH}}}\ \xrightarrow{\ -AlCl_2OH\ }\ H\text{---}\!\!\left[\!H_2C\text{---}\underset{\underset{CH_3}{|}}{\overset{\overset{CH_3}{|}}{C}}\!\right]_{\!n}\!\!CH_2\text{---}\underset{\underset{CH_3}{|}}{\overset{\overset{CH_3}{|}}{C}}\text{---}Cl$$

2.5.2.1 Summary of the Features of Cationic Polymerization

1. Monomers that contain electron-releasing groups (such as isobutylene) can be polymerized by cationic polymerization.
2. The common initiators are protic or Lewis acids.
3. Cationic polymerization can lead to the formation of linear polymers particularly if carried out at low temperatures.
4. Chain-transfer reactions are slow at lower temperatures.

2.6 Ziegler-Natta Catalysis

We have seen that free-radical polymerization of ethylene leads to the formation of low-density polyethylene (LDPE). The reason for the low density of LDPE is the lack of crystallinity owing to branching in the polymer. *Straight-chain* polyethylene can be obtained from Ziegler-Natta polymerization, which involves the use of an organometallic catalytic system [1-5, 18-20].

K. Ziegler, a German chemist, observed an Aufbaureaktion (growth reaction) when ethylene is polymerized in the presence of triethylaluminum [1, 18] (see Eq. 2.24).

$$Et_3Al\ +\ (n\text{-}1)H_2C\!\!=\!\!CH_2\ \xrightarrow[\ 100\ bar\ ]{\ 90\text{-}120\ ^\circ C\ }\ Et_2Al(CH_2CH_2)_nH \tag{2.24}$$

A chain length of up to C_{200} was observed. Surprisingly, in one of the experiments Ziegler observed that only *1-butene* was formed in the reaction. This has led him to investigate the reaction thoroughly and meticulously. He found out that the particular stainless steel reactor where 1-butene was formed had a small impurity of nickel. Ziegler then carried out a systematic investigation on the influence of many transition metal salts and complexes on the reaction. He and his coworkers observed that the combination of $TiCl_4/Et_3Al$ functions as an excellent catalyst for the polymerization of ethylene to polyethylene [18] (see Eq. 2.25).

$$H_2C{=}CH_2 \xrightarrow[\text{25 °C, 1bar}]{\text{TiCl}_4/\text{AlEt}_3} \left[CH_2{-}CH_2 \right]_n \qquad (2.25)$$

polyethylene

The polymerization of ethylene by the combination of triethylaluminum and titanium tetrachloride occurs at room temperature and atmospheric pressure. This is in contrast to the industrial process of free-radical polymerization of ethylene, which involves high temperatures and high pressures (1000-3000 bar pressure and 200 °C). The second important difference is that the polyethylene obtained by using Ziegler's catalyst is completely linear. Because of this it has got better crystallinity and is called High Density Polyethylene (HDPE). While the density of LDPE is 0.91 g/cm^3 that of HDPE is 0.97-0.99 g/cm^3. The melting point of HDPE (130 °C) is also higher than that of LDPE (120 °C) [1]. Because of these differences, LDPE is a soft and low-strength polymer, while HDPE is a tough, high-strength material.

G. Natta, an Italian chemist, made an extremely important extension of the use of Ziegler's catalytic system. He showed the utility of Ziegler catlysts for the room temperature polymerization of propylene. Most importantly, however, Natta observed that the organometallic catalytic system, leads to the preparation of a stereoregular polymer. Thus, polypropylene can have three possible stereochemical orientations [1].

1. **Isotactic**: A stereoregular structure where the sites of steric isomerism in each repeating unit of the polymer has the same configuration. In other words in an isotactic polypropylene all the methyl groups will be located on *one* side of the plane of the polymer chain (Fig. 2.14).

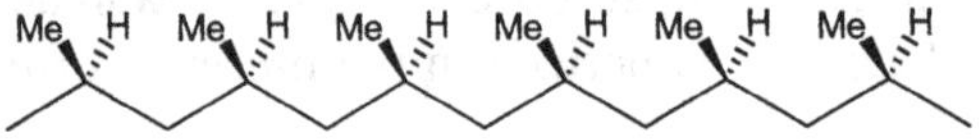

Fig. 2.14. Isotactic polypropylene

2. **Syndiotactic.** A stereoregular polymer where the sites of steric isomerism in each repeating unit are on the opposite side. The methyl groups in alternate repeat units are on the same side and in adjacent repeat units they are on the opposite side (Fig. 2.15).

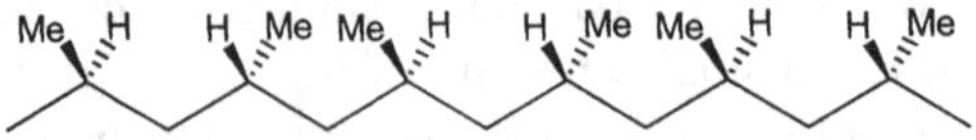

Fig. 2.15. Syndiotactic polypropylene

3. **Atactic.** In this type of polypropylene the orientation of the methyl groups is random.

Remarkably, Natta observed that the Ziegler catalyst system generated predominantly isotactic polypropylene in the polymerization of propylene [1] (see Eq. 2.26).

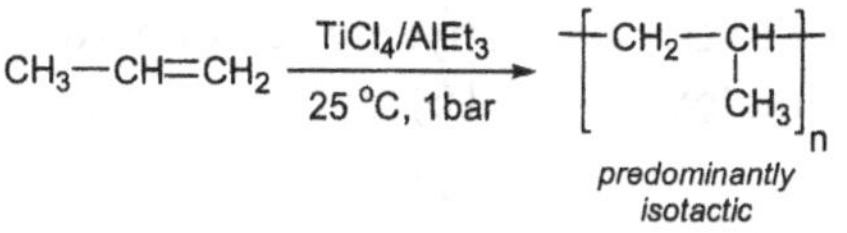

$$CH_3-CH=CH_2 \xrightarrow[25\,°C,\,1bar]{TiCl_4/AlEt_3} \left[CH_2-CH \atop CH_3 \right]_n \qquad (2.26)$$

predominantly
isotactic

Because of the regular disposition of the substituents, isotactic and syndiotactic polymers are highly crystalline and have better mechanical properties than atactic polypropylene, which is amorphous [1].

2.6.1 Mechanism of Polymerization of Olefins Using the Ziegler-Natta Catalysts

Although the exact mechanism of the polymerization of olefins by Ziegler-Natta catalysis is not delineated with the precision that would satisfy every one, it is possible to write a reasonable mechanism [1, 2]. The first of these is a monometallic mechanism. The following mechanism is for a $TiCl_3/AlEt_3$ combination. We have seen above that titanium tetrachloride has been initially used in the polymerization of ethylene. It has been subsequently found that $TiCl_3$ is better than $TiCl_4$.

The first step in the mechanism is an alkylation of the titanium by the alkylaluminum reagent (see Eq. 2.27).

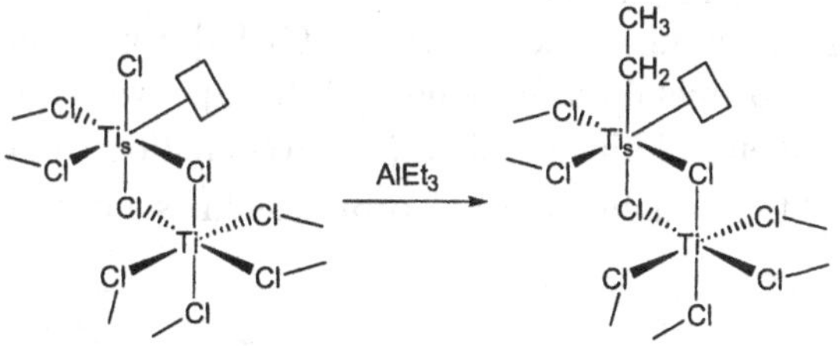

(2.27)

The coordination environment around titanium in $TiCl_3$ (in the solid-state) is six, i.e., each titanium is surrounded by six chlorines. The chlorines are involved as intermolecular bridging ligands between adjacent titanium centers. It is important to realize that this is the solid-state structure and the catalyst system is heterogeneous. Accordingly, only the surface sites will be active and bulk of the material will be not accessible to the reaction. It is reasonable to propose that at the surface several titanium centers will be coordinatively unsaturated. This means that the coordination number will be less than six. A situation with a coordination number of five is depicted above (see Eq. 2.27). Such a titanium is labeled as Ti_s and the vacancy in the coordination sphere is depicted by a square box. Alkylation of such a titanium species leads to the replacement of one of the chlorines by an alkyl group. But the vacancy at titanium is still present. An olefin molecule (such as ethylene or propylene) can coordinate at this site to afford a metal-olefin complex (Fig. 2.16).

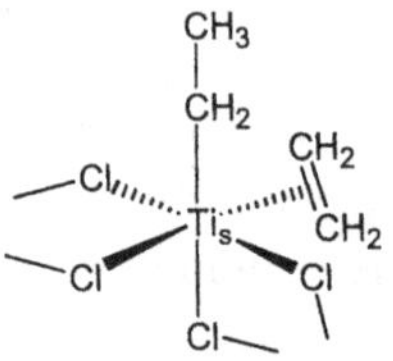

Fig. 2.16. Coordination of ethylene to the vacant site on the surface titanium center

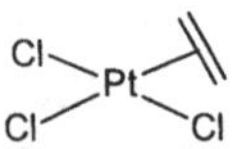

Fig. 2.17. Ethylene coordination to platinum in $K[PtCl_3(C_2H_4)]$

Although the titanium-ethylene complex has not been isolated in the polymerization of ethylene, there are several metal-olefin complexes that have been characterized by many spectroscopic methods and also by X-ray crystallography. The most prominent example is the Zeise's salt $K[PtCl_3(C_2H_4)]$ [18-20] (Fig. 2.17).

From studies on compounds such as $[PtCl_3(C_2H_4)]$ a good deal is now known on the nature of metal-olefin bonding [18-20]. Two types of interactions exist in any metal-olefin complex: (1) A σ-donation from the olefin through its π-orbital to a suitable metal 'd' orbital such as a $d(x^2-y^2)$ orbital.

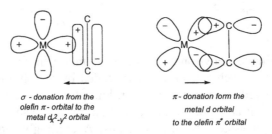

σ - donation from the olefin π - orbital to the metal $d_{x^2-y^2}$ orbital

π - donation form the metal d orbital to the olefin π* orbital

Fig. 2.18. Bonding interactions in metal-olefin complexes

(2) A *back-bonding* involving a π-donation from a metal d orbital to the anti bonding π-orbital of the olefin (Fig. 2.18). These two types of bonding modes act synergistically, i.e., one strengthens the other. The details of the strength of the metal-olefin interactions need not bother us. Suffice it to say that such complexes are prominent members of the organometallic family.

Coming back to the mechanism of polymerization. After the coordination of the olefin the next-step in the sequence of events is termed a *migratory insertion*. This occurs, presumably through the involvement of a four-membered metallacycle intermediate (Fig. 2.19).

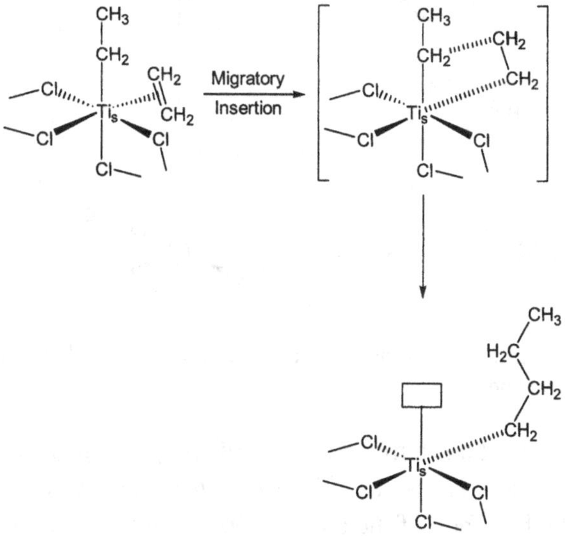

Fig. 2.19. Migratory insertion reaction

At the end of the reaction sequence it *appears* that the olefin is inserted between the Ti-CH$_2$CH$_3$ bonds. Hence, this reaction is known as insertion. However, most mechanistic studies suggest that the migration of the alkyl group (rather than the insertion of the olefin) actually takes place [18-20]. To reconcile these two possibilities the term *migratory insertion* is used. At the end of the migratory insertion the Ti-alkyl chain has grown from Ti-CH$_2$CH$_3$ to Ti-CH$_2$CH$_2$CH$_2$CH$_3$, and the titanium is once again in a coordination number of five, i.e., with a coordinative unsaturation. A fresh olefin molecule can coordinate at this site and by the repetition of the migratory insertion the polymer chain grows. This sequence of events is involved in the propagation of the polymerization (Fig. 2.20).

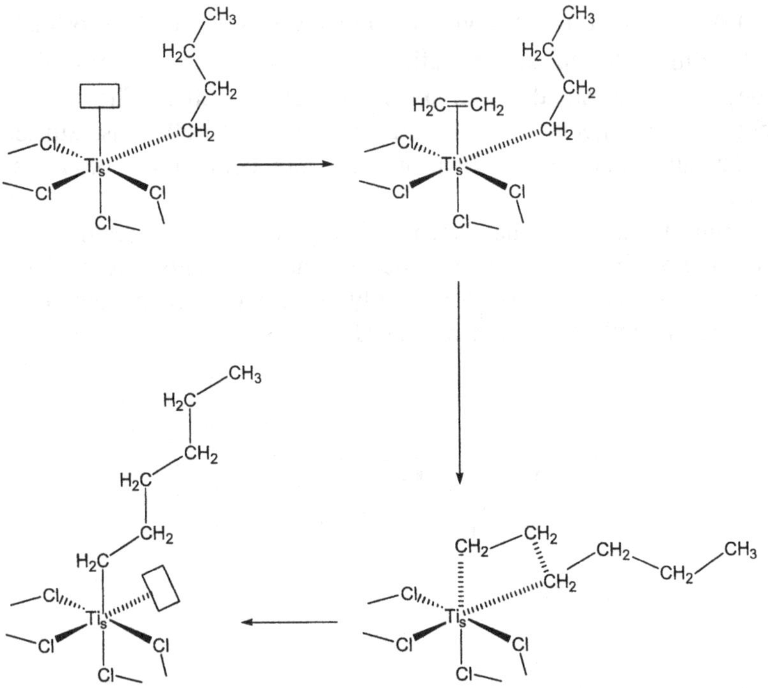

Fig. 2.20. Propagation of the polymer chain in the Ziegler-Natta catalyzed polymerization of ethylene

How does the chain terminate? One of the possible ways this can happen is by a process known as *β-hydrogen transfer*. In this process the hydrogen on the β-carbon of the alkyl chain first interacts at the vacant coordination site of the titanium, and then is finally transferred to the titanium leading to the elimination of the polymer with an unsaturated end group (Fig. 2.21). Evidence for such a mechanism comes from studies on several organometallic compounds. Species such as these have been isolated in other situations showing strong M-H interactions. Such interactions have been labeled as *agostic interactions*.

An alternative mechanism known as the bimetallic mechanism has also been proposed. We start off with a bimetallic intermediate where the aluminum and titanium are bridged to each other (the only difference between this and the earlier proposal is that in the previous instance the role of aluminum was proposed to be limited to alkylating the titanium centers). The remaining steps in the mechanism are nearly similar (Fig. 2.22).

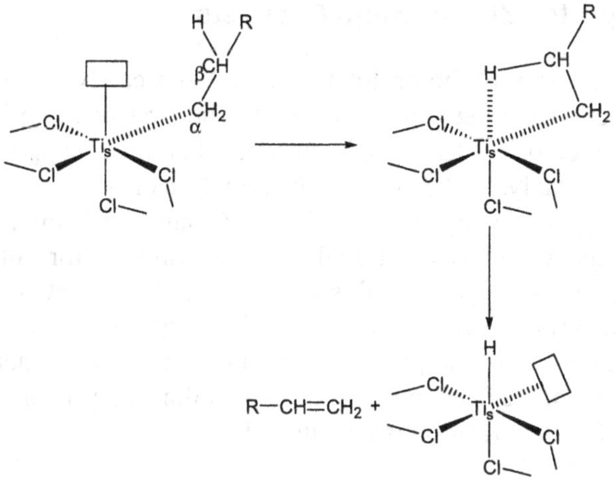

Fig. 2.21. β-Hydrogen transfer in monometallic mechanism

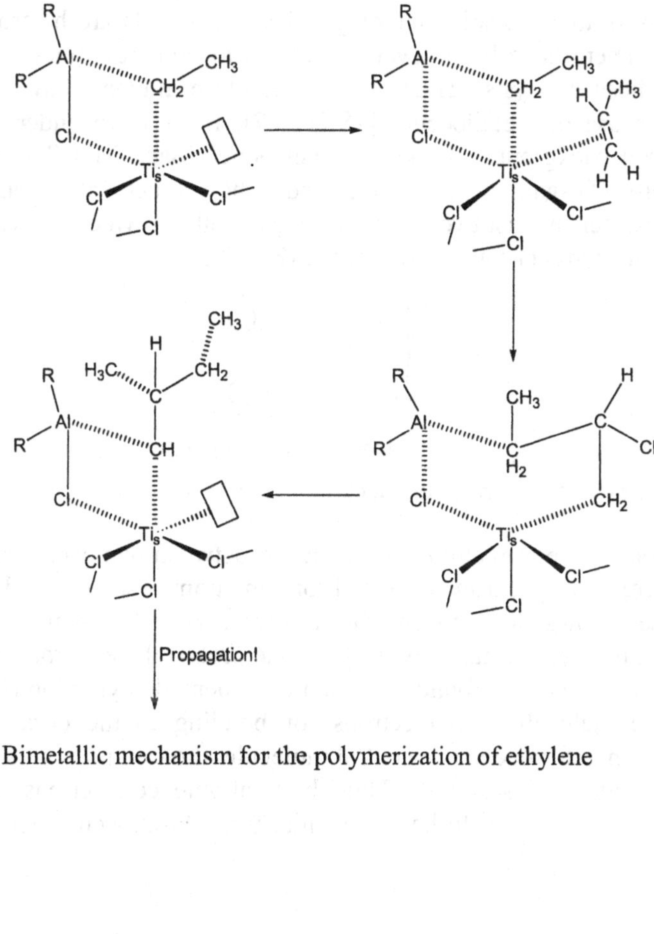

Fig. 2.22. Bimetallic mechanism for the polymerization of ethylene

2.6.2 Supported Ziegler-Natta Catalysts

Since the discovery of the original Ziegler-Natta catalytic system, several new variations have been brought about. The first of these is supporting the titanium chloride on a high surface area inert support. Since the Ziegler-Natta catalyst is heterogeneous in nature only the surface is active. Consequently, dispersing the titanium compound on a high surface area solid will enable more catalytic sites to be available for polymerization. The most popular support for this purpose has been $MgCl_2$. Such $MgCl_2$ supported catalysts have shown major advantages including higher activity, lower catalyst consumption etc. Because only small amounts of catalyst are required, the need to remove the metallic catalyst from the polymer at the end of polymerization is obviated [1].

2.6.3 Recent Developments in Ziegler-Natta Catalysis

As discussed above, traditional Ziegler-Natta catalysts are heterogeneous in nature. There have been several attempts to prepare homogenous versions of these catalysts [21-28]. Many of these attempts have revolved around the use of metallocenes [18-20, 27]. In order to understand this topic a brief background on metallocenes is necessary. Metallocenes are a large family of organometallic compounds where a metal atom/ion is sandwiched between at least two aromatic rings. Well-known examples of metallocenes are ferrocene and chromocene (Fig. 2.23).

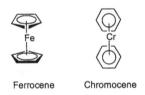

Ferrocene   Chromocene

Fig. 2.23. Structures of ferrocene and chromocene

In ferrocene the iron atom is sandwiched between two cyclopentadienyl rings. In contrast, in chromocene a chromium atom is sandwiched between two benzene rings. In ferrocene the iron atom is in fact bound to all the carbons of the cyclopentadienyl ring. If one probes the electronic configuration of ferrocene it is found that each cyclopentadienyl anion (Cp⁻) contributes six delocalized π–electrons for bonding to the organometallic complex. Since ferrocene is a neutral molecule, the iron present is assigned a formal oxidation state of +2. Thus the total valence electrons present in ferrocene will work out to be 18. Similarly in chromocene each benzene ring contributes 6e and the chromium atom with a zero oxidation state also contributes 6e, making the total number of valence electrons to be 18 (Table 2.3).

Table 2.3. Electron count in ferrocene and chromocene

Electron count in ferrocene	
2Cp⁻	2 x 6 = 12e
1Fe⁺²	1 x 6 = 6e
Total	= 18e
Electron count in chromocene	
2C₆H₆	2 x 6 = 12e
1Cr(o)	1 x 6 = 6e
Total	= 18e

Organometallic compounds having a total valence electron count of 18 are known to be very stable. Following the original discovery of ferrocene, a large number of metallocenes have since been synthesized. The term metallocene now is used in a much more general context and can mean a wide variety of organometallic compounds including those with substituted cyclopentadienyl rings, those with bent sandwich structures and even the half-sandwich compounds. Several other metallocenes with *lesser* number of valence electrons (than eighteen) are known, as for example Cp_2TiCl_2 or Cp_2ZrCl_2. Because of the presence of the additional chlorines these molecules have bent structures (Fig. 2.24). The electron count in these type molecules shows them to be 16e species (Table 2.4).

Fig. 2.24. Bent metallocenes Cp_2ZrCl_2 and Cp_2TiCl_2

Table 2.4. Electron count in Cp_2ZrCl_2 and Cp_2TiCl_2

Ligand or metal	Electrons contributed	Total electrons
2 Cp⁻	2 x 6	12
1 Ti⁺⁴ or Zr⁺⁴	0	0
2 Cl⁻	2 x 2	4
	Total	16

The titanium (or zirconium) present in these complexes is in an oxidation state of +4 and therefore has 0 valence electrons. Each Cp⁻ contributes 6e each, making a total of 12e. Each Cl⁻ contributes 2e each. Thus, Cp_2ZrCl_2 and Cp_2TiCl_2 are 16e compounds and therefore are more reactive than ferrocene. Further the bent structure of metallocenes allows additional ligands to approach the metal ion.

However, initial efforts in using Cp_2TiCl_2 or Cp_2ZrCl_2 as catalysts for ethylene polymerization did not show any promise. A small *accident* changed this. Normally the metallocene halides and the aluminum alkyls are used in the complete absence of O_2 and H_2O. However, it was discovered accidentally that in a combination of $Cp_2ZrMe_2/AlMe_3$ if a small amount of water is introduced it becomes a potent catalyst for ethylene polymerization [25]. Thus, water which is traditionally seen as a foe in Ziegler-Natta catalysis, turns into a friend. The role of water is to convert $AlMe_3$ into an oligomer called methylalumoxane (MAO) $Me_2Al[OAlMe_2]_nOAlMe_2$ where n = 5-20. The precise structure of MAO is still largely unknown.

Indeed, MAO can be deliberately prepared and added to Cp_2ZrMe_2. In a ratio of about 1:1000 (Cp_2ZrMe_2:MAO) the polymerization of ethylene takes place readily producing linear polyethylene. The nature of MAO is quite complex and is believed to be a mixture of linear and cyclic oligomers. Its principal role seems to generate a metal-centered cation by the abstraction of a CH_3^- from Cp_2ZrMe_2 [22-25, 28]. The sequence of steps there after is in keeping with the general principles of organometalic chemistry (Fig. 2.25). This can be summarized as follows.

1. A three-coordinate zirconium cation is generated as a result of the reaction of Cp_2ZrMe_2 with MAO. This is a 14-electron species, which is highly reactive. To some extent the positive nature of the zirconium center is alleviated by the agostic interaction with the methyl group. However, the metal center is quite receptive to any incoming ligand.

2. The reactive three-coordinate zirconium compound is involved in a reaction with ethylene (ethylene coordination). This leads to a metal-olefin complex.

3. The next step is the migratory insertion reaction. This leads to the formation of a metal-alkyl complex. Note that we have ended up again with a three-coordinate zirconium cation, which has a metal-alkyl bond. In effect we have lengthened the alkyl chain by the process of olefin coordination followed by migratory insertion.

4. Propagation of the reaction is facilitated by the rotation of the alkyl group so that steric hindrance is minimized for the incoming

olefin. Olefin coordination and migratory insertion continues the generation of longer alkyl chains. Eventually this process leads to a polymeric chain.

5. The termination reaction is facilitated by the β-hydride elimination. Thus, termination leads back to the reactive three-coordinate zirconium species which can initiate the polymerization again. Recall that this is similar to what we have seen with simple Ziegler-Natta catalysts.

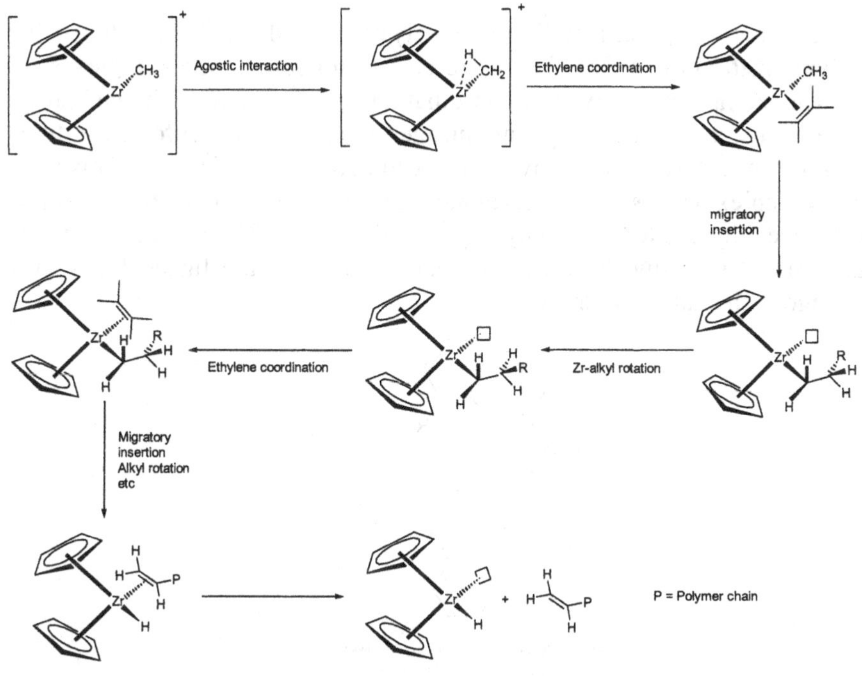

Fig. 2.25. Polymerization of ethylene by Cp_2ZrMe_2/MAO

The most recent developments in this area relate to custom-made zirconocenes for specific applications. The stereo-electronic environment at the zirconium center can be varied in a very subtle manner and this can lead to the formation of specific polymers. Thus, a bridged metallocene in combination with MAO can lead specifically to isotactic polypropylene (Fig. 2.26) [26]. Similarly, variation in the metallocene structure can lead to specific formation of a syndiotactic polypropylene. Even a polypropylene containing alternating isotactic and syndiotactic blocks can be prepared by modifying the metallocene.

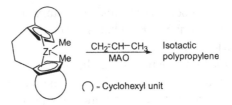

Fig. 2.26. Polymerization of propylene by bridged-zirconocene to afford isotactic polypropylene

More recently, non-metallocene catalysts have also been developed for polymerization of olefins. These are based on coordination complexes. A typical example of the type of coordinating complex used is shown below. It contains a chelating tripodal nitrogen-ligand, which enforces a five-coordinate distorted square-pyramidal geometry around iron(II). This kind of a complex acts as a catalyst (along with MAO as the co-catalyst) to polymerize ethylene to linear polyethylene (Fig. 2.27) [21]. Notice that variation of the substituents on the aromatic groups of the ligand leads to a change in the nature of product.

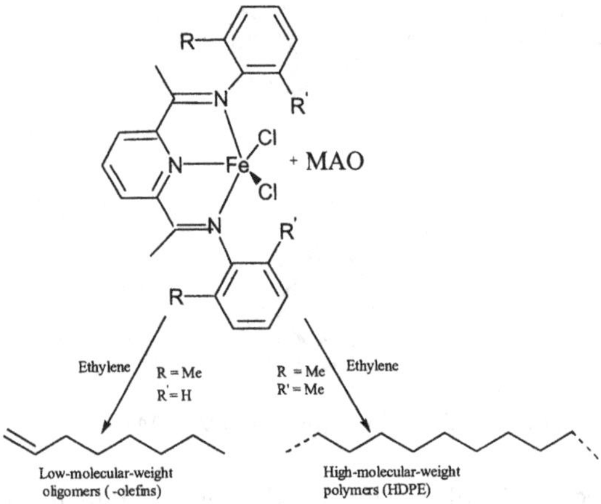

Fig. 2.27. Polymerization of ethylene by transition-metal complexes

The mechanism of polymerization by the transition-metal complexes is similar to what is known for conventional Ziegler-Natta catalyzed reactions (Fig. 2.28). Alkylation of the metal center followed by CH_3^- abstraction leads to a metal-centered cation. Olefin coordination followed by insertion leads to polymer growth. The termination of the polymerization can occur by a β-hydrogen transfer. Clearly the advent of these developments is an indicator that the last word in this area is far from over and the ingenuity of the organometallic chemist and the inorganic chemist can lead to newer catalyst systems.

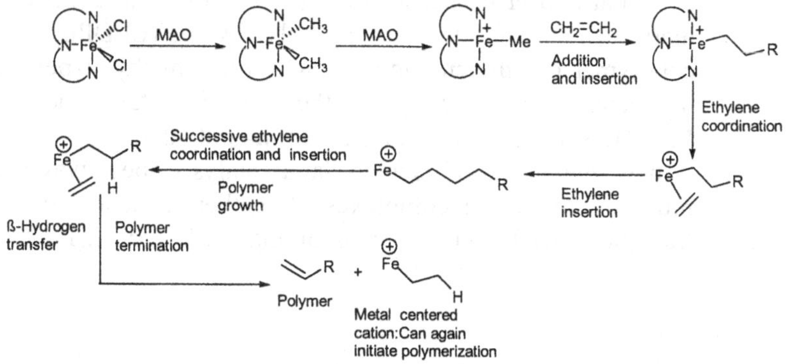

Fig. 2.28. Mechanism of polymerization of ethylene by transition-metal complexes

2.6.4 Summary of the Ziegler-Natta Catalysis

1. A combination of $TiCl_4/Et_3Al$ or $TiCl_3/Et_3Al$ is an excellent catalyst system for the polymerization of olefins. This has been most widely used for preparing linear polyethylene (high-density polyethylene) and isotactic polypropylene. This traditional catalytic system is heterogeneous in nature.

2. The next generations of Ziegler-Natta catalysts have been developed by supporting $TiCl_3$ on inert inorganic supports such as $MgCl_2$. Supported catalysts are more efficient. Higher amounts of polymer are produced per gram of titanium. The mechanism of polymerization involves coordination of the olefin to a vacant coordination site at the surface.

3. New Ziegler-Natta catalysts are homogeneous systems. These are based on a combination of metallocenes and MAO. By appropriate use of these catalysts linear polyethylene, and all the three types of polypropylene (atactic, isotactic and syndiotactic) can be produced in a specific manner. A cationic alkyl-metallocene complex is the active species in the polymerization. MAO helps in alkylating the metallocene (if the metallocene is a metallocene dihalide like Cp_2ZrCl_2) and also in assisting the removal of a CH_3^-. Further it stabilizes the cationic center and also possibly scavenges impurities.

4. Polymers prepared by metallocene catalysts have special properties. Metallocene-polyethylene is an excellent film wrap and has excellent barrier properties to oxygen and moisture. Special polymers such as linear low-density polyethylene (LLDPE) which is a copolymer of ethylene and 1-butene (or 4-methyl-1-pentene) can be manufactured readily by the use of metallocene catalysts. LLDPE is a high strength film-forming polymer.

5. A new generation of Ziegler-Natta catalysts does away with titanium or zirconium complexes altogether. Instead, conventional transition-metal complexes involving Fe(II) or Ni(II) are being used.

2.7 Olefin Metathesis Polymerization

Olefin metathesis is an important polymerization reaction that is particularly useful for cyclic compounds containing a double bond. Before going through this let us look at the olefin metathesis reaction. The latter can be represented by the following simple reaction [1-5] (see Eq. 2.28).

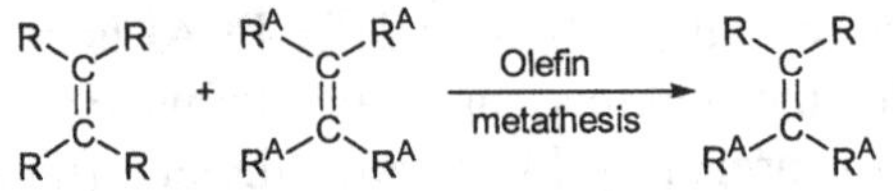

$$
\tag{2.28}
$$

The product in this reaction appears to arise as a result of splitting the reacting olefins exactly into two equal halves and allowing them to join again. This reaction is reversible and several versions are now known.

1. Cross-metathesis (see Eq. 2.29).

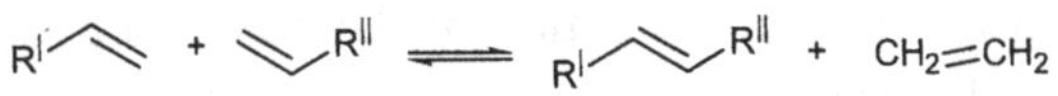

$$
\tag{2.29}
$$

2. Ring-closing metathesis (see Eq. 2.30).

$$
\tag{2.30}
$$

3. Ring-opening metathesis (ROMP) (see Eq. 2.31).

$$
\tag{2.31}
$$

4. Ring-opening metathesis polymerization (see Eq. 2.32).

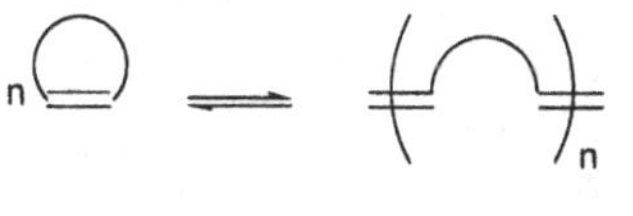

$$n \bigcirc \quad \rightleftharpoons \quad \left(\bigcirc \right)_n \qquad (2.32)$$

5. Acyclic diene metathesis polymerization (see Eq. 2.33)

$$\bigwedge \quad \rightleftharpoons \quad \left(\bigwedge \right)_n \quad + \quad nH_2C=CH_2 \qquad (2.33)$$

The use of ROMP has been widely applied. Thus, cyclopentene can be polymerized by olefin metathesis to a linear polymer. In this reaction the cyclic hydrocarbon is opened up by the catalyst and joined together in a linear fashion (see Eq. 2.34). The catalysts that perform this operation are called ring-opening metathesis catalysts.

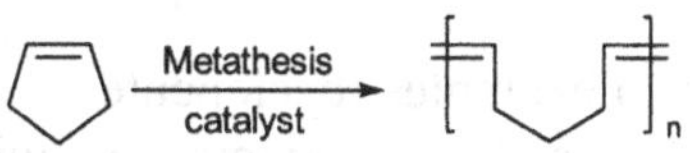

$$\bigcirc \xrightarrow[\text{catalyst}]{\text{Metathesis}} \left(\bigvee \right)_n \qquad (2.34)$$

Representative examples of the recent generation of metathesis catalysts are shown below (Fig. 2.29). Notice that these transition-metal derivatives contain an M=C bond. Such compounds are known as metal carbenes [18-20].

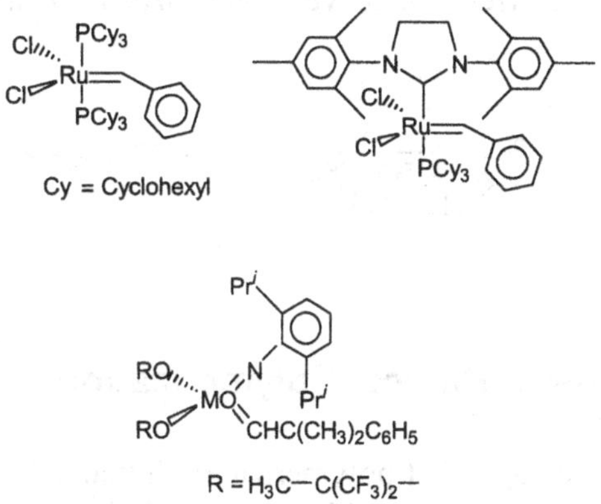

Cy = Cyclohexyl

$$R = H_3C-C(CF_3)_2-$$

Fig. 2.29. Metal carbenes that can be used as metathesis catalysts

The mechanism of the olefin metathesis polymerization involves (a) an initial 2+2 cyclo-addition of a metal carbene with the cyclic olefin. This leads to the formation of a four-membered metallacyclobutane. The next steps (b) (c) and (d) show the polymer propagation reactions (Fig. 2.30).

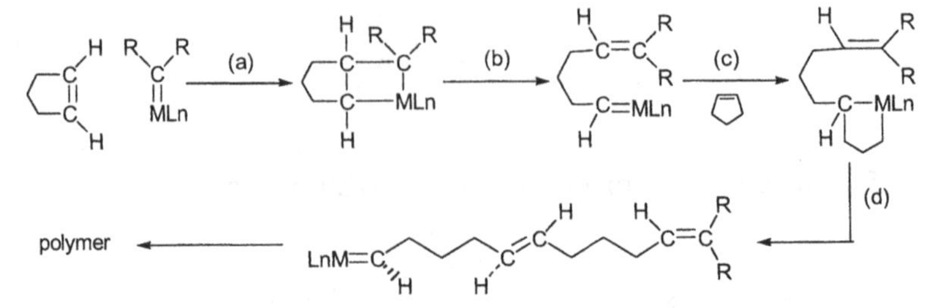

Fig. 2.30. Mechanism of ring-opening metathesis polymerization

Norbornene can also be polymerized by metathesis catalysts to polynorbornene (see Eq. 2.35).

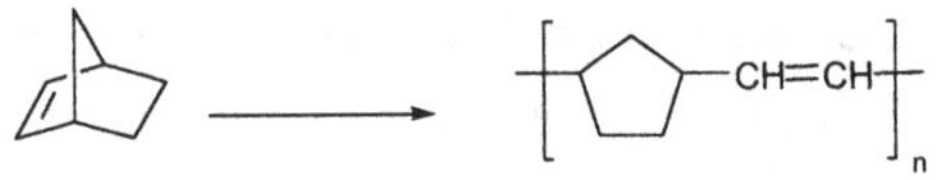

$$(2.35)$$

As can be seen in the two examples (cyclopentene and norbornene) polymerization of cyclic olefins leads to polymers containing double bonds at periodic intervals in their backbone. An interesting application of the metathesis polymerization is the preparation of *trans*-polyacetylene [1]. Metathesis of the tricylic monomer leads to the ring-opening of the cyclobutene ring to afford a polymeric derivative. Heating this polymer eliminates bis-trifluoromethylbenzene to leave behind *trans*-polyacetylene (see Eq. 2.36).

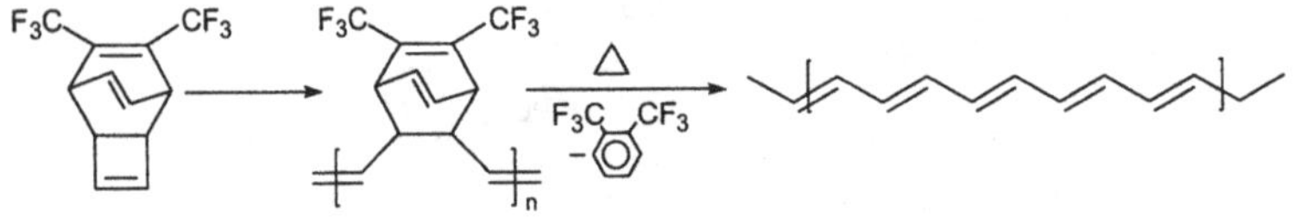

$$(2.36)$$

2.8 Atom Transfer Radical Polymerization

We have seen previously that polymerization initiated by free-radicals suffers from some disadvantages. Mainly, the chain-length cannot be controlled and branching occurs. Some of these disadvantages are overcome in newer methods of radical polymerization. An important new development in this regard is the atom transfer radical polymerization (ATRP) [29-30]. In this process all the chains are initiated essentially at the same time (at the point of catalyst injection) and all the chains grow at the same rate until the monomer is consumed. The important principles of the atom transfer polymerization process are illustrated by the following sequence of reactions in the bulk polymerization of styrene at 130 °C using 1-phenyl ethyl chloride as an initiator and $[Cu(bipy)_2][Cl]^-$ (bipy = 2, 2'-bipyridine) as a chlorine atom transfer promoter (see Eq. 2.37).

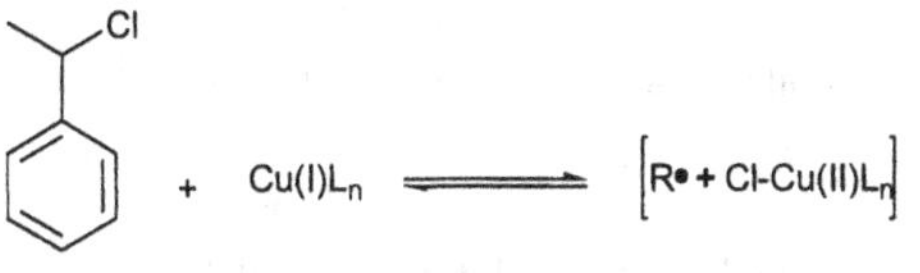

(2.37)

The catalyst being a Cu(I) complex, abstracts a chlorine from the initiator; in this process the metal gets oxidized to Cu(II) generating a free radical R•. The latter can initiate polymerization of a styrene monomer (see Eqs. 2.37-2.38).

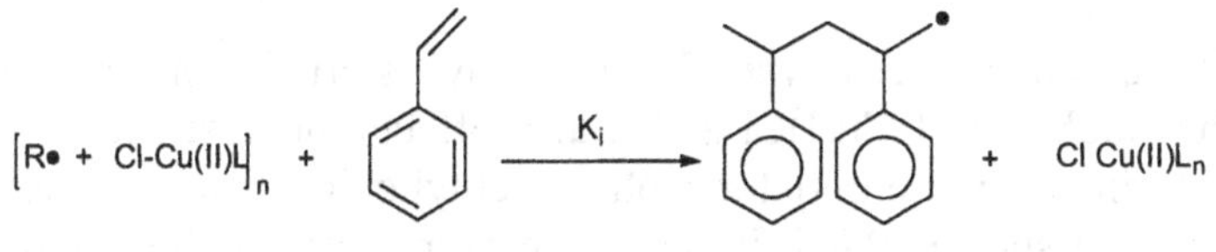

(2.38)

Chlorine transfer can occur to the growing polymer radical (see Eq. 2.39).

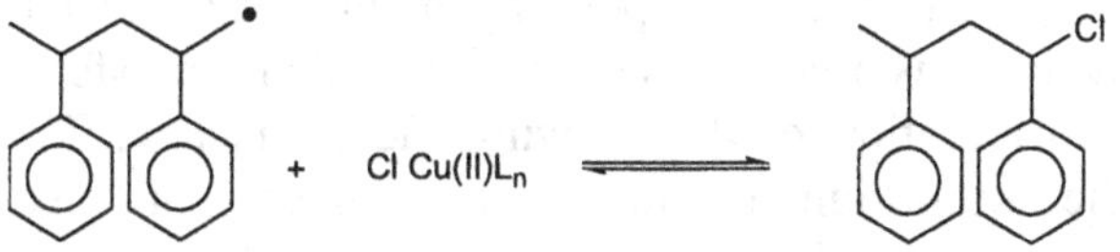

(2.39)

The propagation of the reaction can occur in a similar manner (see Eq. 2.40).

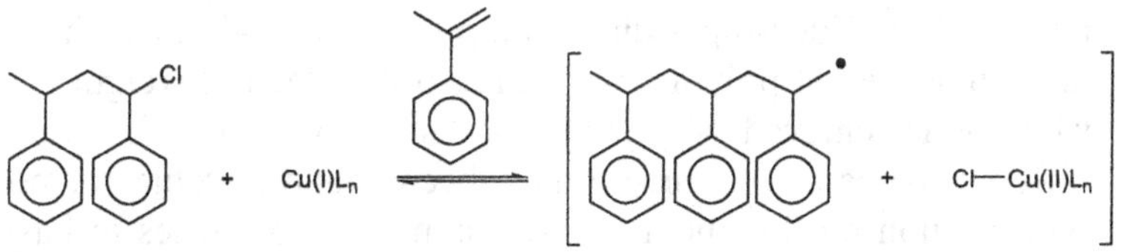

(2.40)

The whole process can be schematically represented as shown in Eqs. 2.41-2.42.

$$R\!-\!Cl \;+\; Cu(I)L_n \;\rightleftharpoons\; \left[R\bullet \;+\; Cl\!-\!Cu(II)L_n\right] \qquad (2.41)$$

$$\downarrow M$$

$$R\!-\!M\!-\!Cl \;+\; Cu(I)L_n \;\rightleftharpoons\; \left[R\!-\!M\bullet \;+\; Cl\!-\!Cu(II)L_x\right]$$

$$R\!-\!Cl \;+\; Cu(I)L_n \;\underset{+M}{\rightleftharpoons}\; P_i\bullet \;+\; Cl\!-\!Cu(II)L_n \qquad (2.42)$$

The following are the key points of the mechanism of atom transfer radical polymerization.

1. The catalyst shuttles between two oxidation states viz., Cu(I) and Cu(II).
2. The bidentate ligand viz., 2,2′-bipyridyl is crucial in increasing the solubility of Cu(I)Cl and it also affects the abstraction of a chlorine from the initiator 1-phenyl ethyl chloride as well as the dormant species P_i-Cl. It is not necessary that this ligand alone be used. Other ligands that can perform in a similar manner are also useful.
3. The alkyl radical R• and the polymer radical P_i are reversible with respect to the corresponding halides R-Cl and P_i-Cl, respectively. This process also involves an atom transfer reaction.
4. The concentration of the growing radical is low. This means that the equilibrium is towards the halides. This also implies that the bimolecular reactions between radicals are low and therefore termination reactions are minimized. This virtually means that a living polymerization is achieved.
5. Thus, in the polymerization of styrene with 1-phenylethyl chloride and a CuCl-bipyridine complex a well-defined high-molecular-weight polymer with a narrow-molecular-weight distribution is achieved. Since the first discovery of ATRP, several different types of monomers have been found suitable for polymerization by this method and this method promises to have important new applications in the future.

2.9 Ring-opening Polymerization

Several cyclic organic molecules can be polymerized by the ring-opening polymerization method [1-5]. Although the metathesis polymerization that we discussed earlier also is a ring-opening polymerization, it was a special class of ring opening that involved a cyclic compound containing a double bond. Other cyclic compounds such as cyclic ethers, lactones, lactams and cyclic imines undergo ring-opening polymerization. A few examples are shown below.

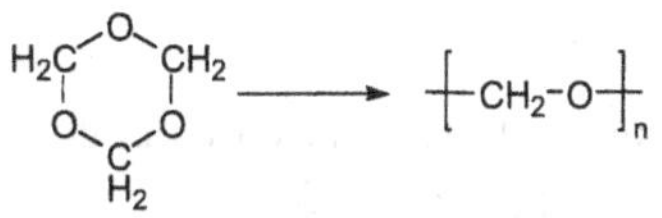

$$H_2C \underset{O-\underset{H_2}{C}-O}{\overset{O}{\diagdown}} CH_2 \longrightarrow \left[CH_2-O \right]_n \qquad (2.43)$$

Trioxane can be polymerized into poly(formaldehyde) by the action of Lewis acids (see Eq. 2.43). Notice that in this method of polymerization there is no change in the chemical composition as the monomer gets converted into the polymer. Caprolactam can be polymerized to nylon-6 by the ring-opening method (see Eq. 2.44).

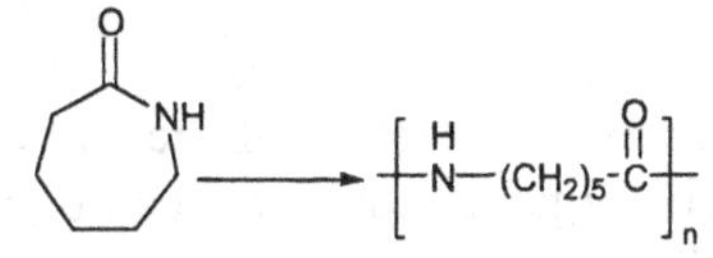

$$\overset{O}{\underset{NH}{\diagdown}} \longrightarrow \left[\underset{H}{N}-(CH_2)_5-\overset{O}{\overset{\|}{C}} \right]_n \qquad (2.44)$$

Caprolactam   Nylon-6

Three-membered rings such as epoxides and aziridines can also be polymerized to their polymeric analogues. The release of ring-strain is an important factor in the polymerization of the three-membered rings. Thus, ethylene oxide can be polymerized to poly(ethylene oxide) by anionic or cationic initiators (see Eq. 2.45).

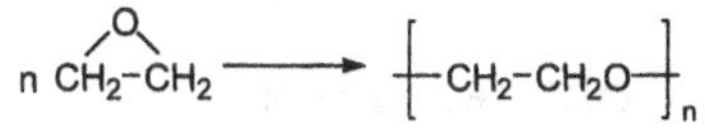

$$n \ CH_2-CH_2 \overset{O}{\diagup\diagdown} \longrightarrow \left[CH_2-CH_2O \right]_n \qquad (2.45)$$

Similarly, ethylene imine polymerizes quite readily in the presence of cationic initiators (see Eq. 2.46).

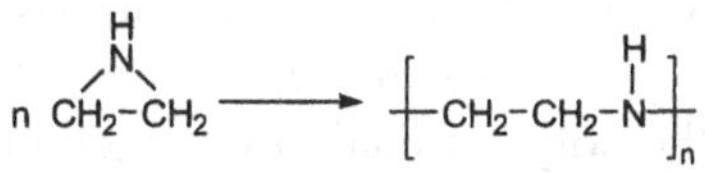

$$n \ CH_2-CH_2 \overset{H}{\underset{N}{\diagup\diagdown}} \longrightarrow \left[CH_2-CH_2-\underset{H}{N} \right]_n \qquad (2.46)$$

Although the mechanism of the ring-opening polymerization would vary depending on the monomer in question it is clear that a reactive species has to be generated which weakens a bond in the ring leading to its scission. This is illustrated by the acid-catalyzed polymerization of trioxane. The first step is the protonation of an oxygen followed by the C-O bond cleavage (see Eq. 2.47) [1, 2].

$$(2.47)$$

Subsequent chain propagation reactions can occur by the electrophilic attack of the reactive carbocation (see Eq. 2.48).

$$(2.48)$$

The chain can terminate by the reaction of the carbocation with an anion. Although several cyclic organic monomers can be polymerized, many more are quite resistant. The most prominent organic rings that cannot be polymerized are benzene and cyclohexane.

To summarize, several cyclic organic compounds can be polymerized by the ring-opening polymerization method. As we will see in the subsequent parts of this book this method is the main route for the preparation of inorganic polymers.

2.10 Group-Transfer Polymerization

Group-transfer polymerization is being thought of as a convenient way for preparing living polymers [1, 2]. However, because of its very nature it cannot be applied to many monomeric systems and has a restricted utility. This polymerization reaction involves the transfer of a trimethylsilyl group from the growing end of the polymer chain to an incoming monomer. Usually this polymerization is carried out on vinyl monomers that contain a carbonyl function. The best examples of such monomers are acrylate esters. The reaction is initiated by a silyl ketene acetal which is activated by a reagent such as F⁻ (see Eq. 2.49). Thus, the addition of HF_2^- to the silyl ketene acetal is believed to form a hypervalent penta-coordinate silicon center. The electrophilicity of the silicon allows the incoming methylmethacrylate to bind to it through the carbonyl oxygen. Subsequently, the trimethylsilyl group is transferred to the next monomer. This process leads to the formation of a linear polymer. Since the chains are all initiated at the same time, the polydispersity index for the polymers prepared by group-transfer polymerization is close to one. Living polymers can be prepared by this method. This reaction can be done at room temperature in contrast to low temperatures that have to be employed for anionic polymerization. Also this polymerization is not sensitive to air (oxygen). However, the polymerization has to be carried out under moisture-free conditions. Another limitation is that this method seems to be mainly applicable to acrylate monomers.

(2.49)

Hypervalent silicon
intermediate

Growing polymer chain

2.11 Condensation Polymers

In all the preceding polymerization methods we have seen how to utilize the double bond in an unsaturated organic compound to link many molecules together into a polymeric chain. Also, in all of these processes the polymer was produced starting from a single monomer. In contrast in this section we will look at polymers that are prepared from the reaction of two *difunctional monomers* with each other. In all the polymerization reactions that we have seen so far there was no *side-product* formation. For example, ethylene was converted into polyethylene; acrylonitrile was converted into polyacrylonitrile and so on. During this conversion the *entire* structural unit of the monomer was incorporated into the polymer without any side-product formation. However, in the preparation of condensation polymers a small molecule (such as water or methanol) is eliminated as the side-product. Another important difference is that condensation polymerization is usually a step-growth polymerization. This means that the polymerization proceeds in a series of steps. To make this point clear let us recall the polymerization of ethylene by the free-radical method. In the free-radical process the polymerization of various chains are initiated by the initiator. At any stage of polymerization, if one analyzes the polymerization mixture one will find a mixture of high polymer along with monomer. No oligomers are found. In contrast, in condensation polymers, the monomers disappear in the initial stage of the reaction itself. However, polymer formation occurs only at the very end. Thus, the molecular weights of the growing chains increase very slowly during most of the course of the reaction, but rapidly increase towards the end. Another feature of the condensation polymerization is that the purity of the monomers is very crucial. In order for high-molecular-weight polymers to be obtained the reacting monomers should be extremely pure. The mechanism of condensation polymerization is illustrated in the next section using polyesters as an example.

2.11.1 Polyesters

Polyesters, as the name implies, are polymers that contain *ester* linkages. As an example of a polyester let us look at poly(ethylene terephthalate). The repeat unit of poly(ethylene terephthalate) is shown in Eq. 2.50.

(2.50)

ester linkage

Before we discuss the preparation of poly(ethylene terephthalate) let us look at the synthesis of simple esters. This will help us to understand polyesters in particular and condensation polymers in general.

Let us consider the reaction of benzoic acid with ethanol. Both are *mono functional*. This reaction leads to the formation of a simple ester. During this reaction water is the side-product (see Eq. 2.51).

(2.51)

carboxylic alcohol   ester + H_2O
acid

Instead of ethanol, let us now use ethylene glycol. The latter is a difunctional reagent containing *two* hydroxyl groups. This reaction leads to the formation of a diester (see Eq. 2.52). Similarly, let us look at the reaction of a dicarboxylic acid with ethanol (see Eq. 2.53). The former is a difunctional reagent containing two carboxylic acid groups.

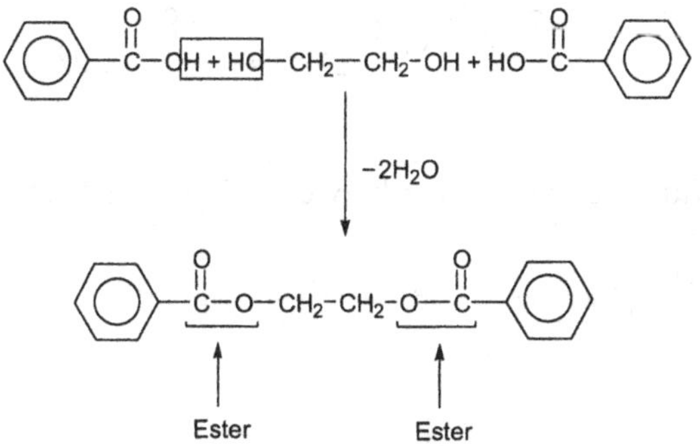

(2.52)

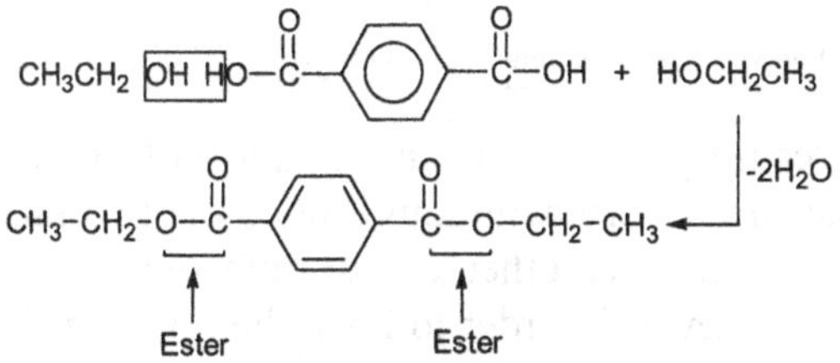

$$-2H_2O$$

Ester   Ester

(2.53)

$$-2H_2O$$

Ester   Ester

Thus, the reaction of ethylene glycol (a diol) with two equivalents of benzoic acid (a mono carboxylic acid) or the reaction of terephthalic acid (a dicarboxylic acid) with ethanol (a mono hydroxy compound) lead to the formation of diesters.

Let us now see what happens if a diol reacts with a dicarboxylic acid (see Eq. 2.54).

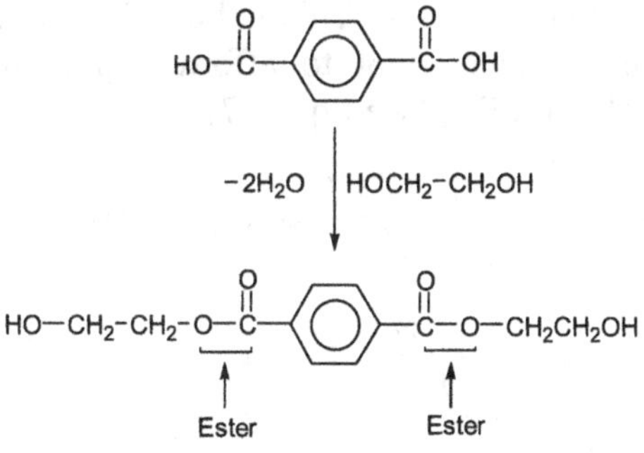

(2.54)

$$-2H_2O \quad | \quad HOCH_2-CH_2OH$$

Ester   Ester

A

Importantly, unlike the diester we have seen earlier, **A** still has two functional groups (-OH groups) for further reaction. These can react now with a dicarboxylic acid to generate a tetra-ester, which has two carboxylic acids at the terminal ends. Thus, alternately the diol and the dicarboxylic acids can react to gradually afford a long-chain polymer (see Eq. 2.55).

$$\text{HO-C} \underset{O}{\overset{O}{\parallel}} \text{-O-CH}_2\text{CH}_2\text{-O-C} \underset{O}{\overset{O}{\parallel}} \text{-C-O-CH}_2\text{CH}_2\text{-O-C} \underset{O}{\overset{O}{\parallel}} \text{-C-OH} \qquad (2.55)$$

Obviously reactions can also occur between oligomers (see Eq. 2.56).

$$\text{HO-C} \underset{O}{\overset{O}{\parallel}} \boxed{\text{Tetramer}} \text{-C-OH} + \text{HO} \boxed{\text{Tetramer}} \text{-OH} \qquad (2.56)$$

$$\downarrow$$

$$\text{HO} \boxed{\text{Tetramer}} \text{-C} \boxed{\text{Tetramer}} \text{-C-O} \boxed{\text{Tetramer}} \text{-OH}$$

Thus, it can be readily seen that in this method of polymerization the monomers disappear in the very beginning. Since the two kinds of difunctional monomers react with each other almost immediately there will be no trace of monomers. However, in order to form the polymer the short-chain oligomers that are formed have to condense with each other to form even longer chains. This process leads to the presence of several medium-molecular-weight oligomers up to the very end. Condensation of such oligomers leads to the formation of high polymers and to the increase in molecular weight during the very end of the reaction.

In practice, one of the ways poly(ethylene terephthalate) is prepared is as follows. First a diester is formed as a result of the reaction of terephthalic acid with ethylene glycol. This is heated at about 270 °C resulting in the loss of ethylene glyol to afford high-molecular-weight poly(ethylene terephthalate) (PET) (see Eq. 2.57).

$$\text{HO-H}_2\text{C-H}_2\text{C-O-C} \underset{O}{\overset{O}{\parallel}} \text{-C-O-CH}_2\text{CH}_2\text{-OH} \qquad (2.57)$$

$$\downarrow \quad 270°C$$

$$\left[\text{O-C} \underset{O}{\overset{O}{\parallel}} \text{-C-O-CH}_2\text{CH}_2 \right]_n$$

Another polyester similar to PET is poly(ethylene naphthalate) or PEN (see Eq. 2.58).

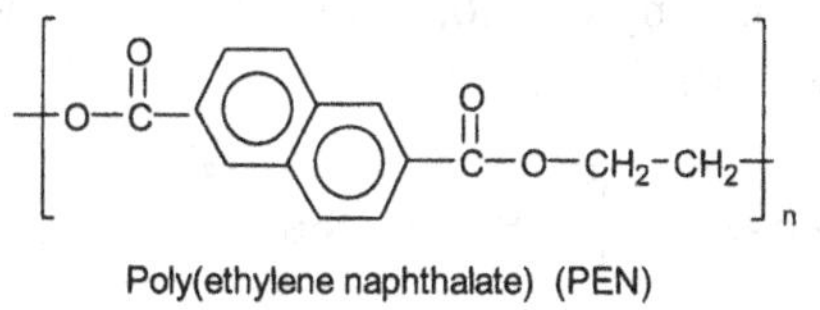

(2.58)

Poly(ethylene naphthalate) (PEN)

The monomers used in the preparation of PEN are ethylene glycol and naphthalene dicarboxylic acid (Fig. 2.31).

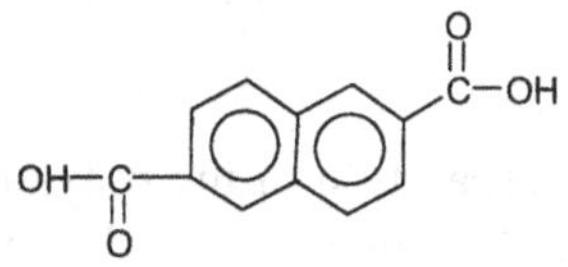

Fig. 2.31. Naphthalene dicarboxylic acid

A variety of polyesters can be prepared by varying the diols and dicarboxylic acids. This principle is quite general and can be applied to any system of difunctional reagents that can react with each other. Some of the other important types of condensation polymers are discussed in the following sections.

2.11.2 Polycarbonates

Polycarbonates are the polyesters of carbonic acid. Thus, the reaction of bisphenol with phosgene ($COCl_2$) or diphenyl carbonate can lead to polycarbonate (see Eq. 2.59).

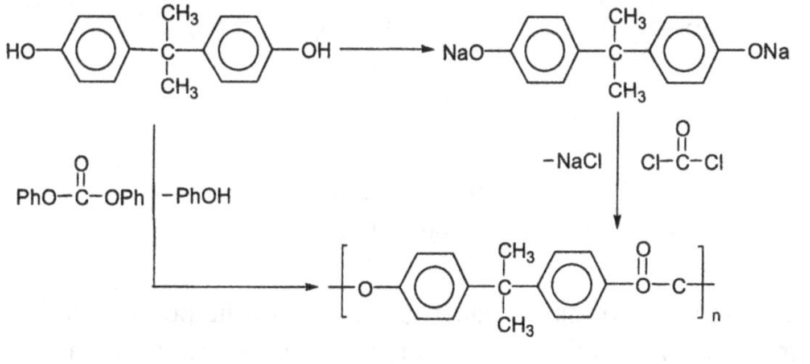

(2.59)

2.11.3 Polyamides

In general polyamides are prepared by a melt polymerization. For example, the reaction of hexamethylene diamine and adipic acid leads to the formation of a salt. The latter is heated at high temperatures (about 220 °C) in a sealed vessel to afford nylon-6,6. Notice the presence of the amide linkage in nylon-6,6 (see Eqs. 2.60, 2.61).

$$\left[O-\overset{O}{\overset{\|}{C}}-(CH_2)_4-\overset{O}{\overset{\|}{C}}-O \right]^{2-} \left[H_3N(CH_2)_6NH_3 \right]^{2+} \tag{2.60}$$

Salt

$$Salt \xrightarrow[-H_2O]{\Delta} \left[\overset{O}{\overset{\|}{C}}-(CH_2)_4-\overset{O}{\overset{\|}{C}}-\overset{H}{\overset{|}{N}}-(CH_2)_6-NH \right]_n \tag{2.61}$$

Nylon-6,6

Another route for the preparation of the polyamides is the interfacial polymerization method. In this method the diamine is dissolved in water (which usually also contains a base such as potassium hydroxide for scavenging the HCl formed in the reaction). The diacid chloride is dissolved in an organic solvent such as dichloromethane or tetrachloroethylene. These two solutions are brought in contact with each other. The polymer is formed at the interface of the two immiscible solvent systems. An example of this polymerization is shown in Eq. 2.62.

$$H_2N-(CH_2)_6-NH_2 + Cl-\overset{O}{\overset{\|}{C}}-(CH_2)_8-\overset{O}{\overset{\|}{C}}-Cl \xrightarrow[\text{polymerization}]{\text{Interfacial}} \left[NH-(CH_2)_6-\overset{H}{\overset{|}{N}}-\overset{O}{\overset{\|}{C}}-(CH_2)_8-\overset{O}{\overset{\|}{C}} \right]_n \tag{2.62}$$

A polyamide, where both the diacid and the diamine contain aromatic units is *Kevlar*. Thus, the dicarboxylic acid is terephthalic acid while the diamine is 1,4-phenylene diamine (see Eq. 2.63).

$$HO-\overset{O}{\overset{\|}{C}}-\text{⬡}-\overset{O}{\overset{\|}{C}}-OH \quad + \quad H_2N-\text{⬡}-NH_2 \tag{2.63}$$

$$\left[HN-\overset{O}{\overset{\|}{C}}-\text{⬡}-\overset{O}{\overset{\|}{C}}-NH-\text{⬡}-NH \right]$$

In Kevlar the rigid aromatic spacer group allows the polymer to be present in an *all-trans* conformation. Such a conformation imparts great strength for the polymer fiber. Further, intermolecular hydrogen bonds make the polymer even more crystalline and rigid and make these polymer fibers strong (Fig. 2.32).

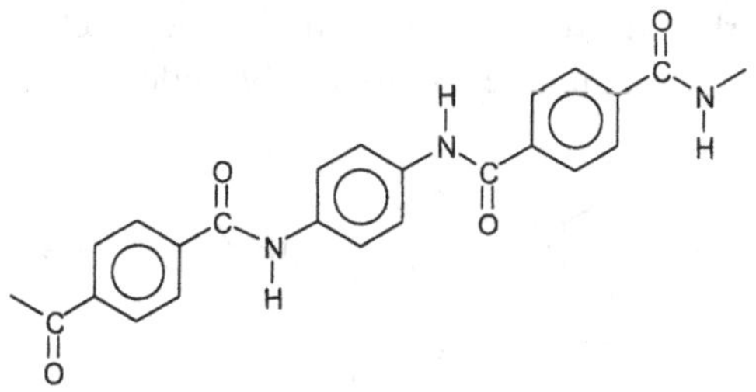

Fig. 2.32. The *all-trans* conformation of Kevlar

A variation of Kevlar is Nomex. In this polyamide the aromatic dicar-boxylic acid is slightly different. Instead of terephthalic acid which con-tains the carboxylic acid functional groups *para* to each other the prepara-tion of Nomex uses a dicarboxylic acid where the two –COOH units are *meta* with respect to each other (Fig. 2.33).

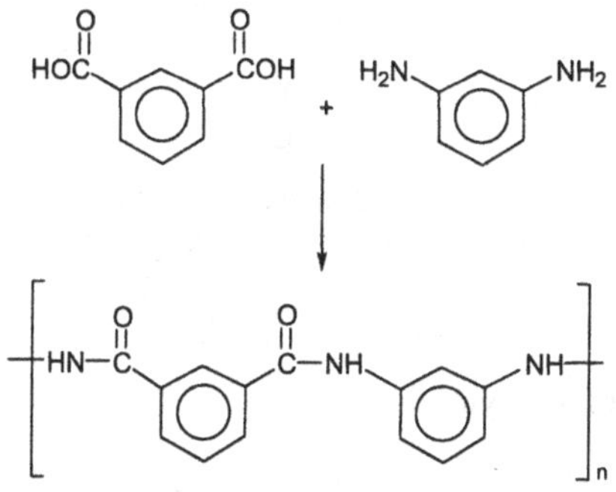

Fig. 2.33. Synthesis of Nomex

This change of the monomer leads to a slight change in the crystallinity of the polymer. Thus, it is possible to vary the polymer structure and hence the properties by small structural changes at the monomer level. By the use of such simple principles polymers with widely different properties can be assembled. Thus, Nomex is slightly less crystalline than Kevlar and can be processed more easily.

2.11.4 Polyimides

Polyimides are a group of extremely strong polymers that are also highly heat and chemical resistant. These polymers contain the *imide* functional-ity. We have seen that the amide functionality $[—C(O)NH—]_n$ is present in synthetic polyamides like nylon-6,6 as well as in natural polyamides like proteins. The imide functional group is created when nitrogen is attached to two carbonyl groups. For example, the reaction between a dian-hydride and a diamine can give rise to a polyimide (Fig. 2.34).

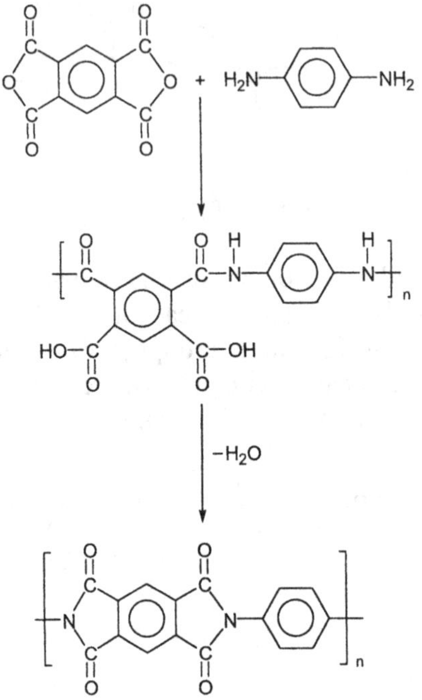

Fig. 2.34. Synthesis of a polyimide

Aromatic polyamides are rigid flat molecules that can stack up on top of each other. Stacking is also aided by the fact that the polyimide chains contain electron-deficient carbonyl groups (acceptors) and electron-rich nitrogens (donors). In successive chains stacking can occur such that a phenylene diimine unit lies on top of the anhydride unit (Fig. 2.35). Because of these structural features aromatic polyimides are extremely crystalline polymers.

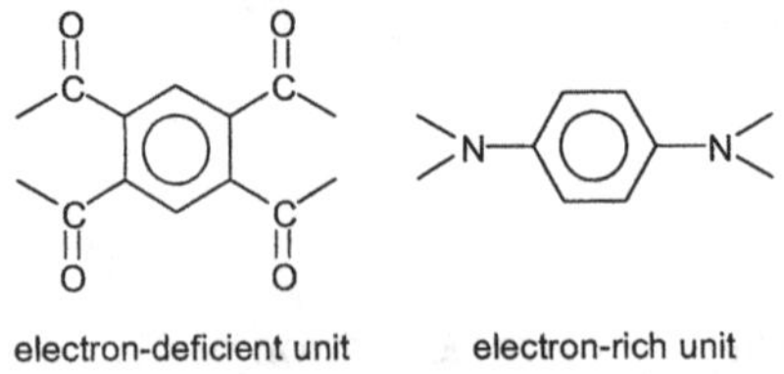

electron-deficient unit   electron-rich unit

Fig. 2.35. Electron-deficient and electron-rich units present in aromatic polyimides

2.11.5 Polysulfones

Polysulfones are a group of polymers that are also heat resistant and thermally stable. A typical repeat unit of a poly sulfone is shown below (Fig. 2.36). These are a special group of polyethers (the ether linkage connects the polymeric chain).

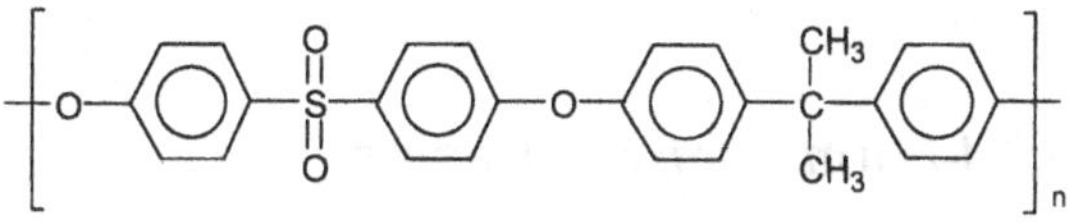

Fig. 2.36. A polysulfone

The polysufone shown above is prepared by the reaction of the disodium salt of bisphenol-A with 4,4'-dichlorodiphenyl sulfone (Fig. 2.37).

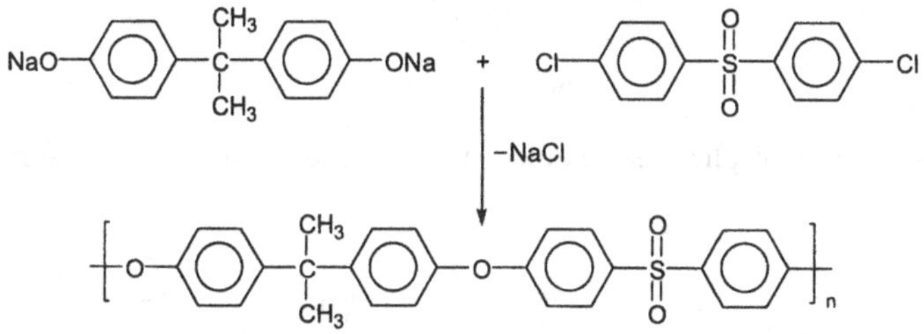

Fig. 2.37. Synthesis of a polysulfone

Other polyethers such as poly(phenylene oxide) are prepared by oxidation reactions using a transition metal complexes containing metal ions such as Cu^+ (see Eq. 2.64).

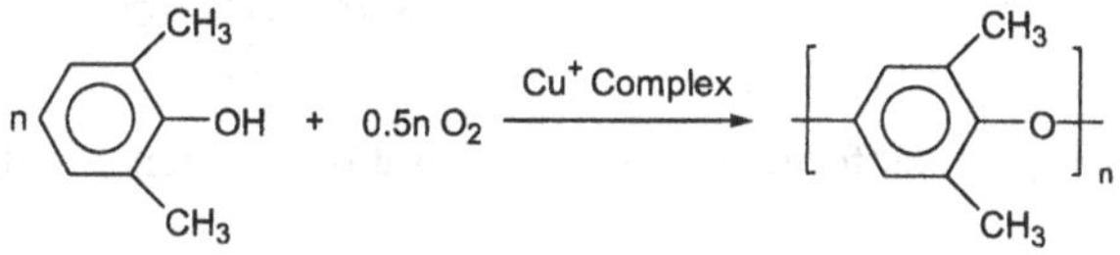

$$(2.64)$$

2.11.6 Thermoset Resins

Thermoset resins are a special family of polymers. There are two stages in the preparation of these polymers. The first is the generation of a *prepolymer*, which is a low-medium-molecular-weight material and which contains many reactive functional groups. The second step is a *curing reaction* of the prepolymer. The curing could be simply a thermal treatment. At a higher temperature the many reactive groups react with each other by extrusion of small molecules such as H_2O or formaldehyde (or both). Alternatively a second chemical reagent is added to react with the reactive groups. Either way, the curing reaction leads to the formation of an intricately networked (crosslinked) material. The final cured polymer is usually thermally very stable. The cured polymer cannot be molded further by thermal treatment. Three typical examples of thermoset resins are given below.

2.11.6.1 Phenol-Formaldehyde Resins

Reaction of a phenol with formaldehyde gives a mixture of compounds containing various methylol phenols (Figs. 2.38 and 2.39).

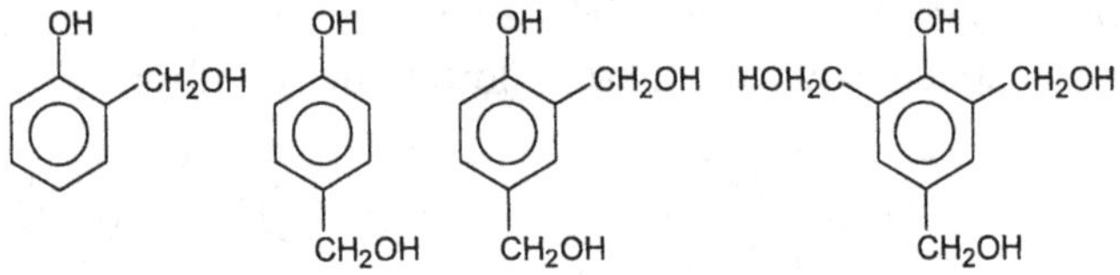

Fig. 2.38. Methylol phenols obtained in the preparation of phenol-formaldehyde resins

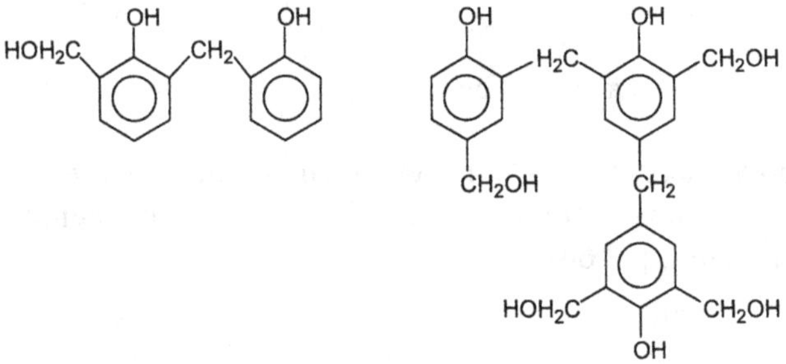

Fig. 2.39. Condensed phenols formed in the condensation of phenol with formaldehyde

The mixture of products shown in Figures 2.38 and 2.39 is called a prepolymer. These products contain a number of reactive sites. Heating these at higher temperatures leads to the formation of highly crosslinked structures. This process of heating the prepolymer, which leads to the final polymer, is called *curing*. The crosslinked structures contain many –CH$_2$– linkages. Because of extensive crosslinking these kinds of polymers are called thermosets. What this means is that after the material is molded into a desirable shape and heated, the resultant highly crosslinked polymer *cannot* be further processed by thermal treatment.

2.11.6.2 Melamine-Formaldehyde Resins

In these resins a multifunctional heterocyclic ring is used as one of the monomers. Thus, melamine, which contains three reactive $-NH_2$ groups, is used as one of the monomers. Condensation of melamine with formaldehyde leads to products that contain the $NH-CH_2OH$ units (Fig. 2.40).

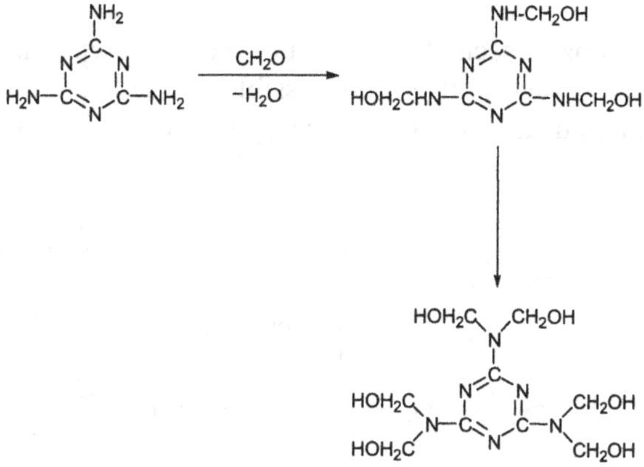

Fig. 2.40. Condensation of melamine with formaldehyde

Crosslink with a -CH$_2$-   Crosslink with a -OCH$_2$-

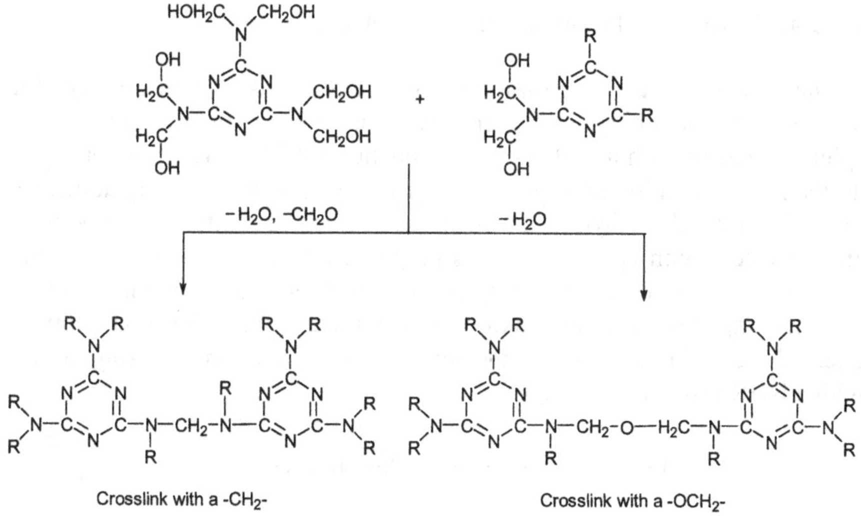

Fig. 2.41. Thermal curing of prepolymers obtained from melamine-formaldehyde condensation reaction

The polyfunctional products shown in Fig. 2.40 are the prepolymers in this instance. Thermal curing of these prepolymers leads to crosslinked products that contain either the CH_2–O linkages or –CH_2– linkages (Fig. 2.41).

2.11.6.3 Epoxy Resins

In the two thermosets seen above the curing of the prepolymer was carried out by a thermal process. It is also possible to use a chemical reagent for the curing reaction. In the case of the epoxy resins this principle is employed.

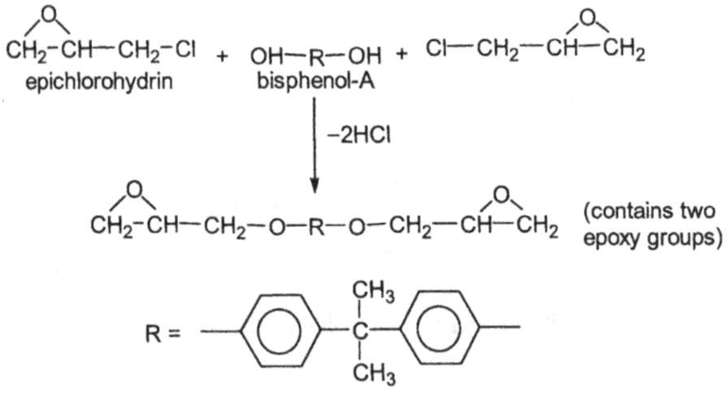

Fig. 2.42. Preparation of epoxy terminated bis-phenol derivatives

The first step in the preparation of the epoxy resins is the assembly of a pre-polymer containing epoxy groups. This is done by the reaction of epichlorohydrin with a diol such as bisphenol-A (Fig. 2.42). Longer chain derivatives containing multiple epoxy groups can be prepared by adjusting the ratio of epichlorohydrin and bisphenol. Thus, a mixture of oligomers and polymers with epoxy groups is prepared by the above reaction. This pre-polymer can be cured rapidly by the reaction with a diamine. Let us see what happens when an epoxide reacts with an amine. A primary amine opens the epoxide to a compound that contains the hydroxyl group along with a NHR group (see Eq. 2.65).

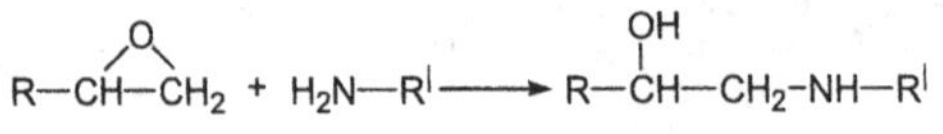

(2.65)

When the prepolymer containing multiple epoxy groups is treated with a diamine, the pre-polymer units are connected (crosslinked) together to generate a thermoset polymer (see Eq. 2.66).

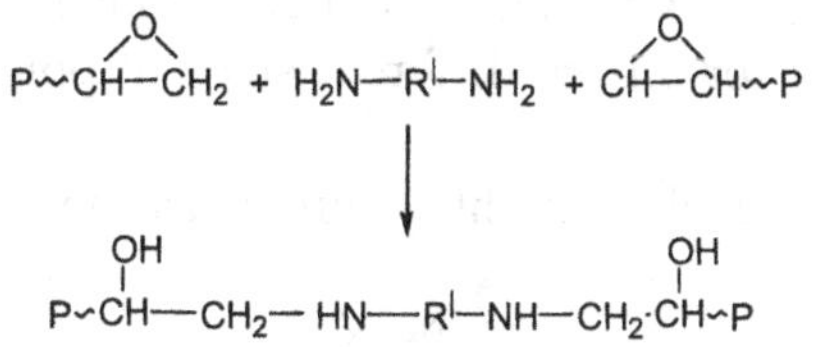

$$P\!\!\sim\!\!\overset{O}{\overset{\diagup\backslash}{CH}}\!\!-\!\!CH_2 \; + \; H_2N\!\!-\!\!R^{\underline{l}}\!\!-\!\!NH_2 \; + \; \overset{O}{\overset{\diagup\backslash}{CH}}\!\!-\!\!CH\!\!\sim\!\!P \qquad (2.66)$$

$$\downarrow$$

$$\underset{\underset{OH}{|}}{P\!\!\sim\!\!CH}\!\!-\!\!CH_2\!\!-\!\!HN\!\!-\!\!R^{\underline{l}}\!\!-\!\!NH\!\!-\!\!CH_2\!\cdot\!\underset{\underset{OH}{|}}{CH}\!\!\sim\!\!P$$

2.12 Some Basic Features About Polymers

We have seen in the above sections a survey of various preparative methods for assembling organic polymers. We will now acquaint ourselves of some basic features that are distinctive of polymers.

2.12.1 Molecular Weights of Polymers

Unlike a small-molecular-weight compound, a polymer has a larger molecular weight. But this is not the only difference. In a sample of a smaller-molecular-weight compound such as benzoic acid, *all* the molecules of benzoic acid will have the *exact* chemical identity. This means that the molecular weight of benzoic acid is a constant quantity. This situation does not apply to a synthetic polymer. For a given polymer the chain-lengths of individual polymer chains will vary from each other (except in naturally occurring polymers such as proteins). Because of this the molecular weights of the polymers will be an *average property*. A distribution of chain lengths and molecular weights is what is found for synthetic polymers. Because of such a distribution we can obtain an average value. Two of the most common molecular weight averages used for polymers are the number average molecular weight M_n and the weight average molecular weight M_w.

The number average molecular weight is the total weight of all the polymer molecules in a sample, divided by the total number of polymer molecules. Thus, the number average molecular weight M_n is defined as shown in Fig. 2.43.

$$M_n \; = \; \frac{\sum\limits_{x=1}^{\infty} N_x \, M_x}{\sum\limits_{x=1}^{\infty} N_x}$$

Fig. 2.43. Definition of the number average molecular weight M_n

In this definition (Fig. 2.43) N_x is the number of moles whose weight is M_x. The summations are for all the different sizes of polymer molecules from $x = 1$ to $x = \infty$.

The weight average molecular weight is defined as shown in Fig. 2.44.

$$M_w = \frac{\sum\limits_{x=1}^{\infty} N_x M_x^2}{\sum\limits_{x=1}^{\infty} N_x M_x}$$

Fig. 2.44. Definition of the number average molecular weight M_w

In general it is seen that M_n is indicative of lower polymer molecular weights, while M_w is indicative of higher polymer molecular weights. It is for this reason that the quantity M_w/M_n (also known as *polydispersity index* or PDI) tells about the range of molecular weight distribution. If this ratio is one (or even close to one) the polymer is a *monodisperse* polymer. In such a polymer all the polymer chains are initiated into polymerization at the same time and further they all grow at the same rate. This leads to polymer chains of (nearly) equal length. Polymerization by anionic or ATRP methods achieve nearly monodisperse polymers. Other methods of polymerization, however, afford polymers with larger PDI' s.

What should be the minimum molecular weights that polymers should have in order to have useful mechanical properties? It is found that the mechanical strength begins to increase slowly with molecular weight. This increase is rapid up to a critical molecular weight (usually about 5,000-10,000). After this the increase in mechanical properties is quite slow before tapering off [1].

2.12.2 Glass Transition Temperature of Polymers

Glass transition temperature (T_g) is a characteristic temperature of amorphous polymers. Since even crystalline polymers contain amorphous domains, T_g is a characteristic thermal signature of any polymer.

Just as a crystalline solid melts into an isotropic liquid at a particular temperature, if a polymer were to be perfectly crystalline it also would similarly melt. This temperature is the melting temperature or T_m. On the other hand, the thermal behavior of an amorphous polymer is slightly different. As a polymer melt is cooled down continuously, at a critical temperature, the polymer motion ceases completely. This temperature is the glass transition temperature (T_g) of a polymer. Below the T_g the polymers behave like inorganic glasses and become hard and brittle. What is the difference between the melting temperature T_m and glass transition temperature T_g? While T_m is a first-order transition, T_g is a second-order transition. Let us look at T_m first. As we heat any crystalline material (such as a crystalline polymer) at a constant rate its temperature increases. The increase of temperature depends upon the heat capacity of the material. (The heat capacity is the amount of heat required to raise the temperature of *one* gram of the material by one degree Celsius). At the melting temperature itself, all the heat supplied goes into melting the solid and not in increasing the temperature. The amount of heat that is required for melting a solid is called the latent heat of melting. It is required to separate the molecules from each other. It is the energy required to be overcome before the solid is converted into the liquid. Thus, there is a break in the *heat supplied vs temperature increase* phenomenon for a crystalline solid at its T_m. After the melting is over, further supply of heat causes the temperature to rise albeit at a slower rate. This is due to the higher heat capacity of the melt in comparison to the solid.

On the other hand, when an amorphous solid such as an amorphous polymer is heated its temperature increases as before. However, when the T_g is reached there is *no break* in the temperature increase. After T_g the rate of increase of temperature with respect to the heat supplied varies in comparison to the situation below the T_g. This means that the glass transition involves a change in heat capacity but does not involve a latent heat change. In contrast, the T_m involves a change in latent heat as well as heat capacity. The former (T_g) is defined as a second-order transition, while the latter (T_m) a first-order transition.

What is the physical significance of T_g? Above the T_g polymers are soft and pliable, because of torsional or segmental motions. These movements could be either in the backbone or the side chain of the polymer.

A polymer with a low T_g has considerable torsional mobility. For example, poly(dimethylsiloxane) $[Si(CH_3)_2O]_n$ has a T_g of -127 $^\circ$C. On the other hand high T_g's mean a stiff and rigid polymer. For example, poly(methylmethacrylate) has a T_g of about 120 $^\circ$C [1].

2.12.3 Elastomers, Fibers and Plastics

Elastomers are a group of polymers that can be stretched to many times their original length. After the force applied for stretching them is removed, these polymers return back to their original shape. This can be represented as shown in Fig. 2.45.

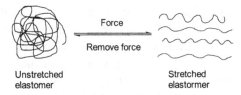

Fig. 2.45. Application of force to stretch elastomers

In general, the elastomer should be lightly crosslinked. These crosslinks prevent the chains from slipping away from each other (when stretched under the influence of a force) (Fig. 2.46).

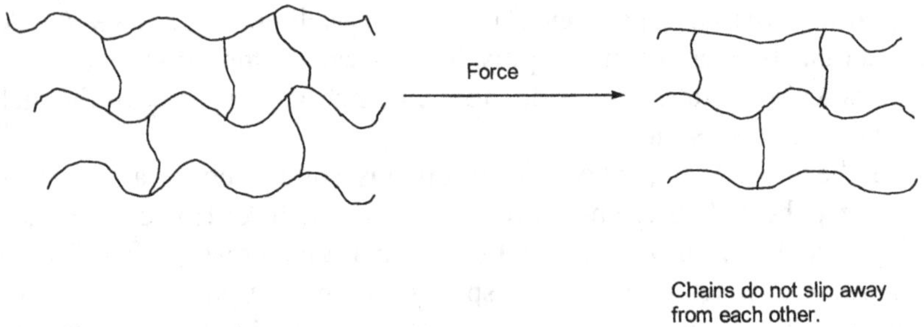

Force ⟶

Chains do not slip away
from each other.

Fig. 2.46. Crosslinking of elastomers

What kind of polymers are elastomers? Polymers that are amorphous and have low T_g's are generally elastomers. Usually the T_g's of amorphous polymers are lower than room temperature. Typical examples of elastomers are polyisoprene, polyisobutylene, polychloroprene, etc (Fig. 2.47).

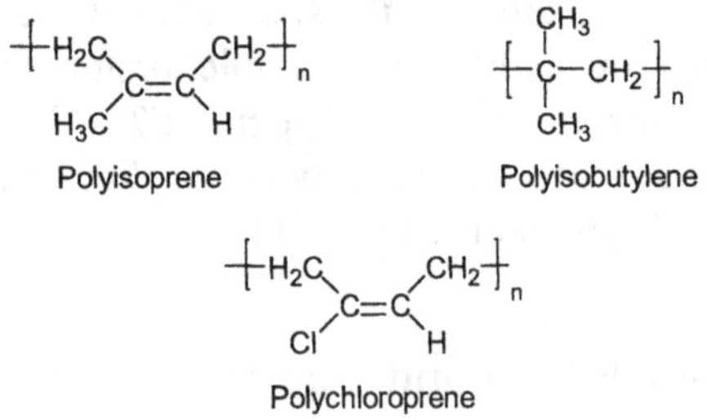

Polyisoprene

Polyisobutylene

Polychloroprene

Fig. 2.47. Examples of elastomers

How is crosslinking achieved in these elastomers? The presence of double bonds in polyisoprene makes the job of crosslinking easier. Heating it with sulfur introduces crosslinks between the chains (Fig. 2.48). Disulfide crosslinks hold different polyisoprene chains together.

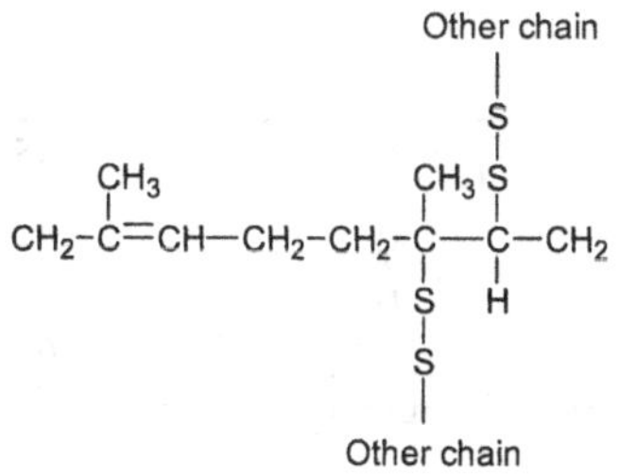

Fig. 2.48. Disulfide crosslinking

Fibers are polymers that are highly crystalline. These polymer chains can be neatly stacked up into a regular arrangement. Usually strong inter-molecular hydrogen bonding helps in this ordered arrangement. Fibers have excellent tensile strength. This means that they can withstand great amounts of strain (any force) applied to deform the polymer, for example, a force to stretch the polymers. However, the tensile strength of fibers is usually in one direction viz., in the fiber direction. They do not have such strength perpendicular to the fiber direction.

Which are the polymers that act as fibers? Polyamides (nylon 6,6, nylon 6, Kevlar), polyesters, polyacrylonitrile, cellulose, etc. are fibers. Highly crystalline polyethylene and polypropylene can also be used as fibers.

Plastics are polymers that are flexible. Their properties are intermediate between those of elastomers and fibers. The glass transition temperatures of many plastics lie above room temperature. Such plastics are processed above their T_g. They are pliable (above their T_g) and so they can be molded to any desired shape. Below their T_g they are hard. This means that the shape is retained. A number of polymers can be used as plastics. These include polymers such as polyethylene, polypropylene etc.

Table 2.5 summarizes the thermal data of some common organic polymers.

Table 2.5. Thermal data of some common organic polymers

Polymer	Repeat unit	T_g(°C)	T_m(°C)	Comments
Nylon 6,6	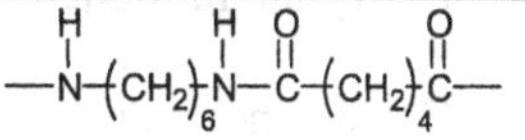	45	267	Fiber
Nylon 6	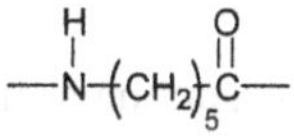	-	233	Fiber
Polyaramide (Nomex)	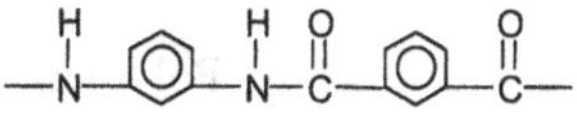	-	390	Fiber

Table 2.5. (contd.)

Polyacrylonitrile	$\left[CH_2-\overset{H}{\underset{\underset{N}{\overset{|||}{C}}}{C}}\right]_n$	85	317	Fiber
Polyethylene terephthalate	$-O(CH_2)_2O-\overset{O}{\overset{||}{C}}-\bigcirc-\overset{O}{\overset{||}{C}}-$	69	270	Thermoplastic fiber
Polypropylene (isotactic)	$\left[CH_2-\overset{H}{\underset{CH_3}{C}}\right]_n$	26	150	Thermoplastic
Polystyrene (isotactic)	$\left[CH_2-\overset{H}{\underset{\bigcirc}{C}}\right]_n$	100	240	Thermoplastic
Polymethyl methacrylate	$\left[CH_2-\overset{CH_3}{\underset{\overset{|}{\underset{OCH_3}{C=O}}}{C}}\right]_n$	105	160	Thermoplastic
Poly(isoprene) (natural rubber)	$\left[\overset{H_3C}{\underset{H_2C}{}}C=C\overset{H}{\underset{CH_2}{}}\right]_n$	-70	36	Elastomer
Polyisobutylene	$\left[CH_2-\overset{CH_3}{\underset{CH_3}{C}}\right]_n$	-70	1.5	Elastomer

References

1. Odian G (2004) Principles of polymerization 4th Ed. Wiley, New York
2. Allcock HR, Lampe F, Mark J (2004) Contemporary polymer chemistry 3rd Ed. Prentice-Hall, Englewood Cliffs
3. Elias, H.-G (1997) An introduction to polymer science. VCH, Weinham
4. Braun D, Cherdron H, Ritter H (2001) Polymer synthesis: theory and practice. Springer-Verlag, Berlin, Heidelberg, New York
5. Stevens MP (1999) Polymer chemistry-an introduction. Oxford University Press, Oxford
6. Mark JE, West R, Allcock HR (1992) Inorganic polymers. Prentice-Hall, Englewood Cliffs
7. Archer RD (2001) Inorganic and organometallic polymers. Wiley-VCH, Weinheim
8. Manners I (2004) Synthetic Metal Containing Polymers. Wiley-VCH, Weinheim
9. Allcock HR (2003) Chemistry and applications of polyphosphazenes. Wiley, Hobolen, New Jersey

10. Jones RG, Ando W, Chojnowski J (eds) (2000) Silicon containing polymers. Kluwer, Dordrecht
11. Clarson SJ, Semlyen JA (eds) (1993) Siloxane polymers. Prentice Hall, Englewood Cliffs
12. Manners I (1996) Angew Chem Int Ed Engl 35:1602
13. Gates DP, Manners I (1997) Dalton Trans 2525
14. Nguyen P, Gomez-Elipe P, Manners I (1999) Chem Rev 99:1515
15. Miller RD, Michl J (1989) Chem Rev 89:1359
16. Neilson RH, Wisian-Neilson P (1988) Chem Rev 88:541
17. DeJaeger R, Gleria M (1998) Prog Polym Sci 23:179
18. Elschenbroich C, Salzer A (1992) Organometallics: a concise introduction 2nd Ed. VCH, Weinheim
19. Crabtree RH (2001) The organometallic chemistry of transition metals 3rd Ed. Wiley, New York
20. Spessard GO, Miessler GL (1997) Organometallic Chemistry. Prentice-Hall, New Jersey
21. Britovsek GJP, Gibson VC, Kimberley, BS, Maddox PJ, McTavish SJ, Solan GA, White AJP,Williams DJ (1998) Chem Commun 849
22. Brintzinger HH, Fischer D, Mülhaupt R., Rieger B, Waymouth RM (1995) Angew Chem Int Ed 34:1143
23. Britovsek GJP, Gibson VC (1999) Angew Chem Int Ed 38:428
24. Fink G, Mühlhaupt R, Brintzinger HH (1995) Ziegler catalysts. Springer-Verlag, Berlin, Heidelberg, New York
25. Kaminsky W, Arndt M (1997) Adv Polym Sci 127:143
26. Scheirs J, Kaminsky W (1999) Metallocene-based polyolefins. Wiley-VCH, Weinheim
27. Togni A, Haltermann RL (1998) Metallocenes. Wiley-VCH, Weinheim
28. Bochmann M (1996) Dalton Trans 255
29. Patten TE, Matyjaszewski K (1999) Acc Chem Res 895.
30. Matyjaszewski K, Xia J (2001) Chem Rev 101:2921

3 Cyclo- and Polyphosphazenes

3.1 Introduction

In the last chapter we have seen that organic polymers could be prepared by a number of ways taking advantage of the rich functional group chemistry of organic molecules. Unfortunately, it is not possible to extend these methodologies completely to inorganic polymers. As we will see in this and the subsequent chapters different types of inorganic polymers require different synthetic strategies. One approach that is common to the preparation of many inorganic polymers is ring-opening polymerization (ROP). This methodology relies on the choice of an appropriate inorganic ring that can be opened to a polymeric material. Obviously, in order for this strategy to succeed suitable inorganic rings have to be discovered and polymerized. In this Chapter we will deal with inorganic polymers known as polyphosphazenes. These polymers contain alternate phosphorus and nitrogen atoms in their backbone (Fig. 3.1).

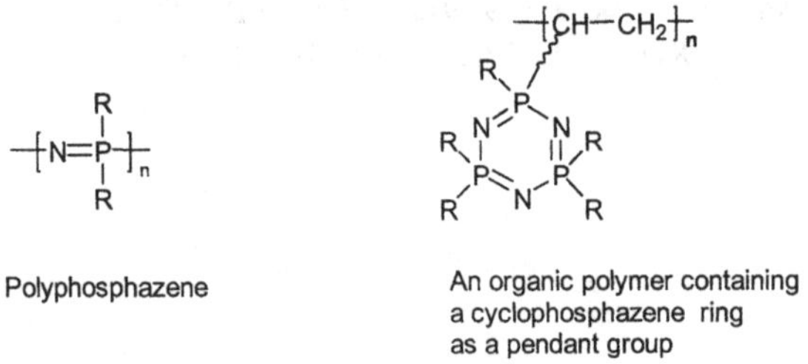

Polyphosphazene

An organic polymer containing
a cyclophosphazene ring
as a pendant group

Fig. 3.1. Polyphosphazene and a cyclophosphazene containing polymer

There are many other polymers that are related to the polyphosphazene family. Thus, there are polymers which are actually made up of an organic backbone but contain a cyclophosphazene ring as a pendant group appended in their side chain. Other polymers that are related to polyphosphazenes are those that contain a third hetero atom in the backbone. Examples of such polymers are shown in Fig. 3.2.

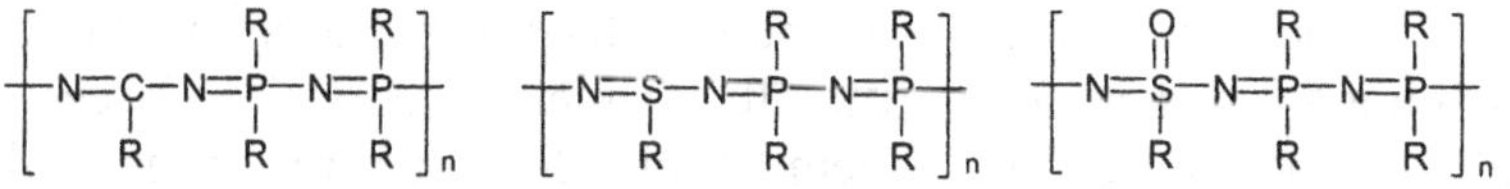

Fig. 3.2. Poly(heterophosphazene)s

Polyphosphazenes have an intimate relationship with their cyclic counterparts. It is essential to obtain an understanding of the structure and reactivity of the cyclic derivatives in order to be able to extend this chemistry to the polymeric analogues. In this chapter, we will therefore, first discuss cyclophosphazenes and then move on to polyphosphazenes. Other polymers related to the polyphosphazene family will be discussed in subsequent chapters.

3.2 Cyclophosphazenes

Cyclophosphazenes are inorganic heterocyclic rings [1-10]. These are among the oldest inorganic rings that have been studied. These were prepared by Liebig and Wöhler as early as 1834. Stokes, an American chemist had also studied these rings in some detail in the 1890's. However, further studies of these intriguing compounds had to await the arrival of powerful spectroscopic methods. The advent of multinuclear NMR and the routine use of single crystal X-ray methods allowed a systematic study of these inorganic rings and consequently a wealth of information is now available for cyclophosphazenes. Currently the research interest in cyclophosphazenes emanates from various reasons as will be seen in the present discussion.

Cyclophosphazenes are made up of a valence unsaturated skeleton containing the $[N=PR_2]$ repeat unit. The ring is composed of alternate phosphorus and nitrogen atoms. Within these rings the phosphorus center is pentavalent and tetracoordinate, while the nitrogen is trivalent and dicoordinate. Thus, each phosphorus is connected to two adjacent ring nitrogen atoms as well as two exocyclic substituents. In contrast, each nitrogen atom is attached to two adjacent phosphorus atoms. The nitrogen atom does not have any exocyclic substituents. The smallest ring that is possible is a four-membered ring and higher rings with an increment of two atoms (phosphorus and nitrogen) are possible. Indeed, cyclophosphazenes form a regular and homologous series. The four-membered ring is not common and stabilization of this ring is possible only by having sterically hindered substituents on phosphorus as is found for $[NP(N\textit{i}Pr)_2]_2$ [2] The most common compounds are those that contain the six- and eight-membered rings. For the series $[NPF_2]_n$, compounds from n = 3 to 40 are known, although not all of them are structurally characterized [8]. The largest structurally characterized ring system is the 24-membered permethyl ring $[NPMe_2]_{12}$. Representative examples of cyclophosphazenes are shown in Figs. 3.3-3.6. It follows from the examples shown in these figures, that cyclophosphazenes containing exocyclic P-N, P-O, P-S, P-C, or even P-M bonds are possible.

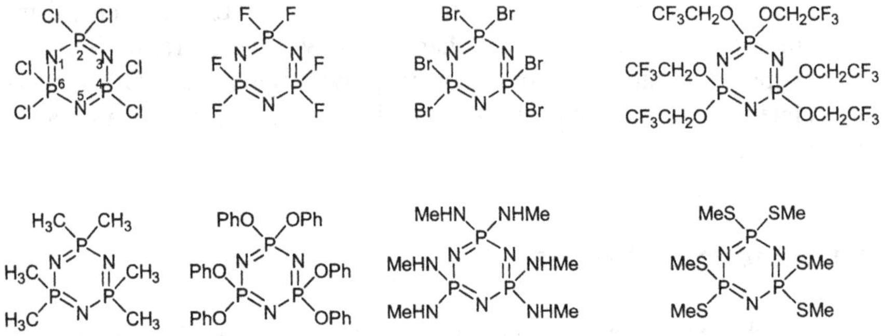

Fig. 3.3. Some examples of cyclotriphosphazenes

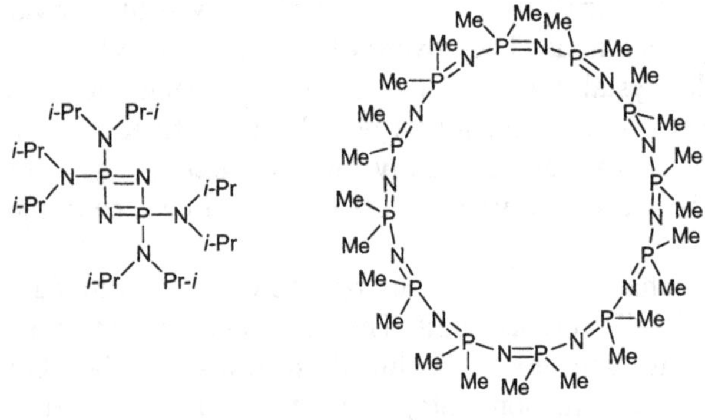

Fig. 3.4. Four- and twentyfour-membered cyclophosphazenes

Further it will be noticed that cyclophosphazenes may be homogeneously (same substituents on all phosphorus centers) or heterogeneously substituted (different substituents). The currently accepted system of nomenclature for these compounds is the one suggested by Shaw [6]. Thus, the six-membered rings are called cyclotriphosphazatrienes and the eight-membered rings are called cyclotetraphosphazatetraenes. The *ene* is to emphasize the valence unsaturated nature of the cyclic skeleton. The numbering scheme starts from nitrogen (the first nitrogen is numbered '1', followed by the phosphorus which is numbered '2' and so on). Thus, $N_3P_3Cl_6$ is called 2,2,4,4,6,6-hexachlorocyclotriphosphazatriene. However, dealing with full names often becomes difficult and in the literature on these compounds it is noticed that simpler ways are devised to address them.

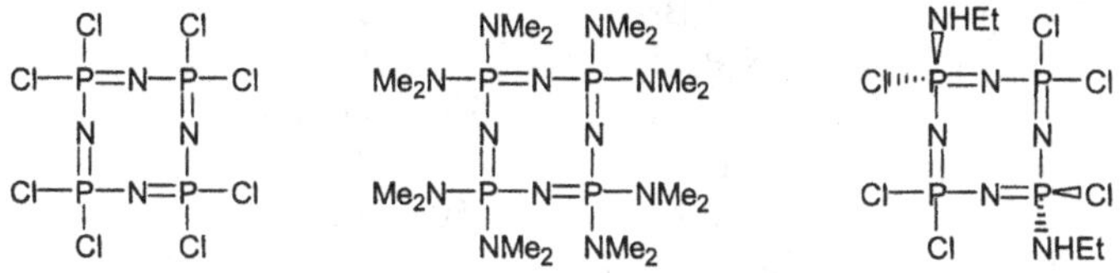

Fig. 3.5. Eight-membered cyclophosphazenes

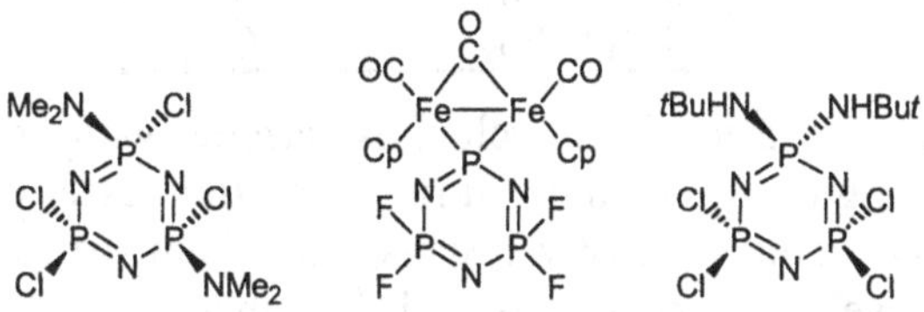

Fig. 3.6. Mixed substituent-containing cyclophosphazenes

Thus, it is much simpler to write the chemical formulas with suitable prefixes to indicate the regio- and stereodisposition of the substituents. For example, 2,4-*nongem-cis*-$N_3P_3Cl_4(NMe_2)_2$ conveys the meaning that the two dimethylamino substituents are present on adjacent phosphorus atoms and are *cis* with respect to each other (Fig. 3.6). Even if one drops '*nongem*' from this prefix the structural information is adequately conveyed. Similarly, the compound 2,2-*gem*-$N_3P_3Cl_4(NHtBu)_2$ can also be written as 2,2-$N_3P_3Cl_4(NHtBu)_2$ or even simply called *gem*-$N_3P_3Cl_4(NHtBu)_2$ (Fig. 3.6). The trivial names of the compounds $N_3P_3Cl_6$ and $N_4P_4Cl_8$ are trimeric chloride and tetrameric chloride, respectively.

Most cyclophosphazenes are relatively stable and have good shelf lives. Some are sensitive to moisture. Even, the chlorocyclophosphazenes, $N_3P_3Cl_6$ and $N_4P_4Cl_8$ can be readily handled in open air (in a well-ventilated hood). Most persubstituted cyclophosphazenes have very low dipole moments and this is reflected in their excellent solubility properties even in nonpolar organic solvents. Table 3.1 summarizes the melting/boiling point data for some of the prominent members of the cyclophosphazene family.

Table 3.1. Melting and boiling points of some cyclophosphazenes

Compound	mp (bp) °C
$N_3P_3Cl_6$	114
$N_3P_3F_6$	27.8 (50.9)
$N_3P_3Br_6$	192
$N_4P_4Cl_8$	124
$N_4P_4F_8$	30.5 (89.7)
$N_5P_5Cl_{10}$	41.3
$N_5P_5F_{10}$	-50.0 (120)
$N_3P_3(NMe_2)_6$	104
$N_4P_4(NMe_2)_8$	238
$N_3P_3(OCH_2CF_3)_6$	38
$N_4P_4(OCH_2CF_3)_8$	65
$N_3P_3(OPh)_6$	111
$N_4P_4(OPh)_8$	86

Although the ring size of the cyclophosphazenes varies considerably we will be examining the six-membered rings and to a lesser extent the eight-membered rings. This is mostly because the chemistry of these rings is much better developed and understood than that of the larger sized ring systems. In the following account we will concern ourselves with the preparation, reactivity and structure of these compounds. Necessarily this description will be brief. Detailed accounts are available in a number of excellent reviews and monographs [1-10].

3.2.1 Preparation of Chlorocyclophosphazenes

Chlorocyclophosphazenes are the most important members of the cyclophosphazene family. These serve as precursors and starting materials for the preparation of a large variety of other cyclophosphazenes. As mentioned before, the four-membered ring $N_2P_2Cl_4$ is *not known*.

The traditional and the still widely used synthesis of chlorocyclophosphazenes consists of the reaction of finely ground ammonium chloride (the source of nitrogen) with phosphorus pentachloride (the source of phosphorus) (see Eq. 3.1). This reaction is carried out in a high boiling chlorinated solvent such as symmetrical tetrachloroethane, $Cl_2CHCHCl_2$ or chlorobenzene. This reaction is quite complex. Ammonium chloride is not soluble in these solvents, while phosphorus pentachloride is. Thus, the overall reaction is heterogeneous in nature. A complex mixture of cyclic and linear products is formed in this reaction. Individual chlorocyclophosphazenes can be separated from each other by various means.

$$nPCl_5 + nNH_4Cl \longrightarrow [NPCl_2]_n + 4n\ HCl \tag{3.1}$$

For example, the cyclic compounds are soluble in petroleum ether and can be extracted into it. The linear polymers are not readily soluble and are left behind. Extraction of the soluble petroleum portion with concentrated H_2SO_4 removes $N_3P_3Cl_6$ because of its greater basicity. However, the higher-membered rings cannot be separated from each other by basicity differences. This has to be accomplished by vacuum distillation.

What is the mechanism of the reaction between NH_4Cl and PCl_5? Although many features of this reaction are not understood with certainty, it is possible to write a probable sequence of events.

The first intermediate that can be isolated in this reaction is the salt $[Cl_3P=N=PCl_3]^+ [PCl_6]^-$. This compound is isolated in about 80% yield in the first hour of the reaction (see Eq. 3.2).

$$3PCl_5 + NH_4Cl \longrightarrow \left[Cl_3P=N=PCl_3\right]^+ \left[PCl_6\right]^- + 4\ HCl \tag{3.2}$$

The initial reaction (see Eq. 3.2) consumes most of the phosphorus pentachloride. The reaction proceeds further by successive chain lengthening events. It is believed that the reagent that causes the increase of chain length is $Cl_3P=NH$. Such a compound is actually a monophosphazene (it is also called a phosphoranimine). Although $Cl_3P=NH$ itself has not been isolated, compounds of this type are now known and have been structurally characterized. The monophosphazene, $Cl_3P=NH$, is presumably formed by the reaction between $[PCl_6]^-$ and $[NH_4]^+$ by the loss of three moles of HCl. The over all chain-lengthening reaction can be shown in a single step (see Eq. 3.3).

$$\left[Cl_3P=N=PCl_3\right]^+\left[PCl_6\right]^- + NH_4Cl \xrightarrow{-4HCl} \left[\begin{array}{ccc} Cl & Cl & Cl \\ | & | & | \\ Cl-P=N-P=N-P-Cl \\ | & | & | \\ Cl & Cl & Cl \end{array} \right]^+ Cl^- \tag{3.3}$$

The above chain which contains five atoms in a series can grow into a larger chain $[Cl_3P=N-P(Cl)_2=N-P(Cl)_2=N-PCl_3]^+$ by further reaction with $Cl_3P=NH$. At this point the chain is sufficiently long that it can close on itself. Thus, ring closure occurs by the loss of $[PCl_4]^+$ and leads to the six-membered ring, $N_3P_3Cl_6$. Eight-, ten- and higher-membered rings are also formed by the ring closure of larger chains. The isolation of short-chain linear phosphazene compounds lends credence to the above mechanism. Why are the cyclic compounds formed in larger yields in the above reaction between ammonium chloride and phosphorus pentachloride at the expense of linear polymeric products? It must be remembered that this reaction is carried out in a solvent, which acts as a diluent. Thus, intermolecular reactions are slower in comparison to intramolecular events. Between the possibilities of chain closure to form a ring versus chain-growth the former is favored because the solvent separates individual molecules from one another. On the contrary, if less dilution is employed or even better, if no solvent is used, linear chain-growth should be favored. As will be shown later this is exactly what happens in the preparation of poly(dichlorophosphazene).

Recently, an alternative method of synthesis of $N_3P_3Cl_6$ has been developed based on the reaction of tris(trimethylsilyl)amine and phosphorus pentachloride [11]. This reaction can be directed by appropriate conditions to result in the formation of cyclo- and linear phosphazenes. The use of the silylamine is guided by the reactivity of the N-Si bond towards chloride and fluoride reagents. The reaction of $N(SiMe_3)_3$ with PCl_5 leads first to the formation of $Cl_3P=NSiMe_3$ by the elimination of Me_3SiCl. In the next step $Cl_3P=NSiMe_3$ reacts with two moles of PCl_5 to afford the intermediate $[Cl_3P=N=PCl_3]^+[PCl_6]^-$ (see Eq. 3.4). Chain propagation occurs by further reaction of the above intermediate with $Cl_3P=NSiMe_3$ (see Eq. 3.5).

$$N(SiMe_3)_3 + PCl_5 \xrightarrow[\text{-2 Me}_3\text{SiCl}]{} Cl_3P=NSiMe_3 \xrightarrow[\text{-Me}_3\text{SiCl}]{2PCl_5} \left[Cl_3P=N=PCl_3\right]\left[PCl_6\right] \quad (3.4)$$

$$\left[Cl_3P=N=PCl_3\right]\left[PCl_6\right] + Me_3SiCl \xrightarrow[\text{-n Me}_3\text{SiCl}]{nCl_3P=NSiMe_3} \left[Cl_3P=N\left(\underset{\underset{Cl}{|}}{\overset{\overset{Cl}{|}}{P}}=N\right)_n PCl_3\right]\left[PCl_6\right] \quad (3.5)$$

The cyclization of the growing chain can occur as a result of the nucleophilic attack of the amine at the phosphonium center followed by an intramolecular cyclization (Fig. 3.7).

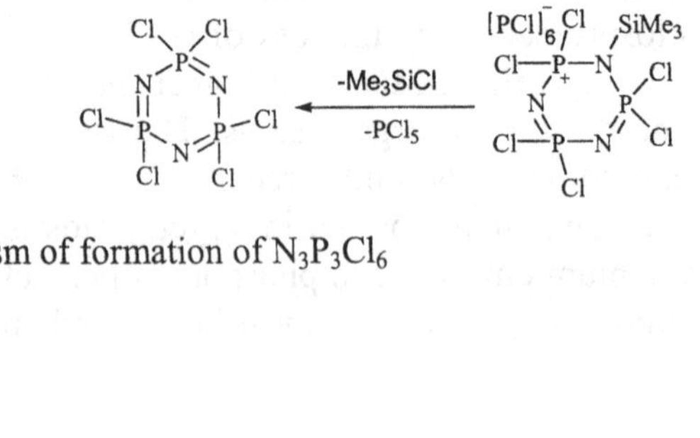

Fig. 3.7. Mechanism of formation of $N_3P_3Cl_6$

3.2.2 Preparation of Fluorocyclophosphazenes and other Halogeno- and Pseudohalogenocyclophosphazenes

The most convenient method of synthesis of fluorocyclophosphazenes consists of metathetical halogen exchange reaction of the chloro analogues effected by an appropriate fluorinating agent (see Eq. 3.6). Although many fluorinating agents are effective the most common and convenient reagents are sodium or potassium fluoride. Usually a polar solvent such as acetonitrile is used as the reaction medium [2, 8].

$$(3.6)$$

Bromocyclophosphazenes are prepared by the reaction of NH_4Br with a mixture of PBr_3 and Br_2 (PBr_5 is not stable) (see Eq. 3.7). This reaction is similar to the NH_4Cl/PCl_5 reaction seen earlier.

$$nPBr_3 + nBr_2 + nNH_4Br \longrightarrow \left[PNBr_2\right]_n + 4nHBr \qquad (3.7)$$

$$n = 3 \text{ or } 4$$

Although iodocyclophosphazenes such as $N_3P_3I_6$ are not known, cyclophosphazenes containing a P-I bond are known and are prepared by the reaction of I_2 with a hydridocyclophosphazene (see Eq. 3.8).

$$(3.8)$$

A hexaisothiocyanatocyclophosphazene can be prepared by the complete replacement of all the six chlorine atoms from $N_3P_3Cl_6$ in a reaction with KNCS. 18-Crown-6-ether is used in this reaction to trap K^+ and to facilitate the liberation of the nucleophile $[NCS]^-$ for further reaction with $N_3P_3Cl_6$ (see Eq. 3.9) [2].

$$(3.9)$$

Among the pseudohalogenocyclophosphazenes, a monocyanophosphazene can be prepared by the replacement of the lone chlorine in $N_3P_3(OPh)_5Cl$ (see Eq. 3.10) [2].

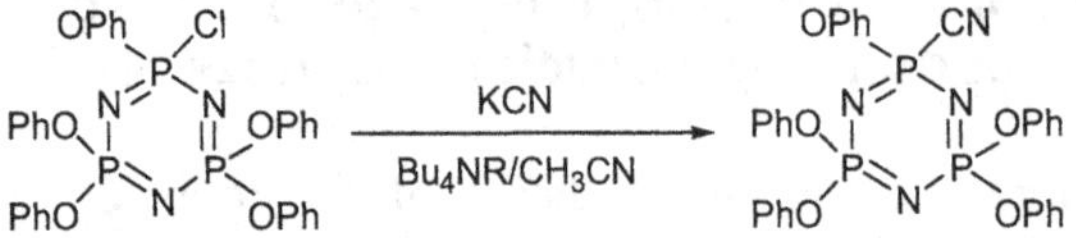

$$(3.10)$$

A similar reaction with sodium azide leads to the isolation of a stable azido derivative (see Eq. 3.11) [2]. Higher azido-containing cyclophosphazenes are explosive [2].

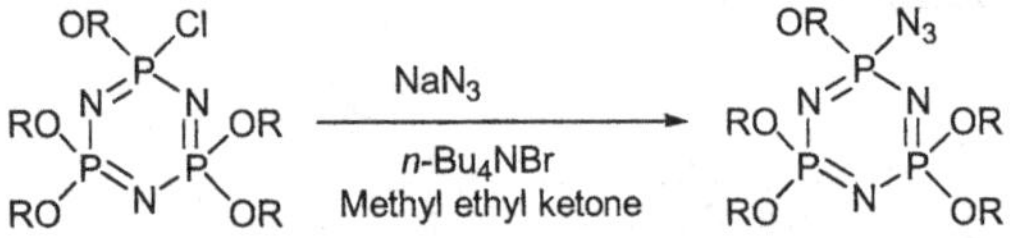

$$(3.11)$$

Hydridocyclophosphazenes are compounds that contain a P-H bond. These can be prepared by the following way. The reaction of $N_3P_3Cl_6$ with a Grignard reagent in conjunction with an organocopper reagent [n-$Bu_3PCuI]_4$ leads to a dimetallic intermediate which upon treatment with isopropanol affords gem-$N_3P_3Cl_4(R)(H)$, a compound that contains a P-H bond (Fig. 3.8). As shown earlier, the hydridophosphazenes are themselves valuable reagents and can be converted to halogenocyclophosphazenes upon treatment with chlorine, bromine or iodine. Other methods of preparing the hydridophosphazenes are also known [2].

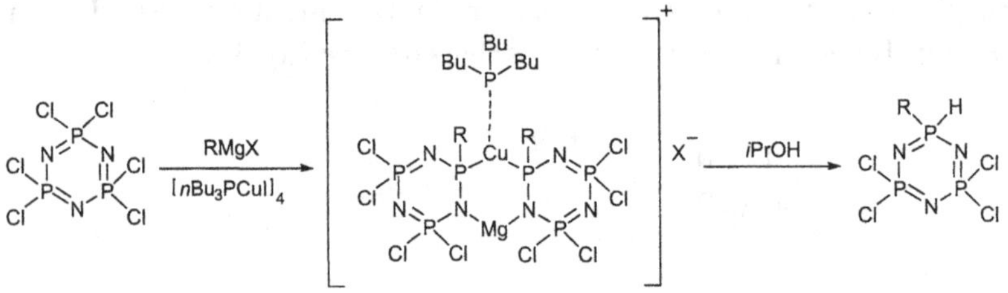

Fig. 3.8. Preparation of hydridocyclophosphazenes

3.2.3 Chlorine Replacement Reactions of Chlorocyclophosphazenes

The replacement of fluorine, chlorine or bromine from the corresponding halogenocyclophosphazenes by another substituent (a nucleophile) is an important aspect of the chemistry of these ring systems [2, 7, 12]. First, this route provides a way of preparing cyclophosphazenes with substituents involving exocyclic P-N, P-O, P-S, P-C, P-H and P-M linkages. Thus, in a way it may be said that the halogen replacement reactions are the most important means to increase the diversity of cyclophosphazenes. Second, as will be explained later, an understanding of these reactions at the small molecule cyclophosphazene level greatly enhances the capability to translate this chemistry to polyphosphazenes. Third, there is a lot of intriguing detail involved in the replacement reactions of halogenocyclophosphazenes. For example, what are the regio- and stereoselectivities in the products formed? What is their dependence?

Among the three rings $N_3P_3F_6$, $N_3P_3Cl_6$ and $N_3P_3Br_6$ the reactivity follows the order P-Br> P-Cl> P-F. This reflects the ease of P-X bond cleavage. Most of our discussion will be confined to chlorocyclophosphazenes because of the wealth of data present for these compounds. In chlorocyclophosphazenes the eight-membered ring $N_4P_4Cl_8$ is much more reactive than the six-membered ring $N_3P_3Cl_6$.

Chlorocyclophosphazenes are quite reactive towards nucleophiles such as amines, alcohols or phenols. Complete replacement of chlorines is readily possible with a number of reagents. For example, the reaction of $N_3P_3Cl_6$ or $N_4P_4Cl_8$ with dimethylamine can lead to the formation of $N_3P_3(NMe_2)_6$ or $N_4P_4(NMe_2)_8$ (see Eqs. 3.12 and 3.13). These reactions also work quite well with primary, secondary and even aromatic amines although individual differences in reactivity are present.

$$N_3P_3Cl_6 + 12HNMe_2 \longrightarrow N_3P_3 (NMe_2)_6 + 6Me_2NH . HCl \qquad (3.12)$$

$$N_4P_4Cl_8 + 16HNMe_2 \longrightarrow N_4P_4 (NMe_2)_8 + 8Me_2NH . HCl \qquad (3.13)$$

It can be noticed that the nucleophile dimethylamine has been used to scavenge the hydrogen chloride formed in the reaction. Instead, a tertiary base such as triethylamine can be used for this purpose.

Although complete substitution of chlorines is possible with many amines, with sterically hindered amines this becomes increasingly difficult for the six-membered ring. For example with adamantylamine ($AdNH_2$) the yields of $N_3P_3(NHAd)_6$ are quite low [2]. In the reactions of $N_3P_3Cl_6$ with even more sterically hindered amines such as dicyclohexylamine or dibenzylamine only two chlorines could be substituted. However, this difficulty does not arise with $N_4P_4Cl_8$ because of its greater reactivity. Fully substituted products $N_4P_4(NHR)_8$ or $N_4P_4(NR_2)_8$ are readily formed.

An interesting feature of the reactions of $N_4P_4Cl_8$ is the isolation of *bicyclic* products in addition to the *normal* fully substituted product, particularly if the reaction is carried out in a solvent such as chloroform (Fig 3.9). The mechanism of formation of the bicyclic product is now fairly well understood [2].

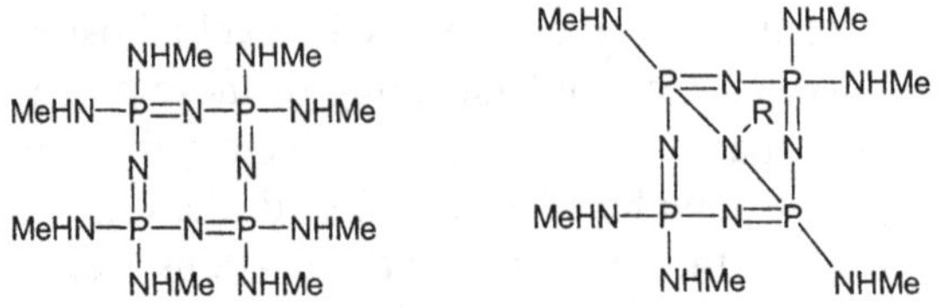
NHMe NHMe
MeHN—P=N—P—NHMe
N N
MeHN—P—N=P—NHMe
NHMe NHMe
MeHN NHMe
P=N—P—NHMe
R
N N N
MeHN—P—N=P
NHMe NHMe

$N_4P_4(NHMe)_8$
Normal product

$N_4P_4(NHMe)_6\ (NMe)$
Bicyclic product

Fig. 3.9. Normal and bicyclic products obtained in the reaction of $N_4P_4Cl_8$ with $MeNH_2$

The most common way to prepare alkoxy- or aryloxycyclophosphazenes containing exocylic P–O bonds is to react the chlorocyclophosphazenes with the sodium salt of the alcohol or phenol (see Eq. 3.14) [2].

$$N_3P_3Cl_6 + 6RONa \longrightarrow N_3P_3(OR)_6 + 6NaCl \tag{3.14}$$

This reaction can also be used for preparing cyclophosphazenes containing P–S bonds. However, there are not many examples of cyclophosphazenes belonging to this category.

Although fully substituted alkoxy- and aryloxycyclophosphazenes are the common products, it is possible to isolate useful penta-substituted derivatives in some instances. Thus, in the reaction between $N_3P_3Cl_6$ and NaOR (R = Ph, -C_6H_4-*p*-CH_3) pentakis aryloxycyclophosphazenes containing one residual P-Cl bond can be isolated. In some instances, such as with $N_3P_3(OCH_2CF_3)_6$, the pentakis derivative $N_3P_3(OCH_2CF_3)_5(Cl)$ can be prepared by a controlled hydrolysis followed by a reaction with PCl_5 (Fig.3.10).

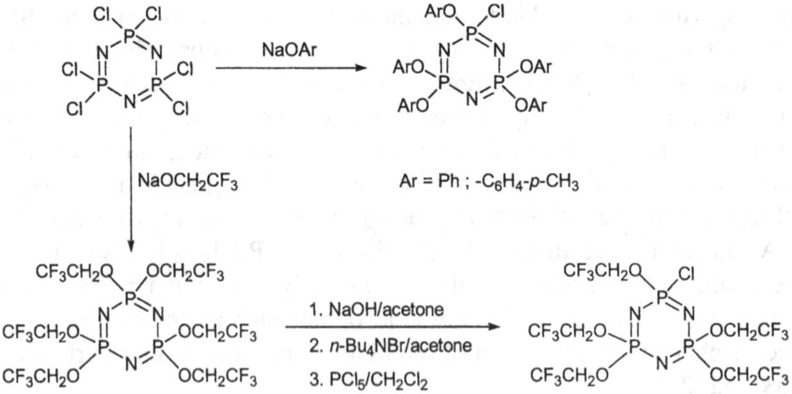
NaOAr
Ar = Ph ; -C6H4-p-CH3
NaOCH2CF3
1. NaOH/acetone
2. n-Bu4NBr/acetone
3. PCl5/CH2Cl2

Fig. 3.10. Preparation of pentakis alkoxy- and aryloxycyclophosphazenes

An interesting aspect of the alkoxycyclophosphazenes $[NP(OMe)_2]_n$, n = 3-6 is that they undergo a thermal rearrangement to cyclophos*phazanes*. This transformation involves a transfer of an alkyl group from the P-O-R unit to a ring nitrogen atom to form the thermodynamically more stable P=O bond (see Eq. 3.15). The mechanism of this rearrangement has been probed by using a mixture of $N_3P_3(OMe)_6$ and $N_3P_3(OCD_3)_6$. Cross-alkylated products are formed indicating that the process of rearrangement involves intermolecular processes. Aryloxycyclophosphazenes or fluoro-alkoxycyclophosphazenes do not undergo the rearrangement reactions [2].

Cyclophosphazene   Cyclophosphazane

(3.15)

The hydrolysis reaction of hexachlorocyclophosphazene, $N_3P_3Cl_6$, can lead to $N_3P_3(OH)_6$. The latter tautomerizes to the metaphosphimic acid $[NHP(O)OH]_3$ which hydrolyzes further to the eventual products, phosphoric acid and ammonia (see Eq. 3.16)

(3.16)

The reactions of $N_3P_3Cl_6$ or $N_4P_4Cl_8$ with organolithium or Grignard reagents are quite complex. Thus, the reactions of an organometallic reagent RM with chlorocyclophosphazenes proceed in two possible ways. In addition to the chlorine replacement reaction by R (elimination of M-Cl) another competing reaction is the elimination of RCl. This latter reaction is believed to occur because of the coordination of the ring nitrogen to the lithium metal ion leading to the formation of an unstable three-coordinate phosphorus-containing intermediate (see Eq. 3.17). Such a species can lead to a variety of products including ring-degraded products [2, 13].

(3.17)

The coordination ability of nitrogen is decreased in fluorocyclo-phosphazenes because of the high electronegativity of the fluorine atoms. This in turn leads to more predictable reactions (elimination of M-Cl and formation of compounds with P-C bonds). For example, the reaction of $N_3P_3F_6$ with dilithium salts of ferrocene or ruthenocene leads to interesting ansa products (see Eq. 3.18) [2, 13].

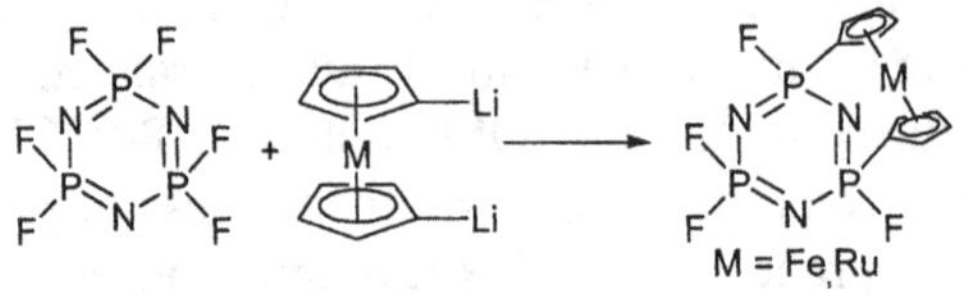

(3.18)

M = Fe, Ru

The reaction of $N_3P_3Cl_6$ with RMgX is also complex and the type of product formed depends on the type of solvent used as well as the nature of the R group in the Grignard reagent. Two prominent products are formed in this reaction: 1) a monoalkylated product and 2) bicyclophosphazene (Fig. 3.11). With sterically less-hindered alkyl groups and also with the phenyl group the bicyclophosphazene is favored. These trends are summarized in Table 3.2.

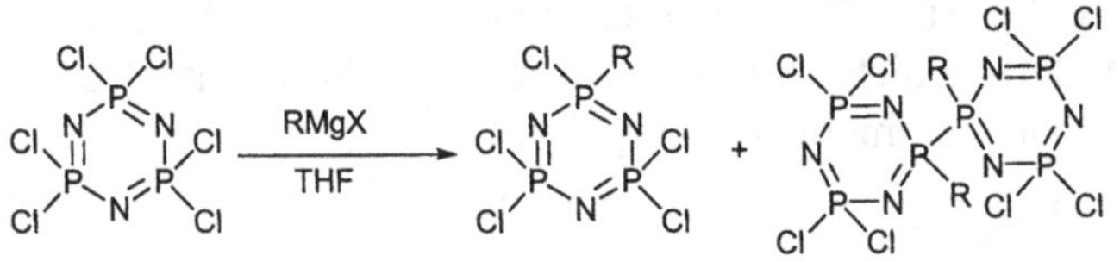

Fig. 3.11. Formation of $N_3P_3Cl_5R$ and $[(N_3P_3Cl_4R)_2]$ in the reactions of Grignard reagents with $N_3P_3Cl_6$

Table 3.2. Product distribution in the reactions of $N_3P_3Cl_6$ with Grignard reagents

S.No.	Grignard reagent	Yield of $N_3P_3Cl_5R$	Yield of $[(N_3P_3Cl_4R)_2]$
1	PhMgCl	0%	100%
2	MeMgCl	15%	85%
3	nBuMgCl	69%	35%
4	Me_3SiCH_2MgCl	100%	0%

3.2.4 Other Methods of Preparing P-C Containing Compounds

There are other methods for preparing cyclophosphazenes containing P-C compounds. Condensation reactions involving the elimination Me_3SiX (X = Br, F) from N-(silyl)-P-(halogeno)phosphoranimines is a good method for the preparation of fully substituted alkyl- or arylcyclophosphazenes (see Eq. 3.19) [2, 14].

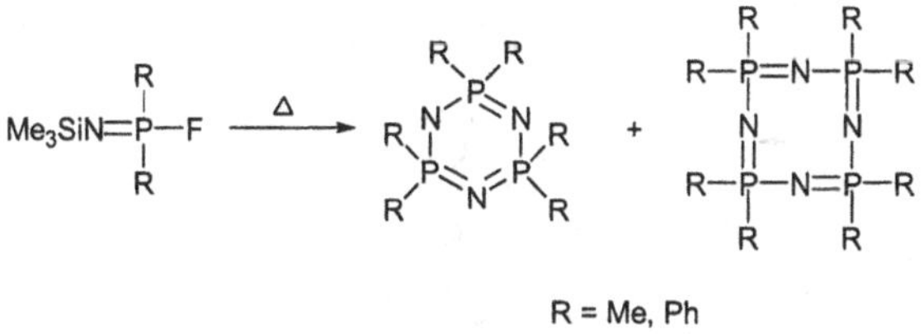

$$\text{Me}_3\text{SiN=P-F} \xrightarrow{\Delta} \text{(3.19)}$$

R = Me, Ph

Cyclization of linear fragments is an effective synthetic method for preparing alkyl- or arylsubstituted cyclophosphazenes. A very good synthon for this purpose is the Bezman's salt $[\text{Ph}_2\text{P(NH}_2)\text{NP(NH}_2)\text{Ph}_2]^+\text{Cl}^-$. This synthon contains a five atom -N-P-N-P-N- fragment and can be readily cyclized with an appropriate reagent to afford the corresponding cyclophosphazene. For example, the reaction of the Bezman's salt with Me_2PCl_3 leads to the formation of gem-$\text{N}_3\text{P}_3\text{Ph}_4\text{Me}_2$ (see Eq. 3.20).

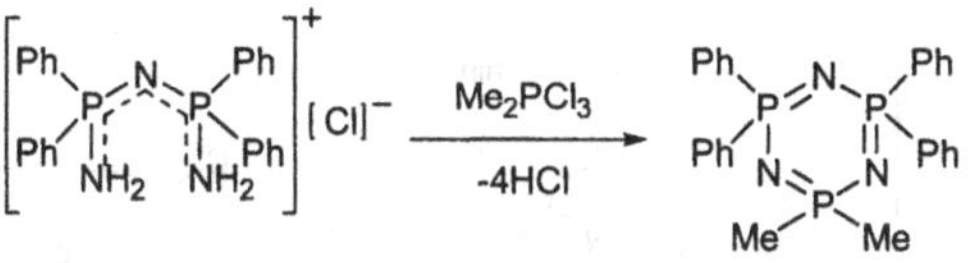

(3.20)

Friedel-Crafts reaction is quite useful to prepare gem-$\text{N}_3\text{P}_3\text{Cl}_4\text{Ph}_2$ and gem-$\text{N}_3\text{P}_3\text{Cl}_2\text{Ph}_4$ (see Eq. 3.21).

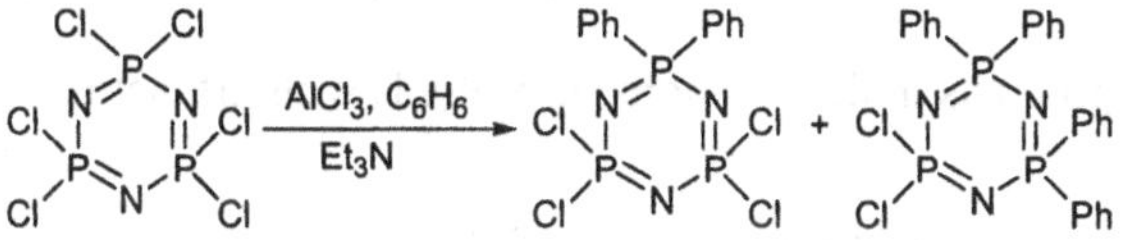

(3.21)

3.2.5 Reactions of Chlorocyclophosphazenes with Difunctional Reagents

So far we have seen the reactions of chlorocyclophosphazenes with reagents that may be termed *monofunctional*. For example, a phenol in its reactions with chlorocyclophosphazenes will react as a nucleophile and form $\text{N}_3\text{P}_3(\text{OPh})_6$ containing P-O linkages. If instead of a phenol we used a biphenol we will have two sites that can react with the phosphorus centers. In this manner there are several *difunctional* reagents such as diamines (ethylenediamine, 1,3-diaminopropane, phenylenediamine), diols (ethyleneglycol, 1,3-propanediol, catechol), and amino alcohols (ethanolamine, propanolamine, *o*-aminophenol) [2, 15].

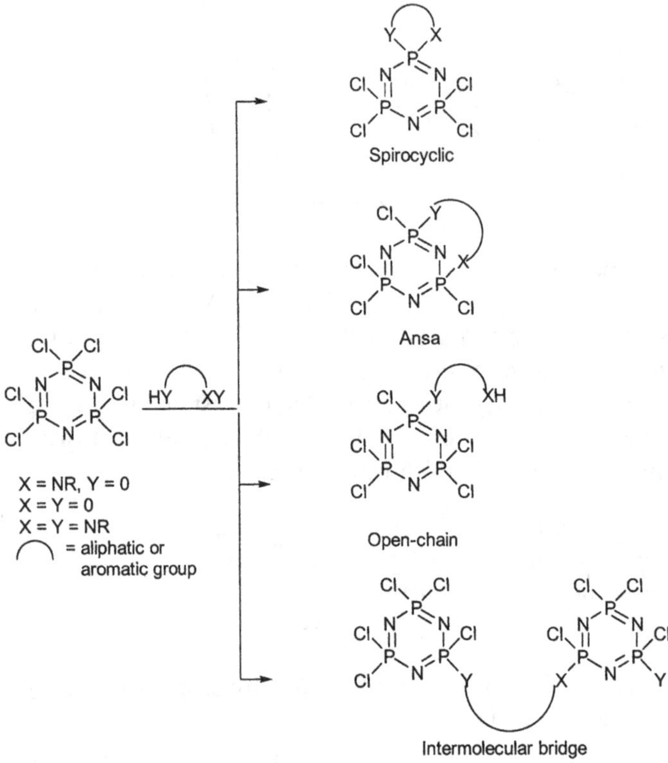

Fig. 3.12. Possible products in the reactions of $N_3P_3Cl_6$ with difunctional reagents

The products formed in the reactions of the difunctional reagents with chlorocyclophosphazenes are slightly different than those that we have encountered so far. Thus, the reactions of a typical difunctional reagent with $N_3P_3Cl_6$ can afford four types of products (Fig. 3.12). The *spirocyclic* product is one where the two ends of the difunctional reagent are attached to the *same* phosphorus center. In the *ansa* product the two ends of the difunctional reagent are attached to *two different* phosphorus centers, but within the same molecule. An *open-chain* compound is one where only one end of the difunctional reagent has reacted leaving the other end free for further reactions. Finally, the *intermolecular-bridged* products are formed as a result of the difunctional reagent linking different cyclophosphazenes. Spirocyclic products are the thermodynamically favorable products. Monospirocyclic compounds can be isolated with many reagents. Trispirocyclic compounds are also possible in many instances. Sequential replacement of pairwise chlorines is possible only in a limited number of examples. A few representative examples of spirocyclic products are shown in Figs 3.13 and 3.14. Among the spirocyclic products the tris catechol derivative is a very interesting compound. This compound forms several inclusion adducts. In other words, this compound in its solid-state has voids in the structure that can accommodate guest molecules. These guests include small molecules such as solvents (example, benzene) or even slightly larger molecules. In fact, in the cavities of the tris catecholate derivative, polymerization of certain monomers such as acrylates could be accomplished in a stereoregular manner. The void sizes, as measured by the diameter of the tunnel formed in the solid-state by trispirocyclotriphosphazenes, can be fine-tuned by the choice of the aromatic diol (Fig. 3.14) [16].

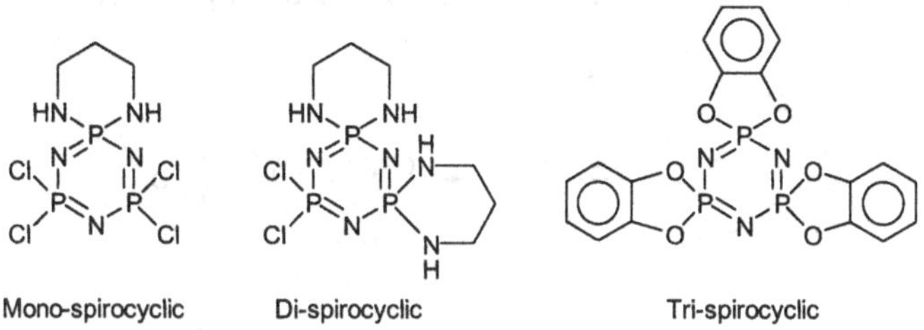

Mono-spirocyclic   Di-spirocyclic   Tri-spirocyclic

Fig. 3.13. Spirocyclic products

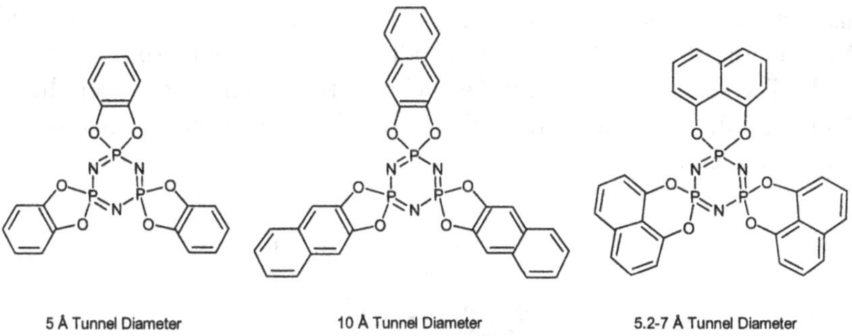

5 Å Tunnel Diameter   10 Å Tunnel Diameter   5.2-7 Å Tunnel Diameter

Fig. 3.14. Trispirocyclotriphosphazenes with varying tunnel diameters in the solid state.

Ansa products are known, although less common than spirocyclic products. Some of these require indirect synthetic procedures. For example, the reaction of $N_3P_3(CH_3)Cl_5$ with 1-aminopropanol affords a product $N_3P_3(CH_3)(NH(CH_2)_3OH)Cl_4$ where the hydroxyl end of the difunctional reagent is free. The latter can be induced to react with the chlorine of the adjacent phosphorus to afford an *ansa* product (see Eq. 3.22).

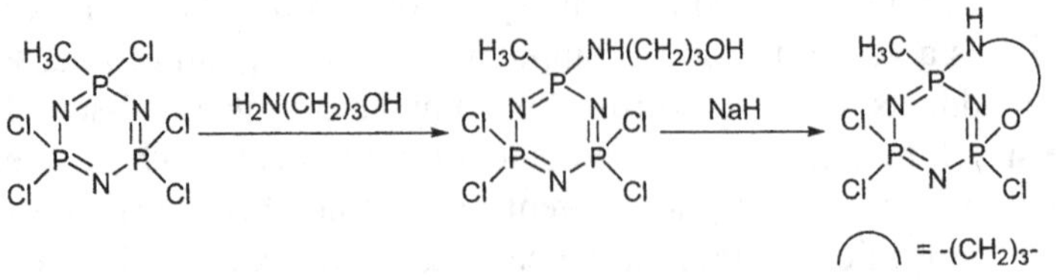

(3.22)

$$\diagup\!\!\frown\!\!\diagdown = \text{-(CH}_2)_3\text{-}$$

Another way of generating an *ansa* product is to prepare a cyclophosphazene precursor that has two *non-geminal* chlorines in a *cis* orientation. Such a compound upon reaction with difunctional reagents can lead to *ansa* products. Thus, the reaction of *non-gem-cis*-$N_3P_3Cl_2(OCH_2CF_3)_4$ with the sodium salt of naphthalene diol affords the *ansa* product (see Eq. 3.23).

(3.23)

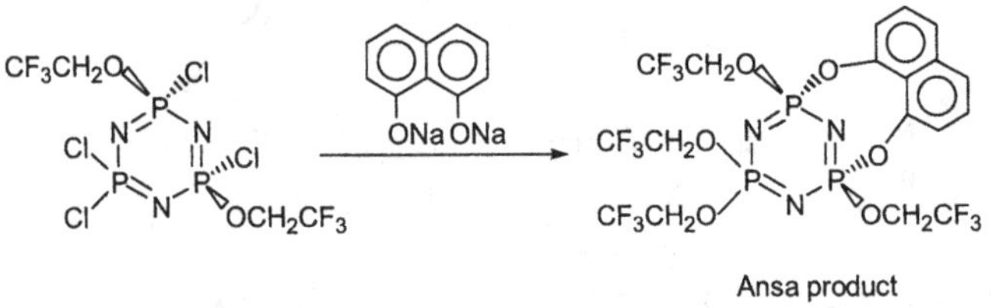

Ansa product

With some diols, *ansa* products are formed in reasonable yields, but have to be separated from the other products formed in the mixture. In this manner, for example, an interesting P-N-P crown-ether can be prepared in the reaction of $N_3P_3Cl_6$ with the disodium salt of an etheroxydiol (see Eq. 3.24). The structural similarity (as well as the difference) between the P-N-P crown-ether and conventional crown-ethers such as 18-crown-6-ether may be noticed [2].

(3.24)

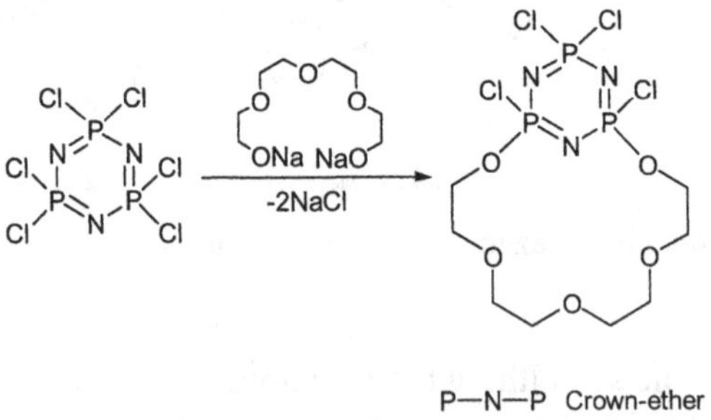

P—N—P Crown-ether

Intermolecular-bridged products are even rare and have only been reported to be formed in the reactions of long-chain aliphatic diamines with $N_3P_3Cl_6$.

3.2.6 Isomerism in Cyclophosphazenes

In the previous section it was mentioned that the reactions of chlorocyclophosphazenes with amines proceed to afford persubstituted (completely substituted) products. However, sequential replacement of chlorine atoms is also possible with many nucleophilic reagents by a simple control of stoichiometry. Various stages of chlorine replacement reactions lead to products ranging from monosubstituted derivatives to persubstituted ones.

After the first replacement of chlorine, since all the phosphorus centers are equivalent, the product $N_3P_3Cl_5R$ has only one possible structure. However, the subsequent replacement of chlorine can occur from a $P(Cl)(R)$ center or a $P(Cl_2)$ center. This can lead to the formation of two regioisomers. The first one will lead to a *geminal* product (the new substituents are on the same phosphorus). If the replacement occurs in the second pathway the compound formed will have a *nongeminal* regiodisposition of the substituents. It can also be seen that the non-geminal product can have two stereoisomers viz., *cis* and *trans* depending on the relative stereodisposition of the substituents. These possibilities are illustrated using the example of dimethylamine as the reagent and $N_3P_3Cl_6$ as the reactant (Fig. 3.15).

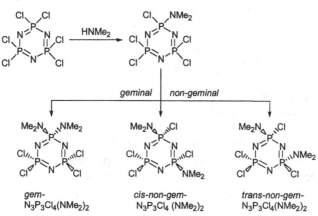

Fig. 3.15. Formation of *geminal* and *nongeminal* products in the reaction of $N_3P_3Cl_6$ with dimethylamine

Similar isomer formation is potentially present at the tris and tetrakis stages of substitution as well. However, at the pentakis and hexakis stages only one isomer can be realized. These possibilities are summarized in Fig. 3.16.

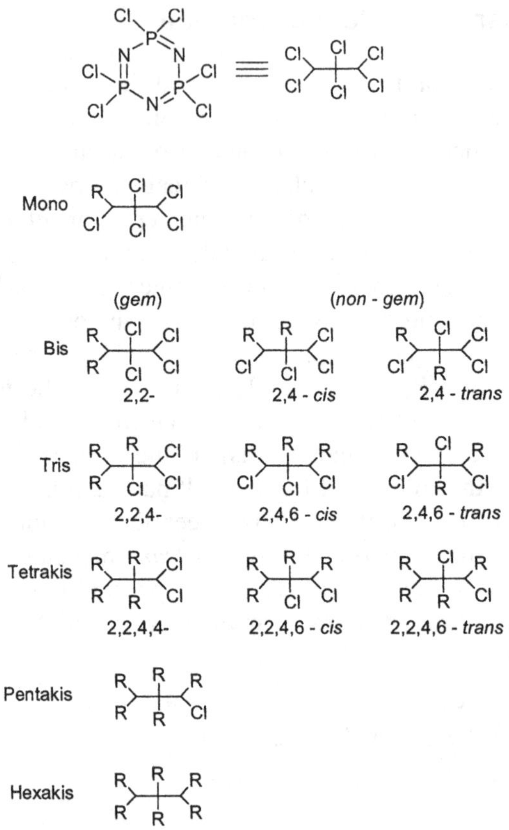

Fig. 3.16. Regio- and stereoisomer possibilities in the sequential chlorine replacement reactions of $N_3P_3Cl_6$

Obviously the number of products that are possible in the replacement reactions of the higher-membered cyclophosphazenes such as $N_4P_4Cl_8$ are even larger. Although studies on chiral cyclophosphazenes are very limited it can be shown that nonsuperimposable mirror images are possible for *trans*-$N_3P_3X_2R_4$ [12].

3.2.7 Mechanism of the Nucleophilic Substitution Reaction

Several mechanistic pathways have been identified in the nucleophilic substitution reactions of chlorocyclophosphazenes [12]. Most of the kinetic experiments have been carried out in the reactions involving chlorocyclophosphazenes with amines. These experiments are suggestive of various mechanistic pathways that can operate in these reactions. The following is a summary of the inference of the kinetic experiments.

3.2.7.1 Preference for Non-geminal Pathway

Preference for *nongeminal* product is *natural*. Imagine the reaction of an amine with a PCl_2 center or a $\equiv P(Cl)NRR'$ center. The phosphorus atom in the former is more *electrophilic* and hence a nucleophile would prefer to attack this center in comparison to the less electrophilic $P(Cl)(NR_2)$ site (Fig. 3.17). This means that *nongeminal* products should *always* be preferred.

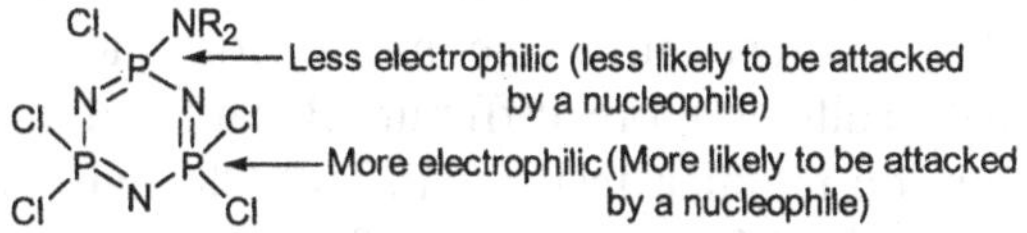

Fig. 3.17. Possible sites of attack of a nucleophile in its reaction with $N_3P_3Cl_5(NR_2)$

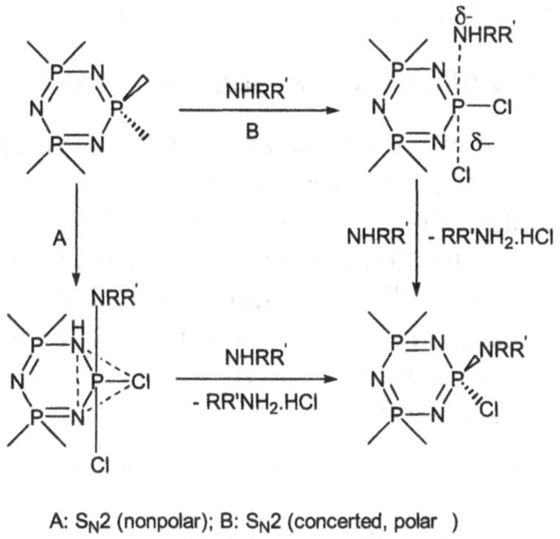

A: S_N2 (nonpolar); B: S_N2 (concerted, polar)

Fig. 3.18. S_N2 reaction pathways in the nucleophilic substitution reactions of cyclotriphosphazene

Although kinetic experiments have established different types of reaction mechanisms one of the most important pathways for the nucleophilic reactions in chlorocyclophosphazenes seems to be a bimolecular associative pathway (S_N2). Two types of transition states have been identified for this mechanism (Fig. 3.18).

(a) Formation of a neutral five-coordinate phosphorus intermediate followed by the expulsion of the leaving group (pathway A, Fig. 3.18).

(b) A concerted S_N2 mechanism involving a polar transition state similar to that found in the carbon system (pathway B, Fig. 3.18).

Either way, the reaction pathway envisages that the four-coordinate phosphorus is converted first into a five-coordinate state before reverting back to the four-coordinate state (Fig. 3.18).

The preponderance of the S_N2 pathway accounts for the faster reactivity of $N_4P_4Cl_8$. The latter is nonplanar and it is easier for the phosphorus center in this molecule to achieve penta-coordination. On the other hand, for the planar $N_3P_3Cl_6$ readjustment of geometry from the four-coordinate to the five-coordinate state is more difficult. Because of these reasons nucleophilic substitution is much faster with $N_4P_4Cl_8$ than with $N_3P_3Cl_6$. For the perfectly planar $N_3P_3F_6$ attainment of the five-coordinate transition state is even more difficult. Consequently, the reactivity of $N_3P_3F_6$ is much lower. Another factor responsible for the slower reactivity of $N_3P_3F_6$ is the higher bond strength of the P-F bond [2, 12].

3.2.7.2 Mechanism of Geminal Product Formation

Dissociative or S_N1 pathways are also known to occur in certain situations. With certain amines that are not very reactive dissociative pathways (S_N1) can also be competitive along with associative S_N2 reactions. In one such mechanism, proton abstraction occurs from a primary amino derivative such as $N_3P_3Cl_5(NHR)$, leading to a three-coordinate P(V) intermediate. Subsequently, geminal products are formed. Thus, for example, reaction of $N_3P_3Cl_6$ with four equivalents of tBuNH$_2$ leads to the formation of *gem*-$N_3P_3Cl_4(HNtBu)_2$ (see Eq. 3.25).

$$N_3P_3Cl_6 + 4tBuNH_2 \longrightarrow \textit{gem-}N_3P_3Cl_4(HNtBu)_2 + 2tBuNH_2 \cdot HCl \qquad (3.25)$$

It is postulated that the mechanism of formation of the geminal product $N_3P_3Cl_4(HNtBu)_2$ involves deprotonation from the monosubstituted product, $N_3P_3Cl_5(HNtBu)$. This leads to the generation of the three-coordinate phophoranimine intermediate (see Eq. 3.26). The three-coordinate phosphorus is sterically strained and is also very electrophilic. Because of this the second nucleophile prefers to attack this center leading to a geminal product. In this mechanism the loss of HCl is the slow process. Hence this mechanism is of the S_N1 type. The abstraction of the proton is accelerated by the addition of a base such as triethylamine [2, 12].

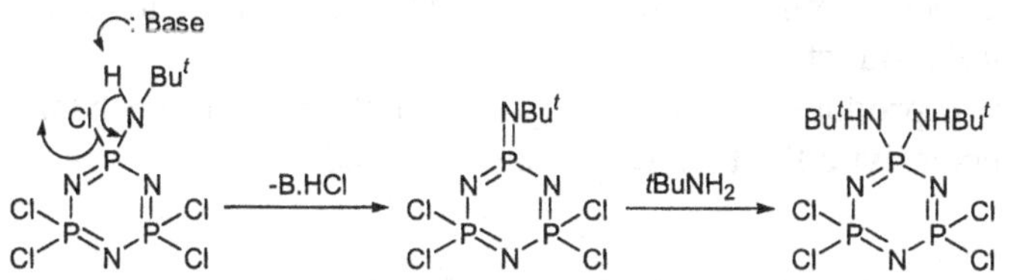

(3.26)

Dissociative pathways are also found in the reactions of compounds such as $N_3P_3(OPh)_5Cl$ with amines. The replacement of the final chlorine from $N_3P_3(OPh)_5Cl$ proceeds by the heterolytic cleavage of the P-Cl bond. This is followed by the attack of the nucleophile, amine, on the phosphonium center to generate $N_3P_3(OPh)_5(NRR')$ (see Eq. 3.27). In this sequence of reactions the heterolytic cleavage of the P-Cl bond is found to be rate determining and hence this pathway belongs to the S_N1 type of reaction [2, 12].

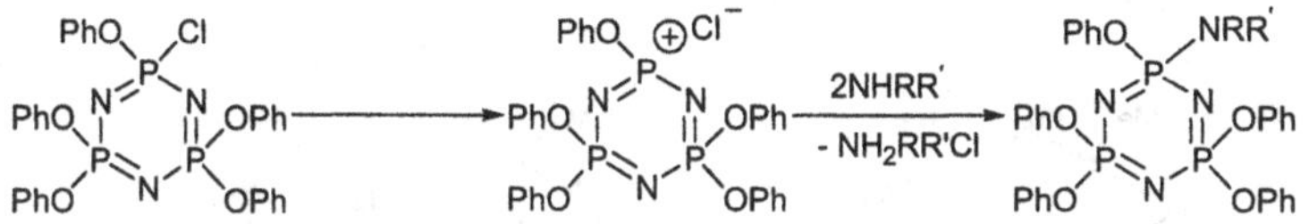

(3.27)

S_N1 pathways also occur in reactions where the chlorine from the P-Cl bond is heterolytically cleaved as a result of a reaction with a Lewis acid such as $AlCl_3$. For example, in the Friedel-Crafts reaction of $N_3P_3Cl_6$ the first step may be the abstraction of the chloride. This is followed by the electrophilic attack of the phosphonium center on the aromatic reagent to afford $N_3P_3Cl_5Ph$ (see Eq. 3.28).

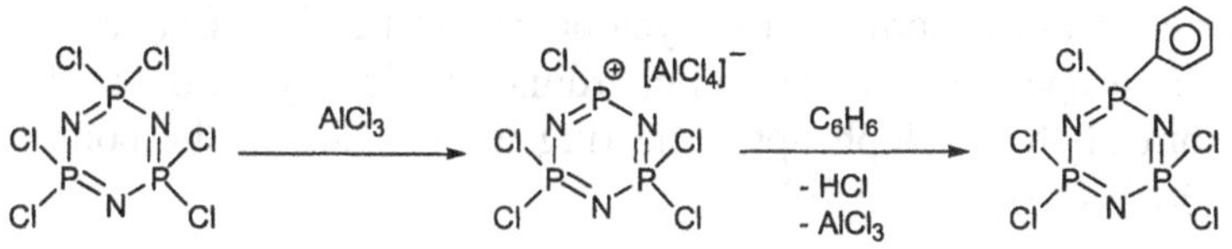

(3.28)

3.2.7.3 Summary of the Mechanisms of the Nucleophilic Substitution Reactions in Chlorocyclophosphazenes

1. The reactivity of cyclophosphazenes with nucleophiles such as amines follows the order $N_4P_4Cl_8 > N_3P_3Cl_6 > N_3P_3F_6$.
2. Many amines react with chlorocyclophosphazenes in a bimolecular process to afford *nongeminal* products. Almost exclusive *nongeminal* product formation is observed in $N_4P_4Cl_8$/amine reactions.
3. Two types of S_N2 mechanisms have been recognized: a nonpolar five-coordinate phosphorus intermediate and a polar concerted

mechanism. Both of these pathways afford the formation of *non-geminal* products.

4. *Geminal* products are formed by proton abstraction mechanism or by a heterolytic P-Cl bond cleavage process.

3.2.8 Cyclophosphazenes as Ligands for Transition Metal Complexes

The skeletal (ring) nitrogen atoms of cyclophosphazenes can act as coordination sites for interacting with transition metal ions [17-19]. This coordination ability increases with the increased presence of electron-releasing substituents on phosphorus and increase in ring size of cyclophosphazenes. For example, larger-membered rings can even function as macrocyclic ligands as seen in $N_6P_6(CH_3)_{12}.MCl_2$ (M=Pt or Pd) and $[N_8P_8(CH_3)_{16}Co]^{2+}[NO_3]_2$ (Fig. 3.19).

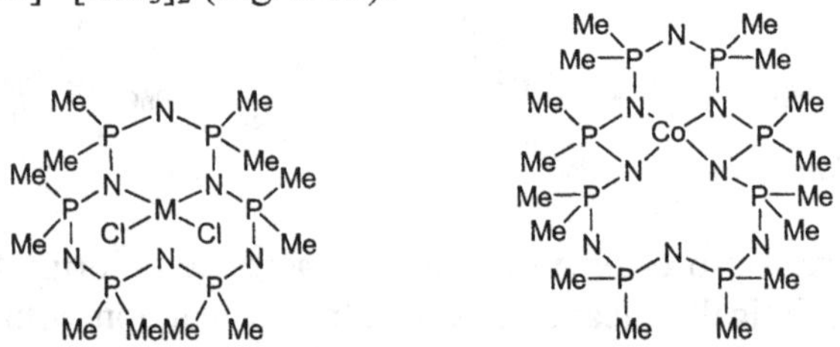

Fig. 3.19. Cyclophosphazenes as macrocyclic ligands

Octakis(methylamino)cyclotetraphosphazene, $N_4P_4(NHMe)_8$, forms the complex $N_4P_4(NHMe)_8.PtCl_2$, where the platinum is coordinated by two opposite ring nitrogen atoms of the cyclophosphazene ring. Examples of an exocyclic nitrogen participating in coordination along with the skeletal nitrogen atom of the cyclophosphazene ring is provided in the complex $N_4P_4(NMe_2)_8.W(CO)_4$ (Fig. 3.20).

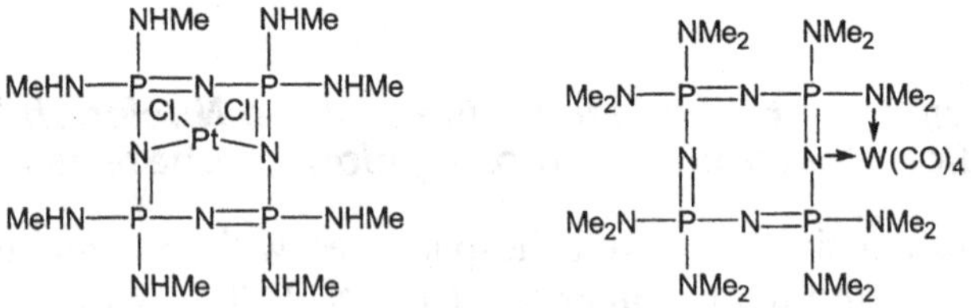

Fig. 3.20. $N_4P_4(NHMe)_8.PtCl_2$ and $N_4P_4(NMe_2)_8.W(CO)_4$

The reaction of $N_4P_4Me_8$ with copper chloride affords the complex $[N_4P_4Me_8HCuCl_3]$ where one ring nitrogen is protonated while the opposite ring nitrogen is involved in coordination to the transition metal. Compounds containing direct P-metal bonds are also possible as shown for the organometallic complex $N_3P_3F_4[\{(CO)(Cp)Fe\}_2(\mu\text{-}CO)]$ (Fig. 3.21).

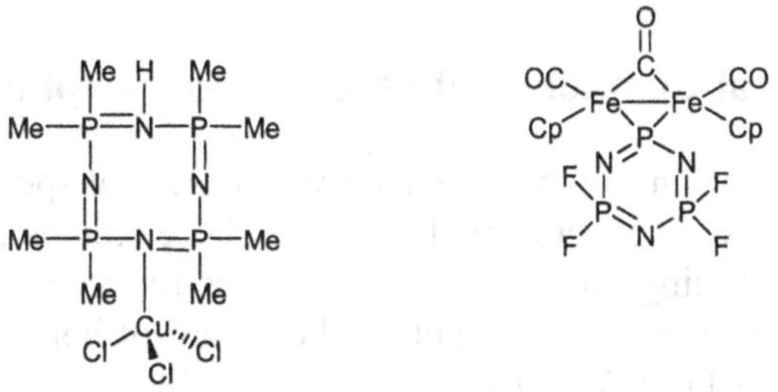

Fig. 3.21. $[N_4P_4(Me)_8H][CuCl_3]$ and $N_3P_3F_4[\{(CO)(Cp)Fe\}_2(\mu\text{-}CO)]$

Another way of using cyclophosphazenes as ligands is to incorporate coordinating units on phosphorus. For example, the reaction of $N_3P_3Cl_6$ with the 3,5-dimethyl pyrazole affords $N_3P_3(3,5\text{-}Me_2Pz)_6$, which is a very versatile multi-site coordination ligand (see Eq. 3.29).

(3.29)

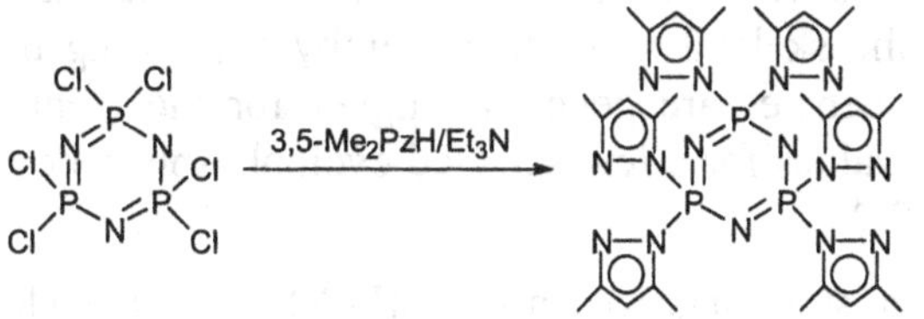

The ligand $N_3P_3(3,5\text{-}Me_2Pz)_6$ can be used to prepare monometallic and bimetallic derivatives (Fig. 3.22).

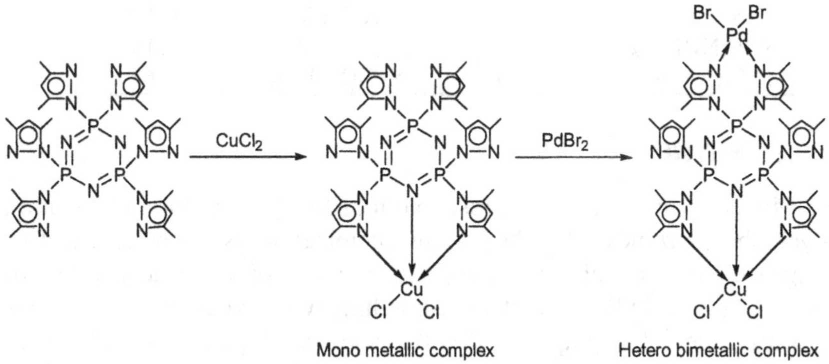

Mono metallic complex   Hetero bimetallic complex

Fig. 3.22. Coordination behavior of $N_3P_3(3,5\text{-}Me_2Pz)_6$

Synthetic approaches such as those shown above can be very useful for assembling a library of coordination ligands using the cyclophosphazene ring as a means for supporting such ligand architectures [17].

3.2.9 Structural Characterization of Cyclophosphazenes

Cyclophosphazenes have been studied by a number of spectroscopic methods as well as by X-ray crystal structures. These studies have helped not only in understanding the molecular structures of cyclophosphazenes but have also shed a great deal of light on the nature of bonding that exists in these compounds [1, 5, 7, 9, 10].

3.2.9.1 Vibrational Spectra

Cyclophosphazenes are characterized by the presence of a strong $\upsilon(P=N)$ stretching frequency in the region of 1150 - 1450 cm^{-1}. The magnitude of $\upsilon(P=N)$ increases with increasing electronegativity of the substituents. In amino derivatives the electron flow from the exocyclic nitrogen atom seems to weaken the skeletal π-bonding thereby depressing the $\upsilon(P=N)$. The ring P=N frequencies are generally higher for the eight-membered rings. The $\upsilon(P=N)$ values for a few selected cyclophosphazene derivatives are given in Table 3.3.

Table 3.3. Vibrational stretching frequencies {$\upsilon(P=N)$} for some selected cyclophosphazenes

Compound	υ (P=N) cm^{-1}	Compound	υ (P=N) cm^{-1}
$N_3P_3F_6$	1300	$N_4P_4F_6$	1436
$N_3P_3Cl_6$	1218	$N_4P_4Cl_8$	1315
$N_3P_3(NHMe)_6$	1175	$N_4P_4(NHMe)_8$	1215
$N_3P_3(NMe_2)_6$	1195	$N_4P_4(NMe_2)_8$	1265

3.2.9.2 ^{31}P NMR

^{31}P NMR is an invaluable spectroscopic tool for the structural elucidation of cyclophosphazenes. The phosphorus chemical shifts are quite sensitive to ring size (six- vs. eight-membered rings), nature of substituents (electronegativity, steric bulk), extent of π-bonding with exocyclic substituents etc., [20]. The ^{1}H-decoupled ^{31}P-NMR spectra of persubstituted cyclophosphazenes show a single peak. In general, the chemical shifts of the eight-membered rings are up-field shifted in comparison to the corresponding six-membered derivatives. Table 3.4 lists the ^{31}P-NMR chemical shifts for a few selected cyclophosphazene derivatives. These chemical shifts are with reference to external 85% H_3PO_4.

Table 3.4. ^{31}P-NMR chemical shifts for a few selected cyclophosphazenes

Compound	δ	Compound	δ
$N_3P_3Cl_6$	+19.3	$N_3P_4Cl_8$	-6.5
$N_3P_3F_6$	+13.9	$N_4P_4F_8$	-17.7
$N_3P_3(NHEt)_6$	+18.0	$N_4P_4(NHEt)_8$	+4.3
$N_3P_3(NMe_2)_6$	+24.6	$N_4P_4(NMe_2)_8$	+9.6
$N_3P_3(OCH_3)_6$	+21.7	$N_4P_4(OCH_3)_8$	+2.8
$N_3P_3(OCH_2CH_3)_6$	+14.3	$N_4P_4(OCH_2CH_3)_8$	-0.6
$N_3P_3(OCH_2CF_3)_6$	+16.7	$N_4P_4(OCH_2CF_3)_8$	-2.0
$N_3P_3(OPh)_6$	+8.3	$N_4P_4(OPh)_8$	-12.6

Proton-decoupled ^{31}P-NMR spectra can also be used for structural elucidation of mixed substituent containing cyclophosphazenes. Different types of three-spin systems such as AB_2, AX_2, ABC, ABX or AMX can be observed. Similarly, the ^{31}P $\{^1H\}$-NMR spectra of cyclotetraphosphazenes can be diagnostic of the structure of the compound. Usually a combination of proton and phosphorus NMR can be quite useful in arriving at the structure of the cyclophosphazene compounds.

3.2.9.3 Electronic Spectra

Most cyclophosphazenes that do not contain other chromophores as substituents do not have prominent absorption in the UV-Visible region of the electromagnetic spectrum [9, 10].

3.2.9.4 X-ray Crystal Structures

The X-ray crystal structures of many cyclophosphazenes are known [7, 9, 10]. A summary of the details of these investigations is as follows.

1. The six-membered rings are generally planar. For example, while $N_3P_3Cl_6$ is very nearly planar, $N_3P_3F_6$ is perfectly planar. Here, planarity is with reference to only the N_3P_3 segment. The substituents on phosphorus lie above and below the plane of the ring. The eight-membered rings are generally nonplanar and have puckered conformations. For example, $N_4P_4Cl_8$ exists in two nonplanar conformations viz., chair and boat forms.

2. The P-N bond lengths in homogeneously persubstituted cyclophosphazenes (where the substituents on the phosphorus are the same) are equal. The observed bond lengths are shorter (about 1.57

– 1.60 Å) than the average P-N single bond distance (1.78 Å). This implies the presence of some type of multiple bonding in these compounds.

3. The bond angles at phosphorus and nitrogen are both quite close to 120°.

4. In heterogeneously substituted compounds the ring P-N bond lengths are unequal. For example, in *gem*-$N_3P_3F_4Ph_2$ three different bond distances are seen: 1.617(5), 1.539(5) and 1.555(4) Å. The longest distance is associated with the N-P-N segment flanking the PPh_2 unit.

X-ray data for a few selected cyclophosphazenes are summarized in Table 3.5.

Table 3.5. X-ray data for a few selected cyclophosphazenes

S.No	Compound	Ring conformation	Bond distance (Å)		Bond angles (°)	
			$P=N_{endo}$	$P-X_{exo}$	NPN	PNP
1	$N_3P_3Cl_6$	planar	1.581	1.98	118.4	121.4
2	$N_3P_3F_6$	planar	1.57(1)	1.52(1)	120(1)	119(1)
3	$N_3P_3(NMe_2)_6$	distorted boat	1.588(3)	1.652(4)	116.7(4)	123.0(4)
4	$N_3P_3(OPh)_6$	nonplanar	1.575(2)	1.582(2)	117.3(3)	121.9(3)
5	$N_3P_3(O_2C_6H_4)_3$	planar	1.59(2)	1.62(1)	117(1)	122(1)
6	$N_3P_3(Me)_6$	distorted chair	1.595(2)	1.788(2)	116.7(1)	122.5(1)
7	$N_3P_3(Ph)_6$	slightly chair	1.597	1.804	117.8	122.1
8	$N_4P_4Cl_8$ (K-form)	boat or tub	1.570(9)	1.991(4)	121.2(5)	131.3(6)
9	$N_4P_4(NMe_2)_8$	saddle	1.58(1)	1.69(1)	120.1(5)	130.0(6)
10	$N_4P_4(OMe)_8$	saddle	1.57	1.58	121.0	132.0
11	$N_4P_4(OPh)_8$	boat	1.560	1.582	121.1	133.9
12	$N_4P_4(Me)_8$	distorted boat	1.591(5)	1.802(6)	119.8(3)	132.0(2)
13	$N_4P_4(Ph)_8$	distorted boat	1.590(5)	1.809(4)	119.8(1)	127.8(2)

3.2.10 Nature of Bonding in Cyclophosphazenes

The nature of bonding in cyclophosphazenes can be understood by examining certain experimental facts and the arguments advanced to explain them. These can be summarized as follows [5, 7, 9, 10].

1. In homogeneously substituted cyclophosphazenes the ring P-N bond lengths are short and equal. In this regard cyclophosphazenes have a similarity with benzene and other aromatic molecules.

2. The P-N bond lengths are also influenced by the nature of substituents on the phosphorus. Thus, presence of more electronegative substituents that are σ-electron withdrawing result in a shortening of the P-N bond distance. Similarly, involvement of ring nitrogens in coordination to metals increases the P-N bond distance in the affected P-N-P segment.

3. However, cyclophosphazenes do not have the other characteristics that are typical of organic aromatic molecules. Thus, the $(4n + 2)$ π rule does not have any special significance in this family of compounds. The six- and eight-membered rings, $N_3P_3R_6$ and $N_4P_4R_8$ are equally stable.

4. Ring-current effects typical of aromatic compounds are absent in cyclophosphazenes. Bathochromic shifts (observed with an increase in π electrons in aromatic compounds) are absent in cyclophosphazenes. In fact, for all practical purposes, cyclophosphazenes may be regarded as transparent in the UV-Visible spectrum.

5. Thus, quite clearly cyclophosphazenes are not aromatic, at least in the context in which this word is used for organic compounds.

However, the shortness and equality of P-N bond lengths need to be accounted for and qualitative bonding models adequately do this. The electronic structure of cyclophosphazenes may be visualized by considering the possible resonance structures as shown in Fig. 3.23.

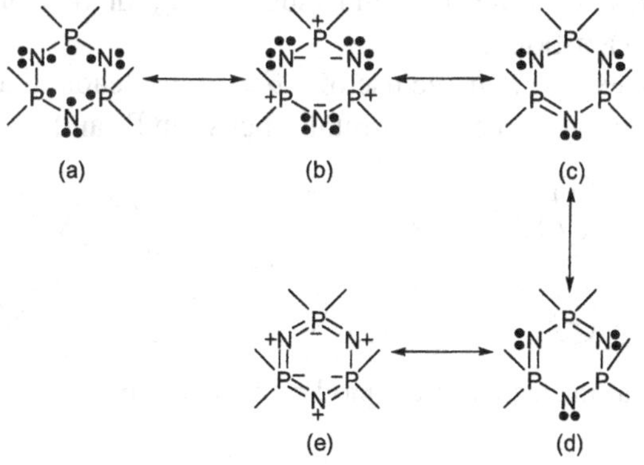

Fig. 3.23. Possible resonance structures for cyclophosphazenes

It is easier to visualize the bonding in cyclophosphazenes by separating the σ-component from the π-component that may be contributing together to the overall bonding picture in these compounds. These arguments can be summarized as follows.

1. If one considers the σ-framework alone, each phosphorus forms four covalent bonds (two exocyclic bonds and two bonds with endocyclic ring nitrogen atoms). This process consumes four of the five valence electrons on phosphorus leaving it with one electron.

2. Which orbitals does phosphorus use for forming these σ-bonds? Recalling the bond angles at phosphorus it is possible that approximate sp^3 hybrid orbitals are utilized.

3. The situation with nitrogen is that it is postulated to form $3sp^2$ hybrid orbitals. One of these lodges a lone pair. The other two are involved in forming a covalent bond each with two adjacent phosphorus centers. Thus the lone pair present on nitrogen is *in-plane* with the N_3P_3 framework.

4. After the σ-bonding requirements are met nitrogen also is left with one electron similar to phosphorus [Fig. 3.23(a)]. Several bonding possibilities are now present. These include ionic contributions as shown in Figure 3.23(b). For a long time there has been the speculation of some kind of π-bonding. Irrespective of the current validity of this model this view is also presented.

5. After the formation of the σ-bonded framework we have an odd electron each on phosphorus and nitrogen. It is speculated that this electron which resides in the p_z orbital on nitrogen overlaps with a suitable d orbital (such as d_{xz} or d_{yz}) on phosphorus to form a π-bond.

6. However, exclusive use of a d_{xz} orbital leads to a sign mis-match in the overlap of the orbitals (between P_6 and N_1) (Fig 3.24).

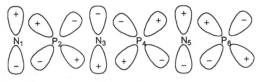

Fig. 3.24. Orbital mismatch in the pπ-dπ interaction between nitrogen and phosphorus.

7. The orbital mismatch can be overcome by the use of a d_{yz} orbital on phosphorus. Dewar suggested that the use of d_{xz} and d_{yz} orbitals in equal proportion leads to limited delocalization over a

three atom (P-N-P) segment. This would account for bond-length reduction as well as bond-length equality, while at the same time explaining the lack of *aromaticity* in cyclophosphazenes.

8. In addition to the above type of π-bonding, other types of π-bonding can also exist although their contribution to the overall multiple bonding may be less. The overlap of the sp^2 orbital on nitrogen (with its lone pair) with an empty d orbital on phosphorus (dx^2-y^2 or dxy) can lead to an *in-plane* π-bonding.

9. The presence of in-plane π-bonding has been suggested by several experimental facts. Thus, if one compares the P-N bond distances between $N_3P_3Cl_2(NHiPr)_4$ and its hydrogen chloride adduct $N_3P_3Cl_2(NHiPr)_4$.HCl one observes significant differences (Fig. 3.25) [5, 7].

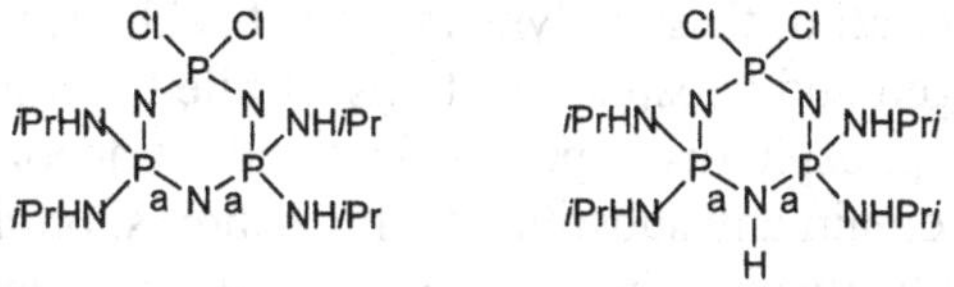

Fig. 3.25. Effect of protonation on P-N bond lengths

10. The P-N bond lengths (Fig.3.25, *a*) are considerably lengthened in compound $N_3P_3Cl_2(NHiPr_4)$.HCl (1.665(5) Å) in comparison to $N_3P_3Cl_2(NHiPr_4)$ (1.589(3) Å). This is possible only if the lone pair of electrons on nitrogen is engaged in bonding to the proton and is consequently not available for being used in π bonding. Similar types of bond lengthening have been noticed in many cyclophosphazene-metal complexes which involve coordination from the ring nitrogen atom. Lastly, a third type of π-bonding is also possible. This involves the exocyclic substituents on phosphorus. Thus, in amino cyclophosphazenes, the lone pair from the exocyclic nitrogen can engage in π-bonding with phosphorus as shown in Fig. 3.26. The exocyclic type of π-bonding is manifested by shorter exocyclic P-N bonds and lengthening of endocyclic P-N bonds.

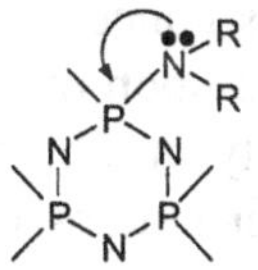

Fig. 3.26. Exocyclic π-bonding in cyclophosphazenes

Clearly in any given cyclophosphazene all of the above types of π-bonding might be contributing to varied extents and the observed structural features are the result of this cumulative effect.

Finally, however, it must be mentioned that considerable theoretical work in main-group compounds suggests that the participation of d-orbitals in any type of bonding may be minimal. Alternative bonding models that avoid the use of d-orbitals and account for all the experimental facts in cyclophosphazenes may soon emerge from such studies.

3.3 Polyphosphazenes

Polyphosphazenes, $[N=PR_2]_n$, are the largest family of inorganic polymers. Over 800 different types of polyphosphazenes are known [16, 21-23]. The common feature of all of these polymers is the backbone which is made up of alternate nitrogen and phosphorus atoms. While the nitrogen is trivalent and dicoordinate, phosphorus is pentavalent and tetracoordinate. Thus, nitrogen does not contain any substituents. Phosphorus, on the other hand, is attached to two substituents. Thus, the basic structural feature of the repeat unit of polyphosphazenes is exactly similar to what is found in cyclophosphazenes.

Although, the repeat unit in all the polyphosphazenes comprises the $[N=PR_2]$ repeating unit, the properties of the polymers can be radically changed by varying the 'R' groups on phosphorus. In this section, we will examine the preparative procedures, the structural features and the applications of polyphosphazenes.

3.3.1 Historical

The American Chemist Stokes discovered, towards the end of the nineteenth century, that heating the six-membered ring, $N_3P_3Cl_6$, to high temperatures leads to the formation of a polymeric material with an empirical formula $[NPCl_2]_n$. This material was called *inorganic rubber* because of its properties. Thus, this was rubbery but had very poor solubility in common organic solvents. Instead of dissolving in solvents, the polymer would swell in them. The most serious concern about the *inorganic rubber* was its sensitivity towards hydrolysis upon exposure to moisture. The polymer was found to degrade completely in ambient atmosphere. These unfavorable properties precluded further studies on this material for a long time [21]. The crucial breakthrough in this area came about by pursuing the following line of thinking.

1. The Stokes's *inorganic rubber* [NPCl$_2$]$_n$ did not dissolve in common organic solvents because it was an extensively crosslinked polymer. This suggestion is also borne out by the swelling of this polymer in solvents. It is known that crosslinked organic polymers also swell in solvents.

2. The degradation of the Stokes's *inorganic rubber* [NPCl$_2$]$_n$ in atmosphere occurs presumably because of the sensitivity of the P-Cl bonds towards the hydrolysis reaction.

Any effort to make the inorganic rubber into a viable material, therefore, needed to concentrate on solving the above two problems.

1. Finding a suitable route to prepare the *linear polymer* [NPCl$_2$]$_n$. Unlike the crosslinked derivative the linear polymer would be expected to be soluble in a large number of organic solvents.

2. The inherent hydrolytic instability of the polymer arising out of the presence of a number of P-Cl bonds can be overcome if the linear polymer can be modified by replacing the P-Cl bonds by other hydrolytically viable linkages.

As will be shown in the following section these strategies resulted in polyphosphazenes emerging as the *largest family* of inorganic polymers [21].

3.3.2 Poly(dichlorophosphazene)

Poly(dichlorophosphazene), [NPCl$_2$]$_n$, can be prepared by several routes. The traditional route is the ring-opening polymerization (ROP) of N$_3$P$_3$Cl$_6$. Although, as discussed above, Stokes as early as in late 1800's had observed the formation of crosslinked poly(dichlorophosphazene), it is to the credit of Allcock that he found the correct recipe for isolation of uncrosslinked linear poly(dichlorophosphazene) [21]. He and his co-workers have made the following important and crucial experimental observations.

1. Heating N$_3$P$_3$Cl$_6$ at 250 °C in vacuum and by allowing the conversion to proceed only up to 70%, linear poly(dichlorophosphazene) could be isolated (see Eq. 3.30). This polymer was soluble in a number of organic solvents such as benzene, toluene, tetrahydrofuran etc., to form clear viscous solutions.

2. Heating beyond 250 °C or allowing the conversion to proceed beyond 70% afforded a crosslinked material similar to what was obtained by Stokes. The crosslinking of the polymer was rapid and the resultant material was totally insoluble.

The linear polymer prepared by Allcock was truly a high polymer with over 15,000 repeat units representing a M_w of about 1.2×10^6. The polydispersity index for the polymer $[NPCl_2]_n$ prepared by the ring-opening of $N_3P_3Cl_6$ is greater than 2.

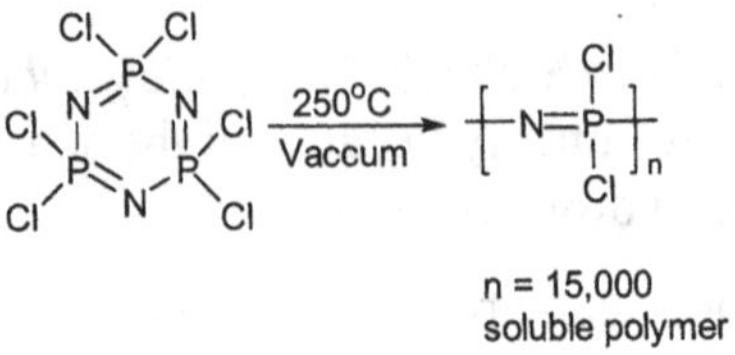

(3.30)

$$Cl_2P_3N_3Cl_6 \xrightarrow[\text{Vaccum}]{250°C} \left[N=P \right]_n$$

n = 15,000
soluble polymer

Hexafluorcyclotriphosphazene, $N_3P_3F_6$, hexabromocyclotriphosphazene, $N_3P_3Br_6$ and the hexathiocyanato cyclotriphosphazene $N_3P_3(NCS)_6$ can also be polymerized by the ring-opening polymerization method to the corresponding linear polymers (Fig. 3.27).

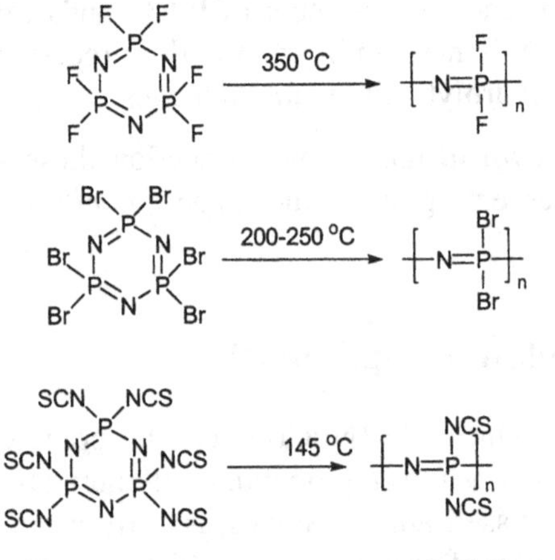

Fig. 3.27. Polymerization of $N_3P_3F_6$, $N_3P_3Br_6$ and $N_3P_3(NCS)_6$

The polymers shown in Fig. 3.27 also are linear polymers and are soluble in organic solvents. However, poly(difluorophosphazene) is not soluble in common organic solvents but only in fluorinated solvents.

The mechanism of the ROP of $N_3P_3Cl_6$ has been elucidated keeping in view the following experimental observations [21].

1. The ionic conductivity of molten $N_3P_3Cl_6$ rises suddenly, coincident with the initiation of polymerization.
2. A number of Lewis acids such as BCl_3 or reagents such as water (in the form of metal hydrates such as $CaSO_4.n\ H_2O$) in catalytic amounts have a beneficial influence in the polymerization reaction.

3. The ring-opening polymerization of the fluoro derivative $N_3P_3F_6$ occurs at a much higher temperature (350 °C) than that of $N_3P_3Cl_6$ (250 °C).

It has been postulated that the initiation of the polymerization occurs by a heterolytic cleavage of the P-Cl bond to generate a phosphazenium ion (see Eq. 3.31).

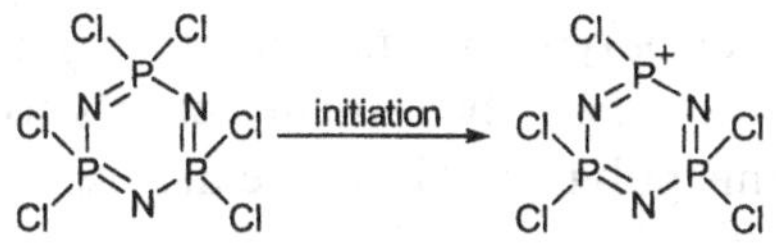

$$(3.31)$$

The propagation reaction which involves the polymer growth occurs as a result of the high electrophilicity of the phosphazenium ion. Since the reaction is carried out in the molten state there is a favorable situation for intermolecular reactions. Thus, the phosphazenium ion can attack a skeletal nitrogen atom of another $N_3P_3Cl_6$, causing ring opening and chain growth (Fig. 3.28).

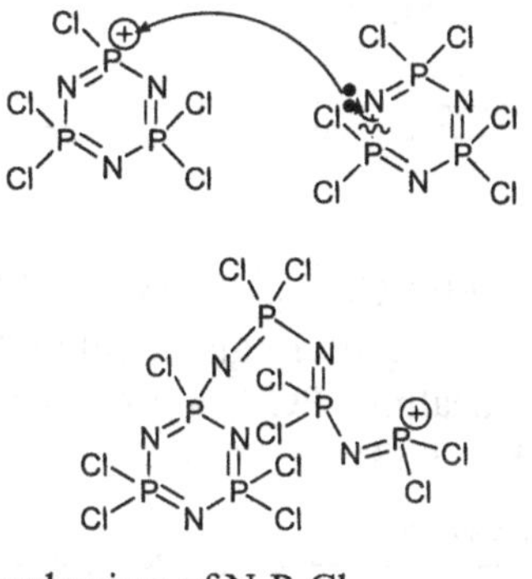

Fig. 3.28. Polymerization mechanism of $N_3P_3Cl_6$

The termination of the chain can occur by the reaction with any nucleophile including traces of moisture. This mechanism accounts for the increase in ionic conductivity at the onset of polymerization of $N_3P_3Cl_6$. Since the P-F bond strength is greater than that of the P-Cl bond, it is also evident why higher temperature is required for the polymerization of $N_3P_3F_6$. Moreover, catalytic amounts of either Lewis acids or water would be expected to facilitate the heterolytic cleavage of the P-Cl bond. Thus, if this polymerization mechanism is correct the presence of at least one P-Cl bond is a must in the cyclic ring in order for polymerization to occur. Indeed, this appears to be true for most cyclophosphazenes. Thus, many organocyclophosphazenes that do not contain a P-Cl bond do not undergo polymerization. But, certain cyclophosphazenes such as $N_3P_3(CH_3)_6$

equilibrate to higher-membered rings such as $N_4P_4(CH_3)_8$. Some other cyclophosphazenes such as the 2-pyridinoxy derivative $N_3P_3(O\text{-}2\text{-}C_6H_4N)_6$ are known to undergo a ring-opening polymerization. Also, strained cyclophosphazenes such as $N_3P_3F_4(\eta^5\text{-}C_5H_4\text{-}Fe\text{-}\eta^5\text{-}C_5H_4)$, that contains a transannularly attached ferrocene moiety undergoes polymerization. This means that other types of mechanisms exist that allow a ring to open [21].

Because of the importance of the linear poly(dichlorophosphazene) towards the synthesis of other poly(organophosphazenes), as we will see in the subsequent sections, there have been many efforts to find alternative synthetic procedures for this polymer. These are discussed in the following sections.

3.3.3 Condensation Polymerization of Cl₃P=N-P(O)Cl₂

De Jaeger and coworkers discovered that the acyclic phosphazene, $Cl_3P=N\text{-}P(O)Cl_2$, can be used to prepare poly(dichlorophosphazene), $[NPCl_2]_n$, either when heated in bulk or in a solvent such as trichlorodiphenyl (see Eq. 3.32) [23-25].

$$n\,PCl_3{=}NP(O)Cl_2 \xrightarrow{\;250\text{ - }280^\circ C\;} Cl_3P{=}N{-}\underset{\underset{Cl}{|}}{\overset{\overset{Cl}{|}}{(P}}{=}N)_{n\text{-}1}{-}POCl_2 \; + \; (n\text{-}1)\,POCl_3 \qquad (3.32)$$

The monomer $Cl_3P=NP(O)Cl_2$ can be prepared by several methods. Two of the best routes of preparing it are given below [23]. The first method of synthesis consists of heating PCl_5 with $(NH_4)_2SO_4$ in boiling *sym*-tetrachloroethane (see Eq. 3.33).

$$4\,PCl_5 \; + \; (NH_4)_2SO_4 \xrightarrow{\;sym\text{-tetrachloroethane}\;} 2\,Cl_3P{=}N{-}P(O)Cl_2 \; + \; 8\,HCl + SO_2 + Cl_2 \qquad (3.33)$$

The second method of synthesis consists of first preparing $[Cl_3P=N=PCl_3]^+\,[PCl_6]^-$ by the reaction of PCl_5 with NH_4Cl (see Eq. 3.34). This is followed by reacting $[Cl_3P=N=PCl_3]^+\,[PCl_6]^-$ with P_4O_{10}. The latter reaction leads to the elimination of $POCl_3$ and formation of the monomer (see Eq. 3.35).

$$3\,PCl_5 \; + \; NH_4Cl \longrightarrow 3\left[Cl_3P{=}N{-}PCl_3\right]^+\left[PCl_6\right]^- \; + \; 4\,HCl \qquad (3.34)$$

$$3\left[Cl_3P{=}N{=}PCl_3\right]^+\left[PCl_6\right]^- \; + \; P_4O_{10} \longrightarrow 3\,Cl_3P{=}N{-}P(O)Cl_2 \; + \; 7\,POCl_3 \qquad (3.35)$$

The reaction of $[Cl_3P=N=PCl_3]^+$ $[PCl_6]^-$ can also be carried out with SO_2 to afford $Cl_3P=NP(O)Cl_2$ (see Eq. 3.36).

$$3\ PCl_5\ +\ NH_4Cl\ +\ 2\ SO_2 \longrightarrow Cl_3P{=}N{-}P(O)Cl_2\ +\ 4\ HCl\ +\ POCl_3\ +\ 2\ SOCl_2 \quad (3.36)$$

$Cl_3P=N\text{-}P(O)Cl_2$ is a low-melting solid (32 $^\circ$C) and can be distilled. Its ^{31}P NMR shows two signals (δ = 0.5 ppm, doublet, $\underline{P}Cl_3$; δ = -12.8 ppm, doublet, $\underline{P}(O)Cl_2$, $^2J(PP)$ = 16.2 Hz).

The bulk condensation (in the absence of solvent) of $Cl_3P=NP(O)Cl_2$ can be carried out between 200-300 $^\circ$C(most often around 240 $^\circ$C). This leads to the elimination of $POCl_3$. After about 65-70% conversion, the reaction becomes very viscous and needs to be stopped. At this stage the average degree of polymerization is about 1000. This corresponds to a molecular weight of about 100,000. It has been reported that the synthesis of poly(dichlorophosphazene) of a molecular weight M_w = 200,000 represents a limit for the bulk condensation reaction. The mechanism of polymerization involves a nucleophilic attack by the nitrogen atom of one monomer molecule on the phosphorus center of another monomer. This leads to the formation of a dimer by the elimination of $POCl_3$. This may be considered as the initiation reaction (see Eq. 3.37).

(3.37)

The initiation reaction (between two monomers) is slower than reaction of the monomer with oligomers (propagation) (see Eq.3.38).

(3.38)

The viscosity of the reaction mixture increases after a certain level of conversion. At this point rapidly growing oligomers can condense with each other (see Eq. 3.39).

(3.39)

Although the bulk condensation leads primarily to the formation of the polymer $[NPCl_2]_n$, side products in low yields such as $N_3P_3Cl_6$, $N_4P_4Cl_8$ and $N_5P_5Cl_{10}$ are also detected [23].

3.3.3.1 Polycondensation in Solution

The most convenient solvent for the polymerization of $Cl_3P=NP(O)Cl_2$ has been found to be molten trichlorobiphenyl. It has been found that in order to have an optimum polymer yield, the polycondensation should be performed first in bulk till about 95% conversion, at which stage the solvent is introduced and polymerization reaction is further continued. Polymers with molecular weights of up to $M_w = 800,000$ were realized by this method [23].

3.3.4 Polymerization of $Cl_3P=NSiMe_3$

Another method of preparation of poly(dichlorophosphazene), $[NPCl_2]_n$, involves an ambient temperature (25 °C) polymerization of the P-trichloro-silyl phosphoranimine $Cl_3P=NSiMe_3$ (see Eq.3.40) [21, 26, 27].

$$n\ Cl_3P{=\!=\!=}NSiMe_3 \xrightarrow[\substack{CH_2Cl_2 \\ 25\,°C}]{2n\ PCl_5} \left[N{=}\underset{\underset{Cl}{|}}{\overset{\overset{Cl}{|}}{P}} \right]_n \tag{3.40}$$

The salient features of this process are as follows:

1. It is a cationic polymerization and it is a living polymerization.
2. This process provides a very good control over the chain length of the polymer.
3. The molecular weights of $[NPCl_2]_n$ produced by this process are about 10^6 with narrow-molecular-weight distribution (PDI: 1.01-1.18).
4. The scope and potential of this polymerization method seem to be quite large.

The monomer $Cl_3P=NSiMe_3$ can be synthesized by two methods. The first method consists of reacting bis(trimethylsilyl)amine $HN(SiMe_3)_2$ with n-butyllithium to afford lithum bis(trimethylsilyl)amide, $LiN(SiMe_3)_2$ (see Eq. 3.41). The latter is reacted with PCl_5 in hexane at -78 °C to afford $Cl_3P=NSiMe_3$ (see Eq. 3.42).

$$HN(SiMe_3)_2\ +\ nBuLi \longrightarrow LiN(SiMe_3)_2 \tag{3.41}$$

$$(\text{Me}_3\text{Si})_2\text{NLi} + \text{PCl}_5 \xrightarrow[\substack{-\text{LiCl} \\ -\text{Me}_3\text{SiCl}}]{-78\,°\text{C}} \text{Me}_3\text{SiN}{=}\text{PCl}_3 \qquad (3.42)$$

The chief drawback of this synthesis (Eqs. 3.41-3.42) is that a side product $(\text{Me}_3\text{Si})_2\text{NCl}$ is also formed in this reaction. This compound is a potent inhibitor for the polymerization of $\text{Cl}_3\text{P}{=}\text{NSiMe}_3$. The close boiling points of $\text{Cl}_3\text{P}{=}\text{NSiMe}_3$ and $(\text{Me}_3\text{Si})_2\text{NCl}$ make it difficult to separate the latter from the former by distillation methods.

An alternative method of synthesis of $\text{Cl}_3\text{P}{=}\text{NSiMe}_3$ overcomes the above problem. Thus, the reaction of tris(trimethylsilyl)amine $(\text{Me}_3\text{Si})_3\text{N}$ with PCl_5 in methylene chloride at 0 °C, immediately followed by quenching with n-hexane affords $\text{Cl}_3\text{P}{=}\text{NSiMe}_3$ in about 40% yield. This compound resonates at -54.0 ppm in its ^{31}P NMR.

The monomer $\text{Cl}_3\text{P}{=}\text{NSiMe}_3$ can be polymerized by the use of various initiators such as PCl_5, or $[(\text{Ph}_3\text{C})(\text{SbCl}_6)]$. The most widely used solvent is methylene chloride although other solvents such as chloroform or toluene have also been found to be effective. The polymerization reaction also occurs in the bulk state (absence of solvent). The polymerization is initiated by two moles of PCl_5 which react with $\text{Cl}_3\text{P}{=}\text{NSiMe}_3$ at room temperature to form $[\text{Cl}_3\text{P}{=}\text{N-PCl}_3]^+[\text{PCl}_6]^-$ (see Eq. 3.43). This cation can react further with $\text{Cl}_3\text{P}{=}\text{NSiMe}_3$ to generate a polymeric species (see Eq. 3.44).

$$\text{Cl}_3\text{P}{=}\text{N-SiMe}_3 + 2\text{PCl}_5 \xrightarrow[-\text{Me}_3\text{SiCl}]{25\,°\text{C}} [\text{Cl}_3\text{P}{=}\text{N-PCl}_3]^+ [\text{PCl}_6]^- \qquad (3.43)$$

$$[\text{Cl}_3\text{P}{=}\text{N-PCl}_3]^+ + n\text{Cl}_3\text{P}{=}\text{N-SiMe}_3 \xrightarrow{-n\text{Me}_3\text{SiCl}} \left[\text{Cl}_3\text{P}{=}\text{N}\left(\overset{\text{Cl}}{\underset{\text{Cl}}{\overset{|}{\underset{|}{\text{P}}}}}{=}\text{N}\right)_n\text{PCl}_3\right]^+ [\text{PCl}_6]^- \qquad (3.44)$$

The reactions shown in Eqs. 3.43-3.44 are facilitated by two factors. 1) The reactivity of the N-Si and P-Cl bonds towards each other to eliminate Me_3SiCl. 2) The spontaneity of the P-N bond formation. The side product Me_3SiCl is quite volatile and is therefore easily removed.

The termination of the polymer chain can be achieved by the addition of N-silylphosphoranimines such as $\text{Me}_3\text{SiN}{=}\text{P(OCH}_2\text{CF}_3)_3$ or $\text{Me}_3\text{SiN}{=}\text{P}(t\text{Bu})(\text{Ph})(\text{F})$ (see Eq. 3.45).

$$\left[\text{Cl}_3\text{P}{=}\text{N}\left(\overset{\text{Cl}}{\underset{\text{Cl}}{\overset{|}{\underset{|}{\text{P}}}}}{=}\text{N}\right)_n\text{PCl}_2\right]^+ [\text{PCl}_6]^- \xrightarrow[\substack{-2\,\text{Me}_3\text{SiCl} \\ -\text{PCl}_5}]{2\,\text{Me}_3\text{SiN}{=}\text{P(OR)}_3} (\text{OR})_3\text{P}{=}\text{N}-\overset{\text{Cl}}{\underset{\text{Cl}}{\overset{|}{\underset{|}{\text{P}}}}}{=}\text{N}\left(\overset{\text{Cl}}{\underset{\text{Cl}}{\overset{|}{\underset{|}{\text{P}}}}}{=}\text{N}\right)_n\overset{\text{Cl}}{\underset{\text{Cl}}{\overset{|}{\underset{|}{\text{P}}}}}-\text{N}{=}\text{P(OR)}_3 \qquad (3.45)$$

3.3.4.1 Block Copolymers

The fact that the cationic polymerization of $Cl_3P=NSiMe_3$ leads to a living polymer makes this process amenable for the synthesis of block copolymers. Thus, after the complete conversion of the first monomer $Cl_3P=NSiMe_3$, addition of a second monomer such as $Me_3SiN=PR_2Cl$ to the living polymer leads to a *diblock* co-polymer (see Eq.3.46). Of course, both blocks contain the P=N back bone but the substituents on phosphorus vary in the two block segments [21, 23, 26].

$$\left[Cl_3P=N\left(\underset{\underset{Cl}{|}}{\overset{\overset{Cl}{|}}{P}}=N\right)_n PCl_3\right]^+ \xrightarrow{n\ Me_3SiN=PR_2Cl} Cl_3P=N\left(\underset{\underset{Cl}{|}}{\overset{\overset{Cl}{|}}{P}}=N\right)_n\left[\underset{\underset{Cl}{|}}{\overset{\overset{Cl}{|}}{P}}=N\right]\left[\underset{\underset{R}{|}}{\overset{\overset{R}{|}}{P}}=N\right]_n PR_2Cl_2 \tag{3.46}$$

$$\underbrace{}_{\text{Block A}} \qquad \underbrace{}_{\text{Block B}}$$

It appears that several types of block polymers can be assembled by this above approach.

3.3.4.2 Polymers with Star Architectures

The cationic polymerization method of $Cl_3P=NSiMe_3$ has many new applications. The preparation of dendrimeric polymers with star-type architecture illustrates the versatility of this method (Fig. 3.29) [21].

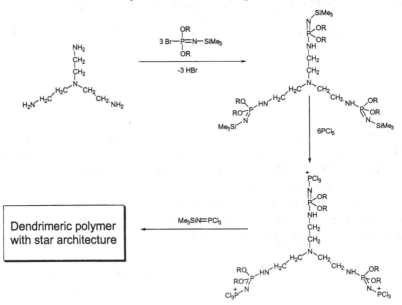

Fig. 3.29. Preparation of dendrimeric polymers

Dendrimeric polymers with three branches emanating from a central core have been prepared (Fig. 3.29). This approach begins with tris(2-aminoethyl)amine which upon reaction with $Me_3SiNP(OR)_2Br$ can be converted to a tris-phosphoranimine. The latter can be converted into the tris-phosphonium cation by reaction with six equivalents of PCl_5. This upon reaction with $Me_3SiN=PCl_3$ allows the polymer to grow in three different directions giving an overall star architecture. It is believed that such a variation of polymer architecture is likely to lead to new properties [21, 23, 26].

3.3.4.3 Polymers Containing Organic Functional End Groups

Another application of the cationic polymerization method of the chloro-phosphoranimines is the preparation of polymers with organic functional groups. Polymers, thus prepared, can be manipulated further with the aid of these terminal functional groups. This methodology is illustrated by the following example. Halogenophosphoranimines such as $BrP(OCH_2CF_3)_2=NSiMe_3$ can be readily modified by replacement of the bromine substituent on phosphorus by an organic substituent that contains vinyl end group to afford $(CH_2=CHCH_2NH)P(OCH_2CF_3)_2=NSiMe_3$ (see Eq. 3.47) [21].

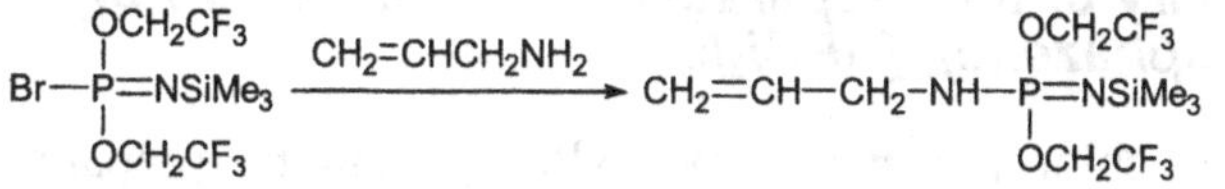

$$ (3.47) $$

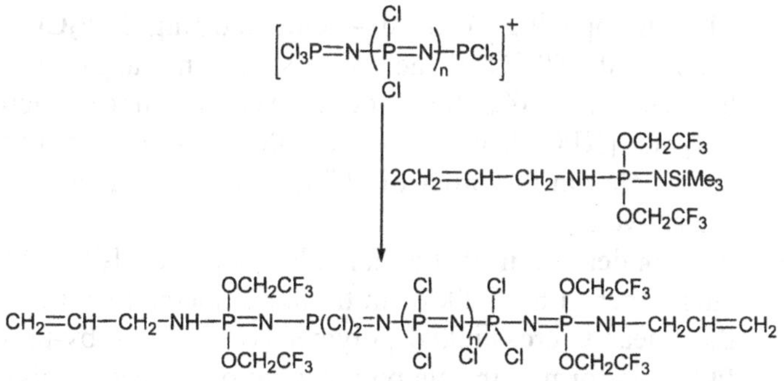

Fig. 3.30. Polymerization to afford polymers that contain terminal vinyl groups

Vinyl end group phosphoranimines can be used to terminate polymer chains. Such polymers would have *two* vinyl end groups which can be utilized further for generating a carbon chain on either end (Fig. 3.30).

Alternatively, the modified phosphoranimine can be used at the beginning of the polymerization reaction as shown in Fig. 3.31.

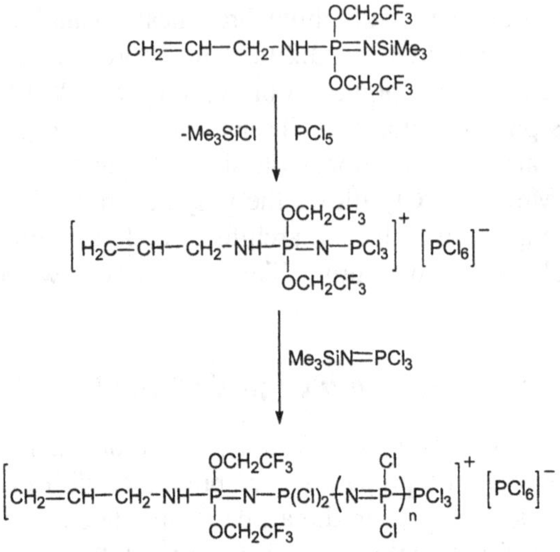

$$CH_2=CH-CH_2-NH-\overset{\overset{\displaystyle OCH_2CF_3}{|}}{\underset{\underset{\displaystyle OCH_2CF_3}{|}}{P}}=NSiMe_3$$

$$\xrightarrow[-Me_3SiCl]{PCl_5}$$

$$\left[H_2C=CH-CH_2\text{-}NH-\overset{\overset{\displaystyle OCH_2CF_3}{|}}{\underset{\underset{\displaystyle OCH_2CF_3}{|}}{P}}=N-PCl_3 \right]^{+} \left[PCl_6 \right]^{-}$$

$$\xrightarrow{Me_3SiN=PCl_3}$$

$$\left[CH_2=CH-CH_2-NH-\overset{\overset{\displaystyle OCH_2CF_3}{|}}{\underset{\underset{\displaystyle OCH_2CF_3}{|}}{P}}=N-P(Cl)_2\!\left(\!N=\overset{\overset{\displaystyle Cl}{|}}{\underset{\underset{\displaystyle Cl}{|}}{P}}\!\right)_{\!n}\!PCl_3 \right]^{+} \left[PCl_6 \right]^{-}$$

Fig. 3.31. Preparation of polyphosphazenes containing one vinyl end-group

These polymers have one vinyl group as the end group which can be utilized for polymer modification by a new chain-growth process.

3.3.4.4 Summary of the Preparative Procedures for Poly-(dichlorophosphazene), [NPCl₂]ₙ

1. Poly(dichlorophosphazene), $[NPCl_2]_n$, can be prepared by at least three different methods of polymerizaton.
2. The ring opening of the six-membered ring, $N_3P_3Cl_6$, occurs in vacuum at 250 °C. If the conversion is not allowed to proceed by more than 70%, this process leads to a linear uncrosslinked polymer $[NPCl_2]_n$ where n = 15,000. This polymer has molecular weights in the order of 10^6 with the polydipersity index being above 2.
3. Polycondensation of the monophosphazene $Cl_3P=NP(O)Cl_2$ occurs either in the bulk or in trichlorobiphenyl solution to afford the linear uncrosslinked polymer $[NPCl_2]_n$. A by-product viz., $POCl_3$ is formed in this polymerization process. Polymers with molecular weights of up to 800,000 can be prepared by this method.
4. Cationic polymerization of the N-silylphosphoranimine $Cl_3P=NSiMe_3$ by initiators such as PCl_5 afford a living polymer. High-molecular-weight $[NPCl_2]_n$ (10^6) with very narrow PDI's (1.01-1.18) are accessible from this approach.

3.3.5 ROP of Substituted Cyclophosphazenes

Substituted chloro- and fluorocyclophosphazenes can also be polymerized to their high polymers by the ROP method. Several alkylcyclophosphazenes such as $N_3P_3Cl_5R$, have been polymerized. These can be prepared from the corresponding hydridophosphazenes (see Eq. 3.48).

$$(3.48)$$

Although cyclophosphazenes containing sterically hindered alkyl groups such as iC_3H_7 or tBu cannot be polymerized to their high polymers, compounds containing sterically unencumbered groups such as Me, Et, nPr or nBu could be readily polymerized (see Eq. 3.49). Other types of cyclophosphazenes, $N_3P_3Cl_5R$ (R = CH_2SiMe_3, Ph, $N=PCl_3$, OCH_2CF_3, carboranyl, metallocenyl) can also be polymerized [28].

$$(3.49)$$

R = Me, Et, $n\text{-}C_3H_7$, $n\text{-}C_4H_9$

Monosubstituted fluorocyclophosphazenes such as $N_3P_3F_5Ph$ and $N_3P_3F_5tBu$ can be polymerized by ring-opening polymerization to the corresponding polymer (see Eq. 3.50).

$$(3.50)$$

Some disubstituted cyclophosphazenes such as [{NPCl_2}_2{NP(R)(CH_2SiMe_3)}]$ have also been polymerized by the ring-opening method to afford the linear polymers (see Eq. 3.51) [29].

$$(3.51)$$

R = CH_3; C_2H_5; $n\text{-}C_4H_9$, C_6H_5

When R = tBu or neoC$_5$H$_{11}$ in the above case (Eq. 3.51) the polymerization is retarded. Some other disubstituted derivatives that have been polymerized include gem-N$_3$P$_3$Cl$_4$(N=PCl$_3$)$_2$, non-gem-N$_3$P$_3$Cl$_4$Me$_2$, non-gem-N$_3$P$_3$Cl$_4$Et$_2$.

Is the presence of chlorine or fluorine substituents on phosphorus necessary for ring-opening polymerization? An overwhelming majority of the compounds that have been polymerized have this feature. A few examples are known that do not have a P-Cl bond but which have been polymerized. For example, the hexapyridinoxycyclotriphosphazene undergoes ROP to afford a polymer with moderate molecular weights (see Eq. 3.52) [30].The mechanism of the polymerization has been suggested to be cationic in nature (heterolytic cleavage of P-O bond in the monomer).

(3.52)

As mentioned previously, heating N$_3$P$_3$Me$_6$ leads to ring expansion to lead to the formation of N$_4$P$_4$Me$_8$. However, polymerization to the corresponding linear polymer is not observed. If there is a $built$-in ring strain in the cyclotriphosphazene structure, the relief of such strain can be a driving force for polymerization. Thus, the reaction of N$_3$P$_3$F$_6$ with 1,1'-dilithioferrocene affords the $ansa$ product N$_3$P$_3$F$_4$(η^5-C$_5$H$_4$-Fe-η^5-C$_5$H$_4$). This compound can be polymerized to afford a polymer where two adjacent phosphorus centers are linked to one ferrocene moiety. Alternatively, the trifluoroethoxy derivative N$_3$P$_3$(OCH$_2$CF$_3$)$_4$(η^5-C$_5$H$_4$-Fe-η^5-C$_5$H$_4$) can also be polymerized in the presence of 1% N$_3$P$_3$Cl$_6$ to afford the corresponding polymer. The latter can also be prepared by the replacement of fluorine atoms from the corresponding poly(fluorophosphazene) (Fig. 3.32) [16,21].

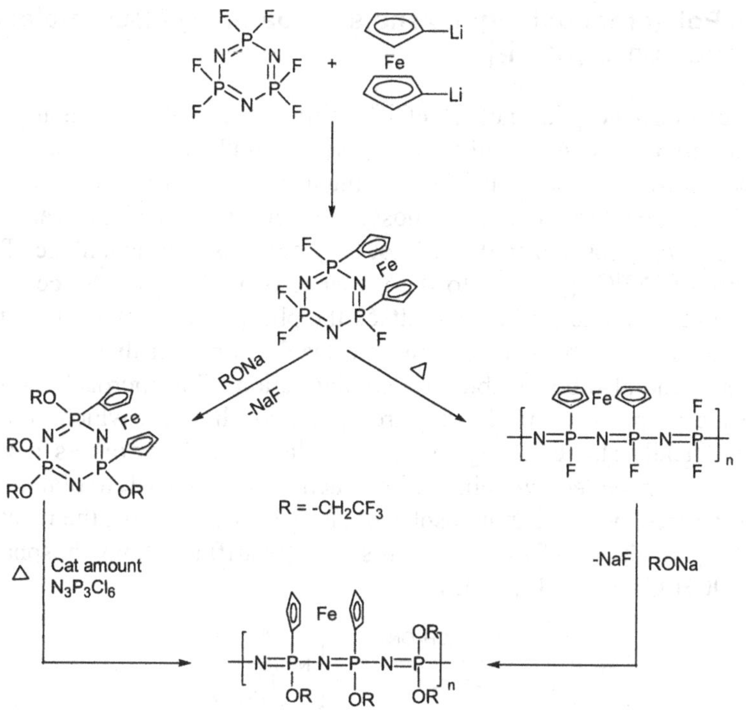

Fig. 3.32. Polymerization of strained cyclophosphazenes

3.3.5.1 Summary of ROP of Organophosphazenes

1. In general, persubstituted cyclotriphosphazenes containing amino, alkoxy or aryloxy substituents cannot be polymerized by ROP. However, a few exceptions exist.
2. Many monosubstituted cyclophosphazenes, $N_3P_3Cl_5R$, with various kinds of R groups can be polymerized by ROP to the corresponding linear polymers. Similarly, many $N_3P_3F_5R$ can also be polymerized to the linear polymers.
3. A few di- and trisubstituted cyclophosphazenes, $N_3P_3Cl_4R_2$ and $N_3P_3Cl_3R_3$, can also be polymerized by ROP.
4. Strained cyclophosphazenes such as non-gem-$N_3P_3F_4(\eta^5$-C_5H_4-Fe-η^5-C_5H_4) and even the alkoxy derivative $N_3P_3(OCH_2CF_3)_4(\eta^5$-$C_5H_4$-Fe-$\eta^5$-$C_5H_4$) can be polymerized by ROP to the linear polymers.

3.3.6 Poly(organophosphazene)s Prepared by Macromolecular Substitution of [NPCl$_2$]$_n$

Poly(dichlorophosphazene), [NPCl$_2$]$_n$, although a high-molecular- weight linear polymer, is not useful for any practical application because of its extreme hydrolytic instability. The eventual products of hydrolysis are ammonia, hydrochloric acid, and phosphoric acid. However, the reactivity of poly(dichlorophosphazene) need not be viewed as a disadvantage. The reactivity of [NPCl$_2$]$_n$ is due to the presence of P-Cl bonds. Since water is able to act as a nucleophile and attack the phosphorus center due to the lability of the P-Cl bond it is quite reasonable to expect that other nucleophiles would also be capable of a similar action. This approach would enable the preparation of chlorine-free polyphosphazenes which should be hydrolytically stable. Using this logic, Allcock and coworkers have performed *macromolecular* substitution reactions on [NPCl$_2$]$_n$ to obtain stable, chlorine-free, poly(organophosphazene)s [16, 21-23]. Thus, the reaction of [NPCl$_2$]$_n$ with CF$_3$CH$_2$ONa affords poly(bistrifluorethoxyphosphazene), [NP(OCH$_2$CF$_3$]$_n$ (see Eq. 3.53).

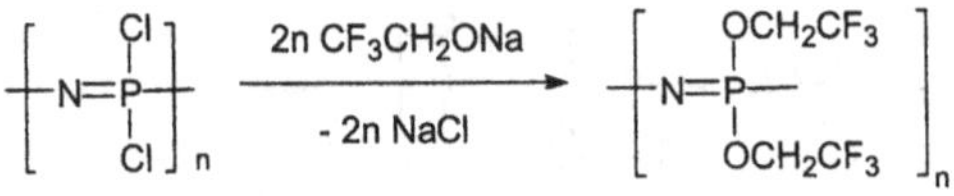

$$
\left[\begin{array}{c} Cl \\ | \\ -N=P- \\ | \\ Cl \end{array} \right]_n \xrightarrow[-\ 2n\ NaCl]{2n\ CF_3CH_2ONa} \left[\begin{array}{c} OCH_2CF_3 \\ | \\ -N=P- \\ | \\ OCH_2CF_3 \end{array} \right]_n \qquad (3.53)
$$

This type of a reaction is quite general. Thus, the reaction of [NPCl$_2$]$_n$ with RONa (R = alkyl or aryl) affords the corresponding polymers [NP(OR)$_2$]$_n$. Similarly, the reactions of [NPCl$_2$]$_n$ with primary or secondary amines affords the corresponding amino derivatives [NP(NHR)$_2$]$_n$ and [NP(NR$_2$)]$_n$ (Fig. 3.33) [21-23].

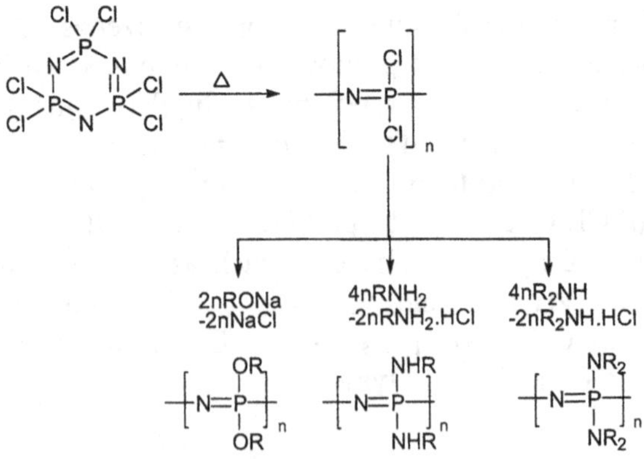

Fig. 3.33. P-Cl Substitution reactions on [NPCl$_2$]$_n$

Even exotic substituents such as those containing metal carbonyl units can be introduced by the nucleophilic substitution reaction (Fig. 3.34).

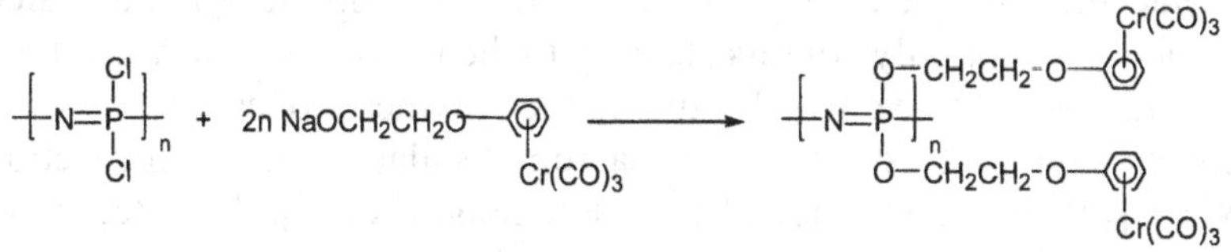

Fig. 3.34. Polyphosphazenes containing organometallic substituents

Other representative examples of polyphosphazenes that have been prepared by the macromolecular substitution of $[NPCl_2]_n$ are shown in Figs. 3.35-3.36.

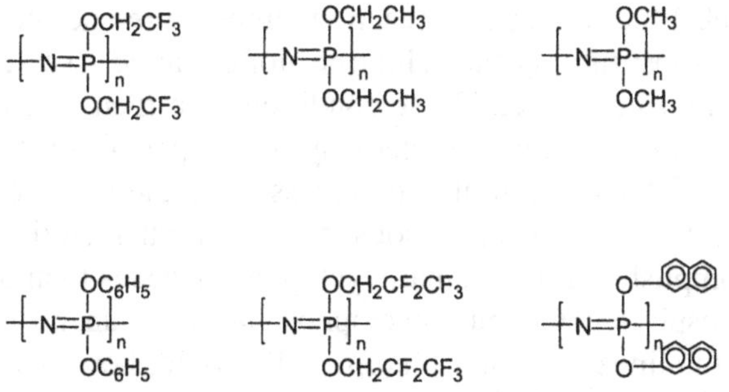

Fig. 3.35. Polyphosphazenes containing alkoxy and aryloxy substituents

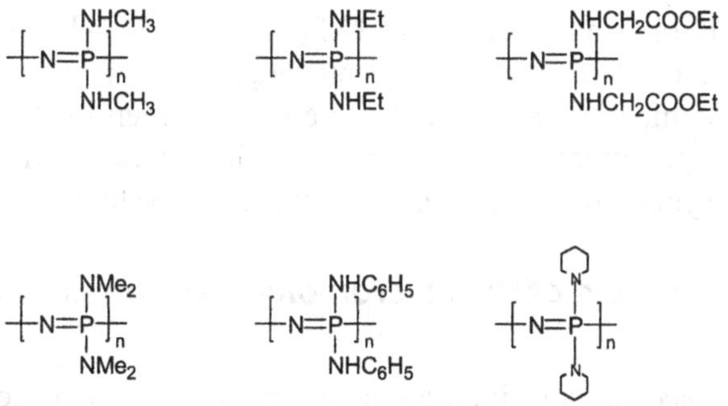

Fig. 3.36. Polyphosphazenes containing alkyl- and arylamino substituents

There are two remarkable aspects of the macromolecular substitution reaction that need to be mentioned. This involves about 30,000 nucleophilic substitution reactions per polymer molecule! This means that for every reacting polymer molecule 30,000 chlorine atoms are being replaced. The obvious *disadvantage* of the parent polymer $[NPCl_2]_n$ is turned around as its major *advantage*. Thus, by this method a number of poly(organophosphazene)s containing P-N or P-O linked side-chains have been assembled. The replacement of the chlorine atoms by the nucleophiles (amines, alcohols or phenols) not only confers stability to the polymers so formed, but it also provides an effective way for tuning the polymer properties.

Another important aspect of the macromolecular substitution reactions on $[NPCl_2]_n$ is the importance of model compounds [31]. As noted above nucleophilic substitution reactions on $[NPCl_2]_n$ are going to be complex because of the large number of such reactions that have to occur per molecule of the polymer. Because of this, often, it is more valuable to carry out a study of nucleophilic substitution reactions on the small molecules viz., the six- or eight-membered rings $N_3P_3Cl_6$ and $N_4P_4Cl_8$ before carrying them out on the polymer. Various factors such as ideal solvent for the nucleophilic substitution reaction, reaction temperature and the time required for complete substitution of the chlorine atoms can be assessed more easily at the small molecule level. These conditions can then be adapted and applied to the more complex polymeric system. Apart from synthetic procedures structural factors can also be assessed by the study of small molecules. Also, full and unambiguous structural characterization is more readily accomplished for the small cyclophosphazenes than the more complex polyphosphazenes. Thus, cyclophosphazenes can be used as *model systems* for the linear polyphosphazenes. It should, however, be noted that polymer behavior in solution is far more complex than those of simple molecules. The conformation and polymer structure in solution is likely to be continuously affected as replacement of chlorines progresses with nucleophiles. Furthermore, unlike in $N_3P_3Cl_6$, in $[NPCl_2]_n$ all the chlorines will not be equally *exposed* to the nucleophiles. It is remarkable that in spite of these inherent and basic differences between the small molecules and the large polymers, the model approach has worked quite well with respect to the cyclophosphazene–polyphosphazene system.

3.3.6.1 Limitations of the Macromolecular Substitution Approach

The macromolecular substitution reaction has been very successful as noted above and several hundred polyphosphazenes have been prepared using this approach. However, there are two important limitations as noted below.

One, *complete* replacement of chlorine atoms from poly(dichlorophosphazene) is not possible with all nucleophiles. Thus, with sterically hindered nucleophiles such as diethylamine, only partial replacement of chlorines occurs. The remaining chlorines have to be replaced with other less hindered substituents (see Eq. 3.54) [21].

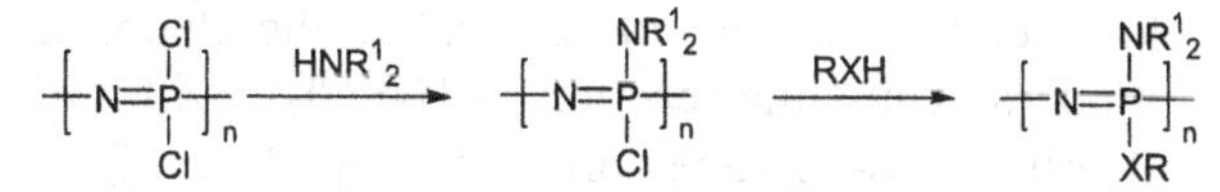

$$
\begin{array}{c}
\left[\!\!-N\!\!=\!\!\overset{\displaystyle Cl}{\underset{\displaystyle Cl}{P}}\!\!-\!\right]_n \xrightarrow{\ HNR^1_2\ } \left[\!\!-N\!\!=\!\!\overset{\displaystyle NR^1_2}{\underset{\displaystyle Cl}{P}}\!\!-\!\right]_n \xrightarrow{\ RXH\ } \left[\!\!-N\!\!=\!\!\overset{\displaystyle NR^1_2}{\underset{\displaystyle XR}{P}}\!\!-\!\right]_n
\end{array}
\qquad (3.54)
$$

R^1 = sterically bulky group
X = O or NH
R = sterically less hindered group

A very good example of the above behavior occurs with adamantyl containing nucleophiles (adamantyl amine, adamantyl alcohol and adamantyl methanol) [32]. Only partial replacement of chlorine atoms occurs even when the adamantly group is separated from the rest of the molecule by a spacer group (Fig. 3.37).

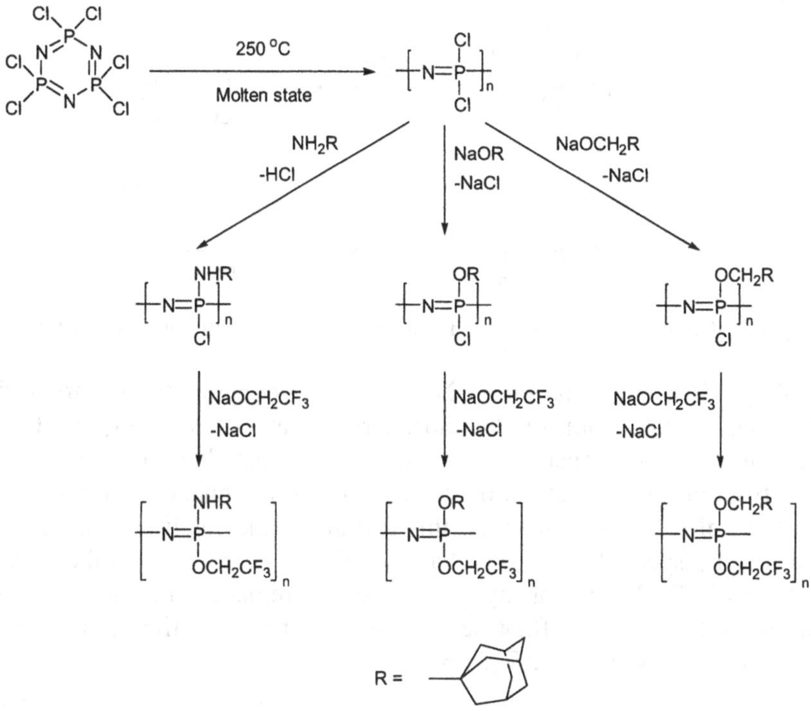

Fig. 3.37. Reactions of $[NPCl_2]_n$ with sterically hindered adamantyl nucleophiles

The second limitation of the macromolecular substitution reaction is that it cannot be applied for the preparation of polyphosphazenes containing P-C bonds.

Reactions of Grignard reagents or alkyllithium reagents with $[NPCl_2]_n$ are not useful in generating $[NPR_2]_n$ because of chain-degradation reactions. The lone pair of electrons at nitrogen interferes in the reactions with RMgX or RLi, by involving in coordination to the metal ion (Fig. 3.38)

[22]. It is speculated that such interactions weaken the chain, causing the formation of oligomers. Thus, the reaction of $[NPCl_2]_n$ with Grignard reagents, RMgX, affords chain-degraded products (Fig. 3.39) [2]. The cyclic rings $N_3P_3Cl_6$ and $N_4P_4Cl_8$ themselves undergo ring degradation in their reactions with Grignard or alkyl(aryl)lithium reagents.

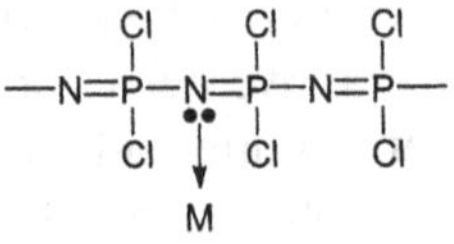

$$\begin{array}{ccc} Cl & Cl & Cl \\ | & | & | \\ -N{=}P{-}N{=}P{-}N{=}P{-} \\ | & | & | \\ Cl & Cl & Cl \\ & M & \end{array}$$

Fig. 3.38. Interaction of the lone pair of electrons on the phosphazene backbone with the metal ion of the Grignard or alkyllithium reagent

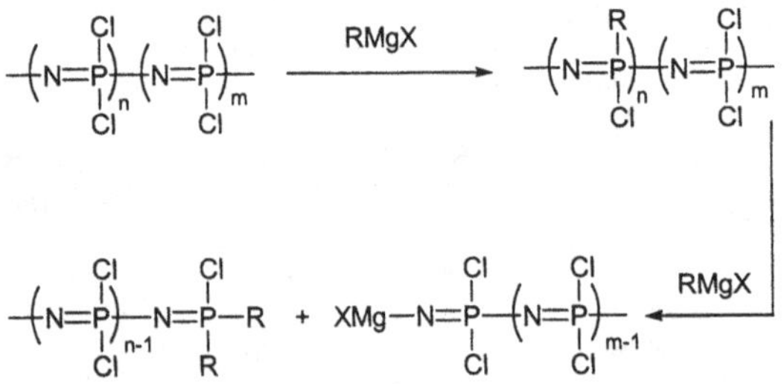

Fig. 3.39. Chain degradation in the reaction of $[NPCl_2]_n$ with Grignard reagents

Poly(difluorophosphazene), $[NPF_2]_n$, which contains the electron-withdrawing fluorine atoms on the phosphorus center would be expected to reduce the electron density at the nitrogen centers. This would inhibit the ability of nitrogen atoms in the polymer chain to engage in an interaction with metal ions. Consequently, this feature should facilitate normal reactions and (especially) prevent chain scission. Accordingly, in the reaction between $[NPF_2]_n$ with phenyllithium normal replacement reactions occur up to about 70-75% of fluorine replacement. It is only after this stage that chain scission occurs (see Eq. 3.55) [21].

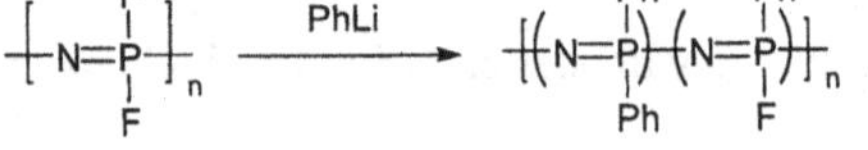

$$\begin{array}{cc} F & \\ | & \\ {+}N{=}P{+}_n & \xrightarrow{PhLi} \end{array} \quad \begin{array}{cc} Ph & Ph \\ | & | \\ {+}(N{=}P){-}(N{=}P){+}_n \\ | & | \\ Ph & F \end{array} \tag{3.55}$$

It is to be noted that a similar reaction of phenyllithium with $[NPCl_2]_n$ leads to chain scission almost as soon as the reaction commences (see Eq. 3.56).

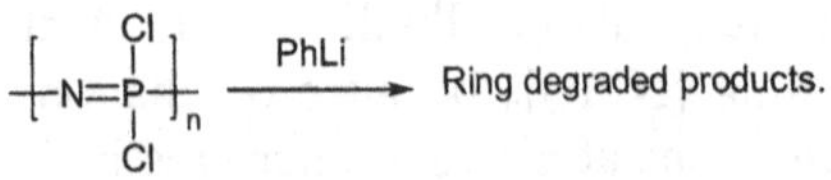

$$\begin{array}{c} Cl \\ | \\ {+}N{=}P{+}_n \end{array} \xrightarrow{PhLi} \text{Ring degraded products.} \tag{3.56}$$

However, $(NPF_2)_n$ is a particularly difficult polymer to be used for routine macromolecular substitution reactions. This is because of the tedious nature of its preparation as well as the fact that it is soluble only in fluorinated solvents. Thus, the preparation of poly(alkylphosphazene)s and poly(arylphosphazene)s requires a completely different approach. This is accomplished by adopting the approaches outlined later.

3.3.6.2 Summary of Macromolecular Substitution Reactions

1. The linear polymer $[NPCl_2]_n$ is very reactive and undergoes hydrolysis in moist air to afford ammonium chloride and phosphoric acid.
2. The reactions of $[NPCl_2]_n$ with many nucleophiles such as amines, alkoxides and aryloxides occurs with replacement of chlorines to afford $[NP(NHR)_2]_n$, $[NP(NR_2)_2]_n$ and $[NP(OR)_2]_n$. Most of these polymers are hydrolytically more stable than the parent polymer. It is advantageous to perform the nucleophilic reactions first on the cyclic model compound $N_3P_3Cl_6$ before implementing these on the polymeric system.
3. Reactions of $[NPCl_2]_n$ with sterically hindered nucleophiles do not lead to complete replacement of chlorines. The partially substituted polymers can be further reacted with an appropriate nucleophile such as trifluoroethoxide to afford *chlorine-free* polymers.
4. Reactions of $[NPCl_2]_n$ with Grignard reagents or with alkyl/aryllithium reagents leads to chain-scission products. On the other hand, reactions of $[NPF_2]_n$ with phenyllithium leads to about 75% substitution of fluorines without chain scission.

3.3.7 Preparation of Polyphosphazenes Containing P-C Bonds by Thermal Treatment of Phosphoranimines

As mentioned in the previous section, poly(dichlorophosphazene) $[NPCl_2]_n$ can not be used as a precursor for the preparation of poly(alkyl-) or poly(arylphosphazene)s. This is due to the chain-scission reaction suffered by $[NPCl_2]_n$ upon reaction with Grignard reagents or alkyl- and aryllithium reagents. The methodology for preparing polyphosphazenes that contain the P-C bonds is dependent on the condensation reactions of the N-silylphosphoranimines, $Me_3SiN=PR_2X$, by the elimination of Me_3SiX. Thus, one of the first examples of the successful implementation of this synthetic route consisted in the preparation of poly(dimethylphosphazene) [14, 33]. The synthetic methodology consists of the following steps:

1. Preparation of the silylaminophosphine derivatives, $(Me_3Si)_2NPR_2$. Note that these compounds already contain the "R" groups that are eventually required in the polymer.
2. Oxidation of $(Me_3Si)_2NPR_2$ to the N-silyl-P-halogeno-phosphoranimines (or simply the monophosphazenes), $Me_3SiN=PR_2Br$. These derivatives themselves were initially found to be unsuitable as *monomers* for the preparation of polymers (as we will see slightly later such compounds containing P-Cl bonds such as $Me_3SiN=PR_2Cl$ can be polymerized by the use of PCl_5 as an initiator).
3. Conversion of $Me_3SiN=PR_2Br$ to the corresponding trifluorethoxy derivative, $Me_3SiN=PR_2(OCH_2CF_3)$.
4. Thermal treatment of $Me_3SiN=PR_2(OCH_2CF_3)$ results in the elimination of $Me_3SiOCH_2CF_3$ and to the formation of the high polymer $[NPMe_2]_n$.

This sequence of reactions is shown in Fig. 3.40.

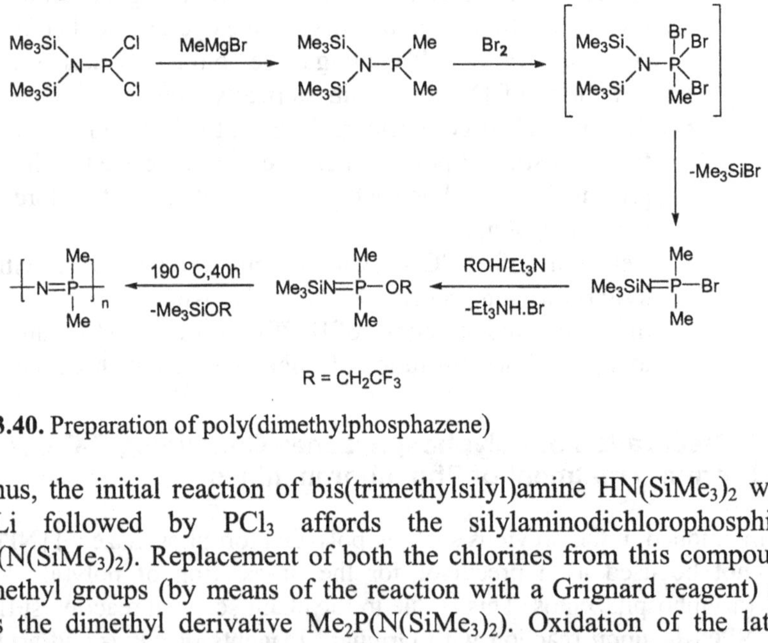

Fig. 3.40. Preparation of poly(dimethylphosphazene)

Thus, the initial reaction of bis(trimethylsilyl)amine $HN(SiMe_3)_2$ with $nBuLi$ followed by PCl_3 affords the silylaminodichlorophosphine $Cl_2P(N(SiMe_3)_2)$. Replacement of both the chlorines from this compound by methyl groups (by means of the reaction with a Grignard reagent) affords the dimethyl derivative $Me_2P(N(SiMe_3)_2)$. Oxidation of the latter with Br_2 affords the compound $Me_3SiN=PR_2Br$ where the $NSiMe_3$ group is formally connected to the phosphorus by a double bond. This compound is converted to the trifluoroethoxy derivative $Me_3SiN=PR_2(OCH_2CF_3)$. Elimination of $Me_3SiOCH_2CF_3$ from $Me_3SiN=PR_2(OCH_2CF_3)$ occurs by a thermal treatment of this monomer at 190 °C to afford the linear polymer $[NPMe_2]_n$. In these polymers there is a direct P-C bond. Interestingly, heat treatment of the *N*-silyl-*P*-fluorophosphoranimines $Me_3SiN=PR_2F$ results in the elimination of Me_3SiF, but leads to the formation of cyclic derivatives. Polymers could not be obtained in this reaction (Fig. 3.41). A similar result has also been found with the chloro and bromo derivatives.

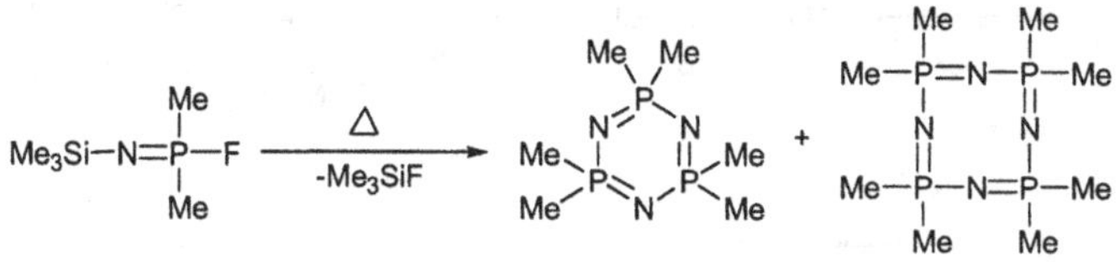

Fig. 3.41. Formation of $N_3P_3Me_6$ and $N_4P_4Me_8$ upon thermolysis of $Me_3SiNP(Me)_2F$

In general the aminophosphines, $N(SiMe_3)_2PR_2$ can be prepared by the reaction of the Grignard reagent with the corresponding dichloro derivative, $N(SiMe_3)_2PCl_2$ (see Eq. 3.57).

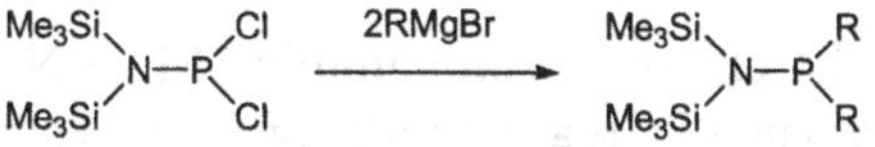

$$(3.57)$$

R = Me, Et, *n*-Pr, *n*-Bu, *n*-hex

Mixed alkyl derivatives containing one alkyl group and one methyl group can be prepared by first reacting $N(SiMe_3)_2PCl_2$ with RMgBr followed by reaction with MeMgBr (see Eq. 3.58).

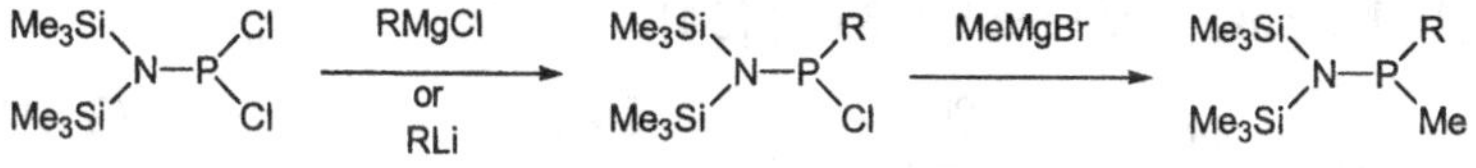

$$(3.58)$$

R = Me, Et, *n*-Pr, *n*-Bu, *n*-hex

In a similar manner the mixed phenyl/methyl derivative can also be prepared (see Eq. 3.59).

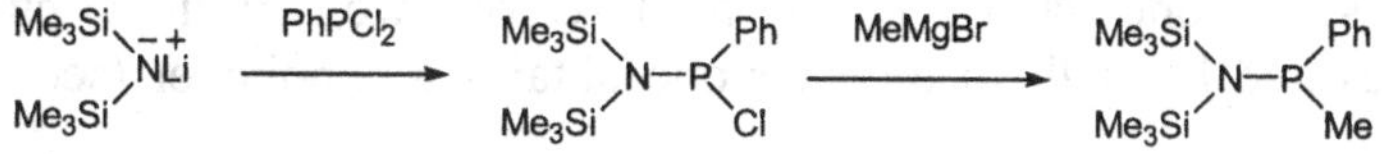

$$(3.59)$$

Recently, the need to use bromine has been avoided [33]. Chlorination of the silylaminophosphines with C_2Cl_6 converts them to the phosphoranimines with P-Cl end groups. These can be converted to the corresponding phenoxy or trifluoroethoxy derivatives which can be polymerized by thermal treatment. With the former, sodium phenoxide is required as an initiator for the polymerization. It should be noted that the thermolysis of other phosphoranimine alkoxides such as Me_3SiNPR_2OMe does not lead to the formation of linear polymers (Fig. 3.42).

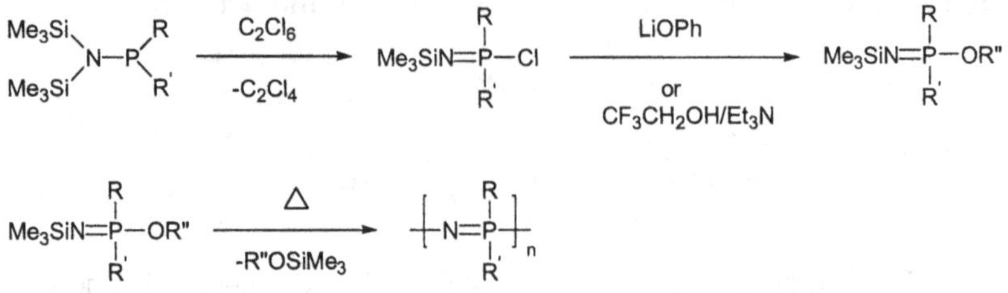

R" = Ph; CH_2CF_3
R = R' = Me, Et, nPr, nBu, $n$hex
R = Me, R' = Ph
R = Me; R' = Et, nPr, nBr, $n$hex

Fig. 3.42. Preparation of $[NPRR']_n$

On the other hand, the chlorophosphoranimine $Me_3SiN=PR_2Cl$ can be polymerized directly by using PCl_5 as an initiator. This is similar to the polymerization of $Me_3SiNPCl_3$ that we had seen earlier (see Eq. 3.60).

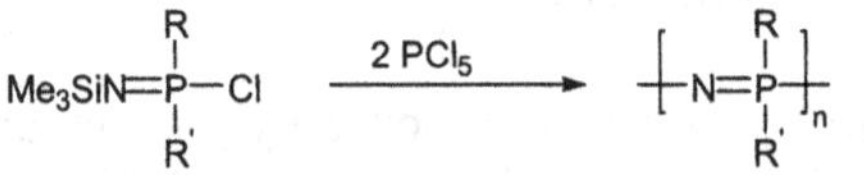

$$ (3.60) $$

The phosphoranimine approach can also be used to prepare poly-(fluoro/aryl)phosphazenes (see Eq. 3.61) [34].

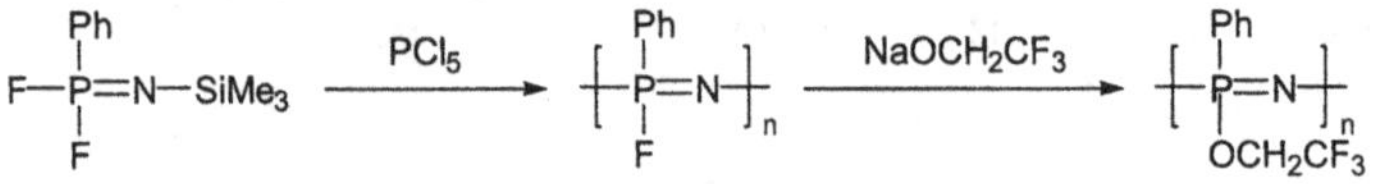

$$ (3.61) $$

Another unique feature of this polymerization is that using this methodology genuine copolymers can also be prepared. Thus, taking advantage of the nearly identical rates of polymerization of various alkyl/aryl phosphoranimines, if one takes a 1:1 mixture of two different phosphoranimines and heats them together one obtains a copolymer (see Eq. 3.62) [33].

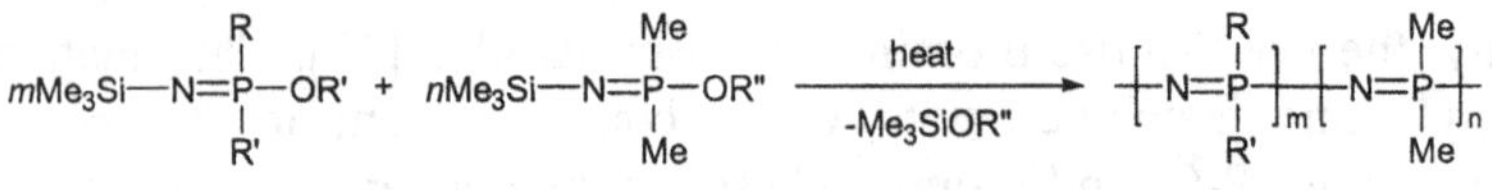

$$ (3.62) $$

R" = Ph, CH_2CF_3

R,R' = alkyl, aryl

The poly(alkyl/arylphosphazene)s prepared by the above method have molecular weights between 50,000 to 200,000 with a polydispersity index of 2.0. The solubility of these polymers seems to depend on the type of substituents present on the phosphorus. Thus, $[NPMe_2]_n$ is soluble in dichloromethane, chloroform, ethanol as well as a 1:1 mixture of THF and water. In contrast, $[NPEt_2]_n$ is virtually insoluble in any organic solvent. However, upon protonation with weak acids $[NPEt_2]_n$ becomes soluble. It has been speculated that protonation destroys the crystallinity of this polymer and enables its solubility. Poly(methylphenylphosphazene), $[NP(Me)(Ph)]_n$ is also soluble in a large number of organic solvents such as chlorinated hydrocarbons as well as THF [33].

Attempts to unravel the mechanism of the thermolysis polymerization of the phosphoranimines have suggested that it does not proceed by a step-growth polymerization. Thus, quenching the reaction at the stage of about 20-50% completion of the reaction and analyzing the products reveals the presence of medium-molecular-weight polymers. It may be recalled that in typical step-growth polymerization processes the molecular weights of the products do not increase dramatically till about 90-99% completion of the reaction. In view of this, a chain-growth mechanism is presumed probably initiated by the heterolytic cleavage of the P-X bond. In this respect the mechanism of polymerization probably is similar to that discussed for the ambient temperature polymerization of $Me_3SiNPCl_3$. The possibility of cyclic intermediates such as $N_3P_3Me_6$ or $N_4P_4Me_8$ is ruled out. It may be recalled that if $N_3P_3Me_6$ is independently thermolyzed, polymerization is not achieved; ring expansion to higher membered rings, however, occurs.

Another type of polymerization of the N-silylphosphoranimines occurs by anionic polymerization. Thus, monomers such as $Me_3SiNP(OR^1)_2(OR^2)$ can be polymerized by the use of initiators such as Bu_4NF to afford polymers $[NP(OR^1)_2]_n$ (see Eq. 3.63) [23]. Polymers with molecular weights (M_n) in excess of 100,000 are achieved.

$$Me_3Si-N{=}P(OR^1)(OR^1)-OR^2 \xrightarrow[-Me_3SiOR^2]{\substack{Bu_4NF \\ 100\ ^\circ C}} \left[N{=}P(OR^1)(OR^1) \right] \qquad (3.63)$$

$OR^1 = OCH_2CH_2OCH_3$ or $OCH_2CH_2OCH_2CH_2OCH_3$
$OR^2 = OCH_2CF_3$

3.3.7.1 Summary of the Thermal Polymerization of N-Silylphosphoranimines

1. A variety of *N*-silylphosphoranimines $Me_3SiNPRR'X$ can be readily prepared and used as monomers for preparing polyphosphazenes.
2. If the leaving group $X = OCH_2CF_3$ or OPh, polymerization can be achieved by heating the monomers between 190-220 °C. The thermolysis reaction leads to the elimination of the silyl ether Me_3SiX. If X = F, Cl or Br, thermolysis reaction does not lead to polymerization. However, if X = Cl, by the use of PCl_5 as the initiator, ambient temperature cationic polymerization can be achieved.
3. The poly(alkyl/arylphosphazene)s prepared by the thermolysis of $Me_3SiNPRR'X$ have high molecular weights between 50,000-200,000, a PDI of 2.0 and in general have good solubility in common organic solvents.
4. The mechanism of the polymerization is believed to proceed by a chain-growth mechanism which is initiated by the heterolysis of the P-X bond.
5. Copolymers of the type $[\{NP(R^1)(R^2)\}_x\{NP(R^3)(R^4)\}_y]_n$ can be prepared by thermolyzing a 1:1 mixture of $Me_3SiNP(R^1)(R^2)X$ and $Me_3SiNP(R^3)(R^4)X$.
6. Monomers of the type $Me_3SiNP(OR^1)_2(OR^3)$ can be polymerized by using anionic initiators such as Bu_4NF.

3.3.8 Modification of Poly(organophosphazene)s

Many poly(organophosphazene)s can be modified without chain scission. Thus, it is possible to carry out conventional organic reactions on the side groups (substituents) of the polyphosphazene backbone. This is an important feature that allows elaboration of polyphosphazenes into many new types of polymers. A few examples are given below to illustrate this principle.

3.3.8.1 Polyphosphazenes Containing Phosphino Ligands

Polyphosphazenes containing phosphino ligands can be prepared as shown in Fig. 3.43.

Reaction of $[NPCl_2]_n$ with the sodium salt of *p*-bromophenoxide affords $[NP(OC_6H_4\text{-}p\text{-}Br)_2]_n$. Lithiation of this polymer followed by reaction with PPh_2Cl affords the phosphine-containing polyphosphazene $[NP(OC_6H_4\text{-}p\text{-}PPh_2)_2]_n$ which can be considered as a polymeric ligand. A modification of this approach involves the substitution of chlorines on [NPCl$_2$]$_n$ by sodium phenoxide followed by reaction with the sodium salt of p-bromophenoxide to afford the mixed polymer [NP(OPh)(OC$_6$H$_4$-p-Br)]$_n$. This can be treated as above with n-butyllithium. Only the bromo substituents are affected. Both the polyphosphazene phosphino ligands as well as metal-containing polyphosphazenes can be made by this route. The ratio of phenoxide/p-bromophenoxide can be modulated to afford the desired loading of phosphino ligands. The phosphine ligands can be used to link transition metals to generate coordination/organometallic complexes. Such derivatives can be used in catalysis.

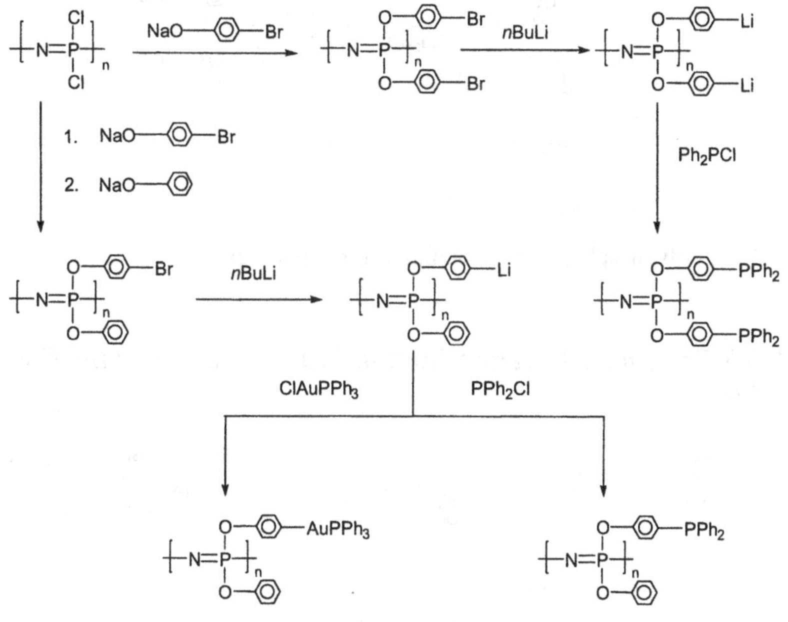

Fig. 3.43. Polyphosphazenes containing phosphino ligands

3.3.8.2 Polyphosphazenes Containing Carboxy Terminal Groups

Polyphosphazenes can be modified to incorporate new functional groups. Thus, for example polyphosphazenes containing carboxy groups can be obtained by the methodology shown in Fig. 3.44.

The reaction of polydichlorophosphazene, [NPCl$_2$]$_n$, with NaO-C$_6$H$_4$-p-COOEt affords [NP(O-C$_6$H$_4$-p-COOEt)$_2$]$_n$. Hydrolysis of this polymer with tBuOK does not cause chain scission but affords stable polymers containing COOH end groups, [NP(O-C$_6$H$_4$-p-COOH)$_2$]$_n$. Interestingly, such a polymer upon treatment with metal salts can lead to a crosslinked hydrogel.

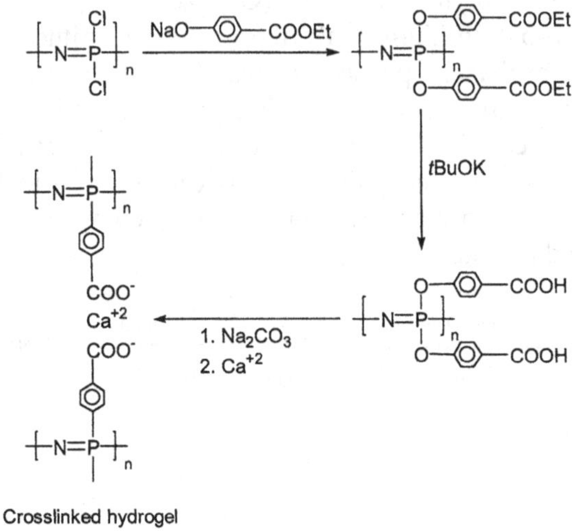

Fig. 3.44. Polyphosphazenes containing carboxyl end-groups

3.3.8.3 Polyphosphazenes Containing Amino or Formyl End Groups

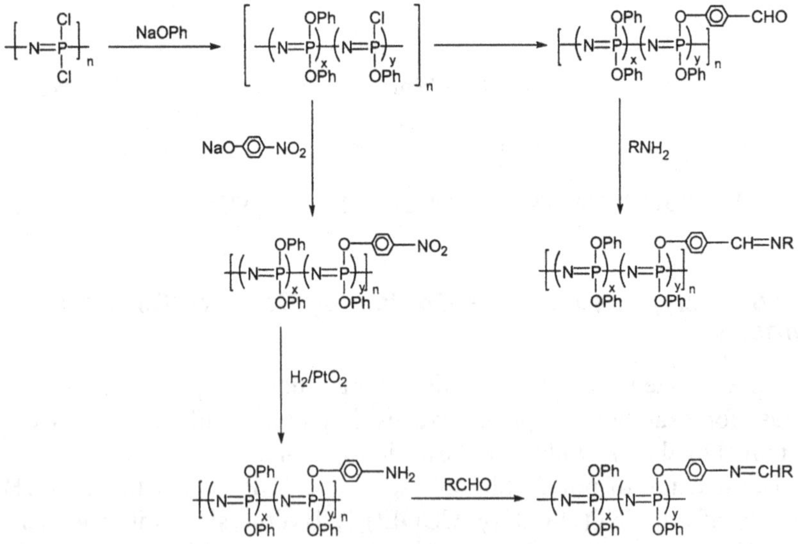

Fig. 3.45. Preparation of polyphosphazenes with Schiff base linkages

Another example of the derivative chemistry of polyphosphazenes is illustrated by the preparation of polyphosphazenes containing reactive functional end groups such as the amino or the formyl functional groups (Fig. 3.45). It can be readily appreciated that such functional groups can be further elaborated. Thus, polyphosphazenes containing Schiff base linkages can be prepared. The protocol consists of reacting poly(dichlorophosphazene), $[NPCl_2]_n$, with sodium phenoxide followed by sodium salt of p-nitrophenoxide to afford a polyphosphazene containing some nitro end groups. The nitro groups can be reduced to amino end groups by standard organic reaction procedures. The amino groups thus generated are now available for reaction with aldehydes to afford polymers with Schiff base linkages. This procedure can also be carried out by first putting the formyl groups on the polymer backbone followed by utilizing these functional groups for condensation with amines to afford polymers with Schiff base linkages (Fig. 3.45).

3.3.8.4 Polyphosphazenes Containing Glucosyl Side Groups

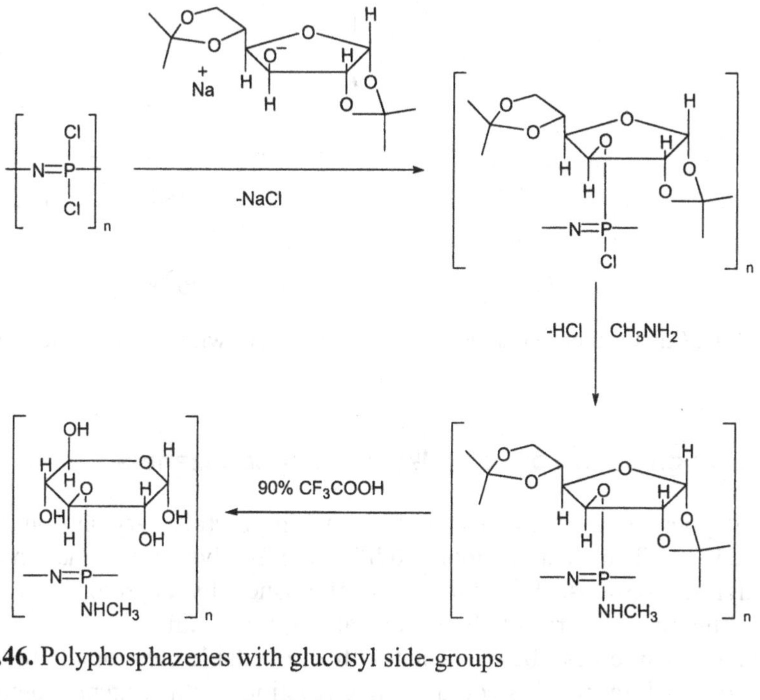

Fig. 3.46. Polyphosphazenes with glucosyl side-groups

Complex polyphosphazenes containing glucosyl side-groups have been prepared (Figs. 3.46-3.47) [35]. The multifunctional nature of the glucose reagents implies that some of these reactive groups have to be protected before a reaction is attempted. Thus, the sodium salt of diacetone glucose is allowed to react with poly(dichlorophosphazene) to replace the chlorine atoms partially. The partially substituted polymer is fully relieved of its remaining chlorines by reaction with methylamine (or even with an aryloxide such as phenoxide). The resulting polymer can be treated with 90% CF_3COOH to afford the deprotected glucosyl units as side-chains (Fig. 3.46). Crosslinking of the glucosyl groups either by radiation or by reaction with hexamethylisocyanate affords a crosslinked hydrogel (Fig. 3.47). It is important to once again note the robust nature of the polyphosphazene backbone that survives so many reactions carried out on the side chains.

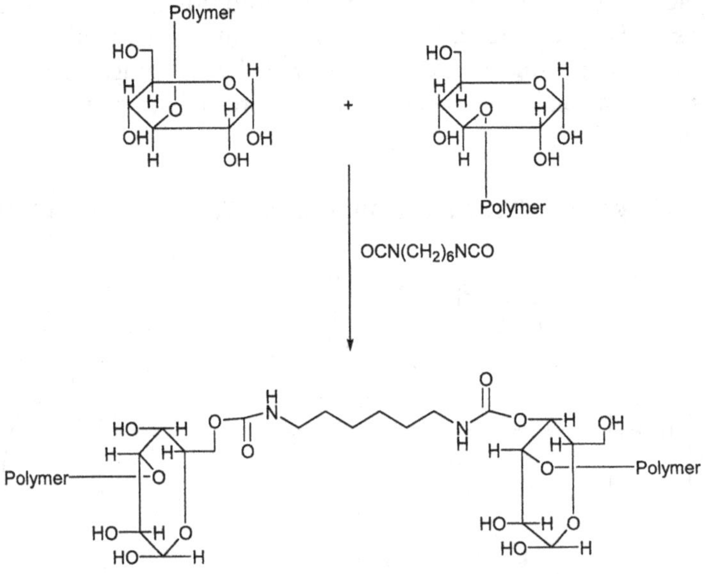

Fig. 3.47. Chemical crosslinking of polyphosphazenes with glucosyl side-groups

3.3.9 Modification of Poly(alkyl/arylphosphazene)s

Many types of modifications can be done on poly(alky/arylphosphazene)s [14, 23, 33]. The most common modification involves deprotonation of the methyl side-groups. This can be accomplished by deprotonation by *n*-butyllithium to afford a polymer containing a carbanion in its side-chain. Such a carbanion can be quenched with any suitable electrophile to afford modified polymers. This could include reactions with reagents containing chloro end-groups such as RMe_2SiCl, aldehydes or esters. These reactions are shown in Fig. 3.48.

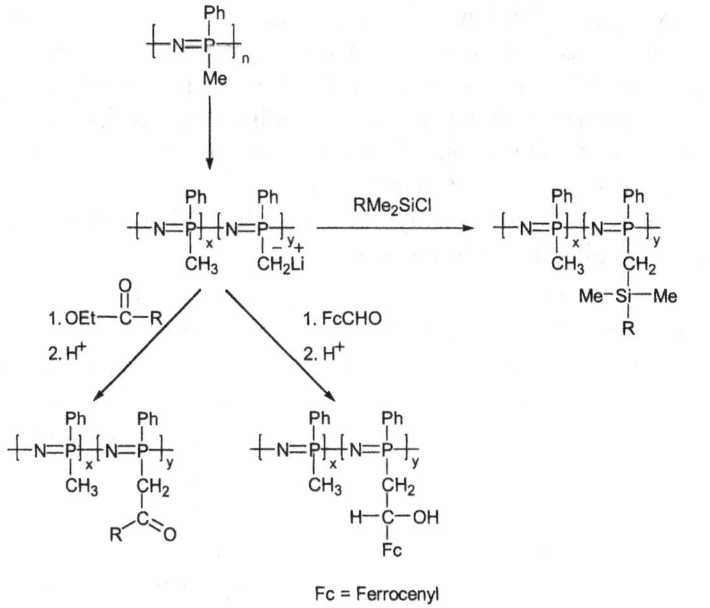

Fig. 3.48. Modification of [NP(Ph)(Me)]ₙ

The polymer containing the carbanion can also be used to polymerize the six-membered hexamethylsiloxane by ROP. The result is the generation of a graft copolymer where the siloxane polymer chain is present as a graft on the main polyphosphazene side-chain (Fig. 3.49).

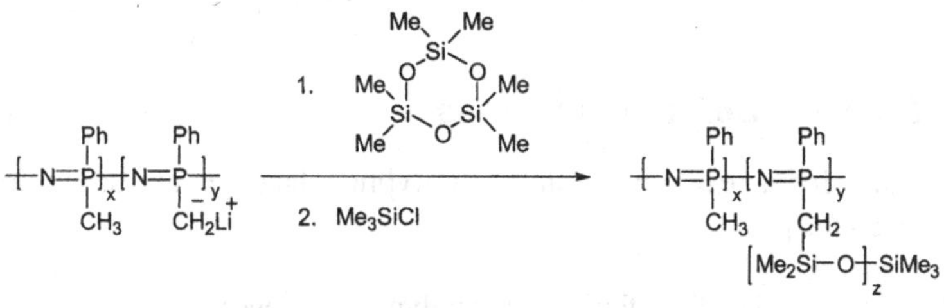

Fig. 3.49. Grafting of polysiloxane on the side-chain of a polyphosphazene by the reaction of a carbanion with [Me₂SiO]₃

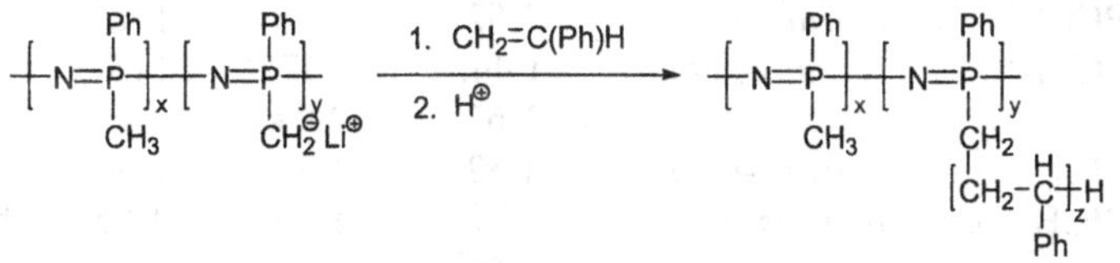

Fig. 3.50. Grafting of polystyrene on polyphosphazenes

The polymer carbanion can also be used as an anionic initiator for polymerizing a conventional organic monomer to generate a poly-phosphazene-*co*-organic polymer (Fig. 3.50). The polystyrene graft copolymer prepared as shown above is actually separated into the individual polystyrene and polyphosphazene phases. This is found by the detection of two separate T_g's at +100 °C and +37 °C.

A more homogeneous methyl methacrylate graft copolymer can be prepared by a similar methodology (Fig. 3.51).

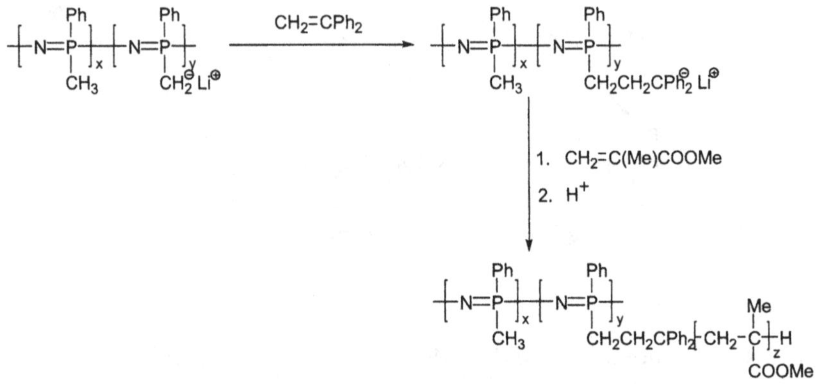

Fig. 3.51. Grafting methyl methacrylate on polyphosphazene

3.3.10 Structure and Properties of Polyphosphazenes

3.3.10.1 X-ray Diffraction Studies

X-ray data for a few representative polyphosphazenes are summarized in Table 3.6 [1-2].

Table 3.6. X-ray data for a few selected polyphosphazenes

Polymer	Polymer conformation	Bond distance (Å)		Bond angles (°)	
		P=N	P-X	NPN	PNP
[NPCl$_2$]$_n$	cis-trans-planar	1.52	1.96	118	141.5
[NPCl$_2$]$_n$	cis-trans-planar	1.44, 1.67	1.97, 2.04	115	131
[NPF$_2$]$_n$	cis-trans-planar	1.52	1.47	119	136
[NP(CH$_3$)$_2$]$_n$	cis-trans-planar	1.59, 1.56	1.80	112.5	135.9
[NP(nPr)$_2$]$_n$	cis-trans-planar	1.59	1.80	114.2	132.6

The number of polyphosphazenes that have been studied by X-ray diffraction are not very large. However, these studies have been quite informative. Two different measurements on [NPCl₂]ₙ revealed difference in metric parameters, although the main structural features are similar. Most polyphosphazenes seem to adopt the *cis-trans-planar* conformation over the *trans-trans-planar* conformation. Steric repulsion between the substituents present on phosphorus is minimized in the *cis-trans* planar conformation. Thus, if one considers a P2-N3 bond (bond labeled A, Fig. 3.52) as a reference, the disposition of the immediate back-bone substituents atoms N1 and P4 are *cis* with respect to each other. On the other hand, with respect to the adjacent N3-P4 bond (bond labeled B, Fig. 3.52), the disposition of the backbone substituents P2 and N5 are *trans* with respect to each other. Thus, the stereochemical orientation of the polymer chain alternately varies from *cis* to *trans*.

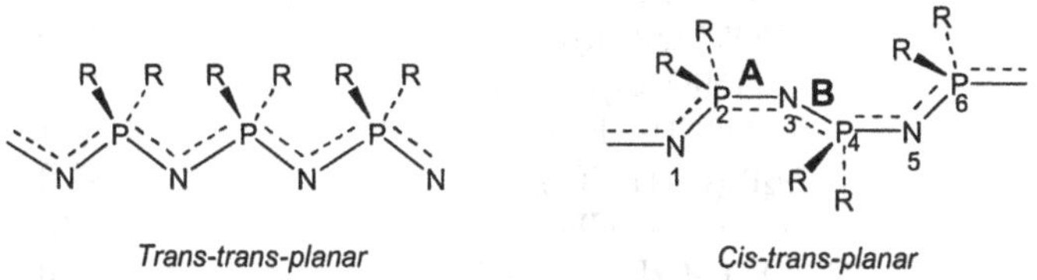

Trans-trans-planar   Cis-trans-planar

Fig. 3.52. *Trans-trans-planar* and *cis-trans-planar* conformations of polyphosphazene chains

In polymers such as [NP(Me)(Ph)]ₙ where the possibility of stereoregular polymers exists (recall polypropylene) atactic structure has been found. Most poly(alkyl/phenylphoshazenes) are amorphous which is consistent with their atactic structures.

The P-N bond distances seem to be constant along the polymer chain although in the case of [NPCl₂]ₙ both constant and alternate P-N bond distances have been found in different studies. All the polymers are characterized by wide bond angles at nitrogen.

3.3.10.2 ³¹P NMR

The phosphorus NMR chemical shifts of a few selected polyphosphazenes are given in Table 3.7. Usually slightly broad signals are seen in the NMR spectrum of the polymers because of the increase in viscosity when polymers are dissolved in solution. The ³¹P-NMR chemical shifts of polyphosphazenes indicate that these are about 20-30 ppm upfield shifted with respect to the cyclic trimers. The cyclic tetramers have intermediate chemical shifts (entries 1, 2, 5, 8, 11 in Table 3.7). In polyphosphazenes containing mixed substituents the ³¹P-NMR chemical shifts are diagnostic of the type of groups present. For example, in polymers $[(NPCl_2)_2(NP(Cl)Me)]$ the δPCl_2 resonates at -19.8 ppm while another resonance at +10.2 ppm is seen which is assigned to the phosphorus that contains one methyl and one chlorine substituent ($\delta P(Cl)(Me)$), respectively.

Table 3.7. ^{31}P-NMR chemical shifts for some selected polyphosphazenes

S.No	Compound	ppm
1	$[NPCl_2]_n$	-20.0
	$N_4P_4Cl_8$	-6.5
	$N_3P_3Cl_6$	+19.3
2	$[NP(OCH_3)_2]_n$	-6.0
	$N_4P_4(OCH_3)_8$	+2.8
	$N_3P_3(OCH_3)_6$	+21.7
3	$[NP(Me)(Ph)]_n$	+1.8
4	$[NPMe_2]_n$	+8.3
5	$[NP(OCH_2CF_3)_2]_n$	-10.0
	$N_4P_4(OCH_2CF_3)_8$	-2.0
	$N_3P_3(OCH_2CF_3)_6$	+16.7
6	$[NP(OCH_2CH_2\ OCH_3)_2]_n$	-8.0
7	$[NP(OCH_2CH_2OCH_2CH_2OCH_3)_2]_n$	-7.7
8	$[NP(OPh)_2]_n$	-19.7
	$N_4P_4(OPh)_8$	-12.6
	$N_3P_3(OPh)_6$	+8.3
9	$[NP(OC_6H_4\text{-}p\text{-}Br)_2]_n$	-19.7
10	$[NP(O\text{-}C_6H_4\text{-}\ p\text{-}COOH)_2]_n$	-19.4
11	$[NP(NHMe)_2]_n$	+3.9
	$N_4P_4(NHMe)_8$	+12.2
	$N_3P_3(NHMe)_6$	+23.0
12	$[(NPCl_2)_2(NP(Cl)Me)]_n$	-21.0(δPCl_2)
		+10.2[$\delta P(Cl)(Me)$]
13	$[(NPCl_2)_2(NP(Cl)Et)]_n$	-19.8(δPCl_2)
		+17.4[$\delta P(Cl)(Et)$]
14	$[NP(OCH_2CF_3)_2\ (NP(OCH_2CF_3)(Ph)]_n$	-8.2[$P(OR)_2$]
		+5.6[$P(R)(OR')$]

3.3.10.3 Molecular Weights

As mentioned in previous sections, the molecular weights determined for $[NPCl_2]_n$ obtained by ROP of $N_3P_3Cl_6$ are in the range of 10^6 with a PDI greater than 2. The PDI's become better for $[NPCl_2]_n$ obtained from ambient temperature cationic polymerization of Me$_3$SiN=PCl$_3$ and achieve a value of nearly 1.01 [21]. It has been observed that poly(organophosphazene)s obtained by macromolecular substitution retain more or less the same chain lengths as the parent polymer. This indicates that the polymer does not suffer chain degradation during the process of macromolecular substitution. This, of course, is not true in the reactions of [NPCl$_2$]$_n$ with Grignard reagents or alkyl/aryllithium reagents. Poly(alkyl/arylphosphazene)s have molecular weights in the range of 50-200,000 with a molecular-weight distribution of 2.0.

3.3.10.4 Solubility Properties

The solubility properties of polyphosphazenes depend on the type of substituents present on the phosphorus centers. Poly(dichlorophosphazene), [NPCl$_2$]$_n$ is soluble in a variety of organic solvents including THF, benzene, toluene etc. On the other hand, poly(difluorophosphazene), [NPF$_2$]$_n$ is soluble only in fluorinated solvents such as perfluorodecalin or perfluoro-2-butyl-THF. While polymers such as [NP(OCH$_2$CF$_3$)$_2$]$_n$ are extremely hydrophobic and are only soluble in organic solvents, polymers with etheroxy side-groups such as [NP(OCH$_2$CH$_2$OCH$_2$CH$_2$OCH$_3$)$_2$]$_n$ are soluble in organic solvents as well as water. Thus, placing hydrophobic or hydrophilic groups on the side-chain of the polyphosphazene transfers these properties to the polymer.

Among poly(aminophosphazene)s, [NP(NHCH$_3$)$_2$]$_n$ is soluble in organic solvents as well as water. Among poly(alkyl/arylphosphazene)s, while [NPEt$_2$]$_n$ is totally insoluble and can be solubilized only by the use of initial protonation, other polymers such as [NPMe$_2$]$_n$ and [NP(Me)(Ph)]$_n$ are soluble in organic solvents. Poly(dimethylphosphazene) [NPMe$_2$]$_n$ is also soluble in a 1:1 mixture of THF and water. It is however, insoluble in water [21, 33].

3.3.10.5 Hydrophobic and Hydrophilic Properties

Since the parent polymer poly(dichlorophosphazene) is hydrolytically unstable, it is difficult to assess its properties in terms of hydrophobicity or hydrophilicity. The presence of nitrogen atoms (with lone pair of electrons) in the backbone of polyphosphazenes suggests that intrinscally polyphosphazenes would be inclined to be hydrophilic polymers. The lone pair on the nitrogen atoms assists polymer solvation by polar solvents particularly by those that can participate in hydrogen bonding (solvents such as water or alcohols). However, the actual situation would depend on the substituents on phosphorus also. If the substituents can augment the intrinsic nature of polyphosphazenes the polymers would be hydrophilic. On the other hand, if the substituents have opposite effects the results could be different. Thus, polyphosphazenes containing various side-groups have varying degrees of affinity towards water (Fig. 3.53).

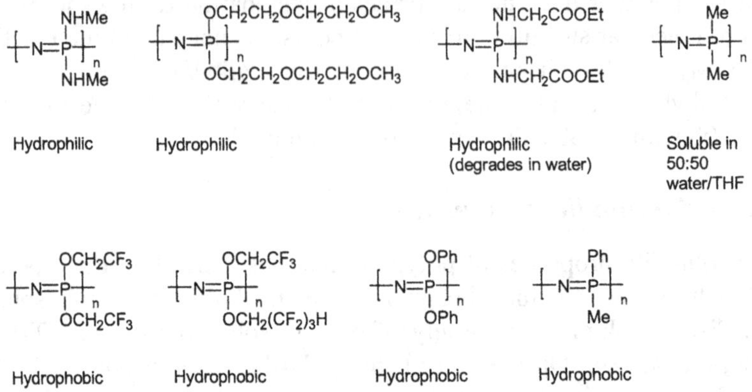

NHMe	OCH$_2$CH$_2$OCH$_2$CH$_2$OCH$_3$	NHCH$_2$COOEt	Me
Hydrophilic	Hydrophilic	Hydrophilic (degrades in water)	Soluble in 50:50 water/THF

OCH$_2$CF$_3$	OCH$_2$CF$_3$	OPh	Ph
Hydrophobic	Hydrophobic	Hydrophobic	Hydrophobic

Fig. 3.53. Hydrophilic and hydrophobic polyphosphazenes

Polymers such as [NP(NHMe)$_2$]$_n$, and [NP(OCH$_2$CH$_2$OCH$_2$CH$_2$-OCH$_3$)$_2$]$_n$ are water-soluble and their hydrophilic properties clearly stem from the presence of side-groups that can hydrogen bond in aqueous solutions. The intrinsic hydrophilicity of the NH moieties in poly(aminophosphazenes) [NP(NHR)$_2$]$_n$ is overcome by the hydrophobicity of the R groups, particularly when R becomes a large hydrocarbon entity.

The hydrophobicity of polyphosphazenes can be assessed by contact angle measurement experiments. Thus, this experiment measures the contact angle made by a drop of water on a polymer film. In a hydrophilic polymer film the water drop would very quickly smear out. For a typical hydrophobic polymer such as Teflon [CF$_2$CF$_2$]$_n$, a polymer formulation used in nonstick kitchenware, the contact angle is 108°. In polyethylene this value is 94°. On the other hand, the contact angle for [NP(OCH$_2$CF$_3$)$_2$]$_n$ is 108°. This means that [NP(OCH$_2$CF$_3$)$_2$]$_n$ is as hydrophobic as Teflon! The contact angles measured for [NP(Me)(Ph)]$_n$ is 73° indicating that the alkyl or aryl groups do not shield the backbone nitrogens sufficiently well and allow their interaction with water.

3.3.10.6 Thermal Properties

Polyphosphazenes have diverse thermal properties, and as one would anticipate these would also depend upon the nature of side-groups. Many poly(organophosphazene)s containing P-O and P-N linkages are known to decompose or depolymerize at high temperatures around 300 °C. On the other hand, poly(alky/arylphosphazene)s which contain P-C bonds are more stable towards thermal treatment.

3.3.10.7 Glass Transitions and Skeletal Flexibility of Polyphosphazenes

Polyphosphazenes are among the most flexible polymers known. This is reflected in low glass transition temperatures (T_g). Only below the glass transition temperatures all conformational mobility is frozen and polymer becomes a glass. The skeletal flexibility of polyphosphazenes seems to arise from many factors.

1. The P-N bond distance in polyphosphazenes (av. 1.59 Å) is longer than C-C bond distances (1.54 Å) found in organic polymers. The longer bond distance would allow a greater torsional freedom around it which would make the polymers flexible.
2. Secondly, in polyphosphazenes only the *alternate* backbone atoms (phosphorus) have substituents unlike in organic polymers where every atom in the backbone has substituents. More intermolecular space (free volume) is therefore available for the polyphosphazene chains.
3. The bond angles at nitrogen in the polymer chain are quite wide.
4. The P=N *double bond* in the polymer chain is not similar to a C=C double bond and it has been pointed out that even if the postulate of pπ-dπ bonding is valid, the torsion of the P-N bonds would allow several suitable d orbitals on phosphorus to participate in π-bonding.

The glass transition temperatures for a few representative polyphosphazenes are given in Table 3.8.

The parent polyphosphazenes $[NPF_2]_n$ and $[NPCl_2]_n$, have extremely low glass transition temperatures of -95 °C and -66 °C respectively. This data suggests that these polymers have a lot of torsional freedom or skeletal flexibility. Only below the glass transition temperatures all the torsional motion of the polymers is arrested.

Substituting the chlorines in $[NPCl_2]_n$ by alkoxy groups lowers the T_g's: $[NP(OCH_3)_2]_n$ (-74 °C). This trend continues up to the *n*-butoxy derivative $[NP(OnBu)_2]_n$ (-105 °C). The presence of trifluoroethoxy groups or etheroxy groups also helps to keep the T_g lower: $[NP(OCH_2CF_3)_2]_n$ (-66 °C); $[NP(OCH_2CH_2OCH_2CH_2OCH_3)_2]_n$ (-84 °C).

Table 3.8. Glass-transition temperatures of various types of polyphosphazenes

S.No	Compound	T_g (°C)
1	$[NPCl_2]_n$	-66
2	$[NPF_2]_n$	-95
3	$[NP(NCS)_2]_n$	-57
4	$[NP(OCH_3)_2]_n$	-74
5	$[NP(OCH_2CH_3)_2]_n$	-84
6	$[NP(OCH_2CH_2CH_3)_2]_n$	-100
7	$[NP(OCH_2CH_2CH_2CH_3)_2]_n$	-105
8	$[NP(OCH_2Ph)_2]_n$	-31
9	$[NP(OCH_2CH_2Ph)_2]_n$	-33
10	$[NP(OCH_2CH_2CH_2Ph)_2]_n$	-47
11	$[NP(OCH_2CH_2\,OCH_3)_2]_n$	-75
12	$[NP(OCH_2CH_2OCH_2CH_2OCH_3)_2]_n$	-84
13	$[NP(OCH_2CH_2\text{-}O\text{-}C_6H_4\text{-}p\text{-}C_6H_4\text{-}p\text{-}I)_2]_n$	+111
14	$[NP(OCH_2CF_3)_2]_n$	-66
15	$[NP(OCH_2CF_3CF_3)_2]_n$	-25
16	$[NP(OCH_2CH_2O\text{-}C_6H_5\text{-}\eta^6\text{-}Cr(CO)_3)_2]_n$	+43
17	$[NP(OC_6H_5)_2]_n$	-8
18	$[NP(O\text{-}C_6H_4\text{-}p\text{-}C_6H_5)_2]_n$	+93
19	$[NP(O\text{-}C_6H_4\text{-}SiMe_3\text{-}p)_2]_n$	+27
20	$[NP(O\text{-}C_6H_4\text{-}Br\text{-}p)_2]_n$	+44
21	$[NP(NHCH_3)_2]_n$	+14
22	$[NP(NHEt)_2]_n$	+30
23	$[NP(NHCH_2COOEt)_2]_n$	-23
24	$[NP(NH(CH_2)_3SiMe_2OSiMe_3)_2]_n$	-67
25	$[NP(NHC_6H_5)_2]_n$	+91
26	$[NP(NHAd)_x(OCH_2CF_3)_y]_n$ x = 1.06; y = 0.94	+180
27	$[NP(NHAd)_x(OCH_2CF_3)_y]_n$ x = 0.42; y = 1.58	+40
28	$[NP(OAd)_x(OCH_2CF_3)]_y$ x = 0.8; y = 1.2	+46
29	$[(NPCl_2)_2NP(N{=}PCl_3)_2]_n$	-37
30	$[(NPCl_2)_2NP(Cl)(N{=}PCl_3)]_n$	-41
31	$[(NPF_2)_2\,(NP(F)(Ph))]_n$	-69
32	$[NP(CH_3)_2]_n$	-46
33	$[NP(Ph)(Me)]_n$	+37
34	$[\{NP(Me)_2\}\{NP(Me)(Ph)\}]_n$	-3

The T_g's of poly(aminophosphazenes) are higher than $[NPCl_2]_n$. Thus, $[NP(NHCH_3)_2]_n$ has a T_g of +14 °C. In the aniline derivative $[NP(NHPh)_2]_n$

the glass transition temperature increases to +91 °C. While full replacement of chlorines from $[NPCl_2]_n$ by the sterically hindered adamantyl amine does not occur, even the partially substituted polymer $[NP(NHAd)_x(OCH_2CF_3)_y]_n$ (x = 1.06; y = 0.94) has a glass transition temperature of +180 °C! This is one of the highest T_g's known for a polyphosphazene.

The polyphosphazene $[NP(Me)_2]_n$ has a T_g of -46 °C. Replacing the methyl groups gradually by phenyl groups increases the T_g: $[\{NP(Me)_2\}\{NP(Me)(Ph)\}]_n$ (-3 °C); $[NP(Me)(Ph)]_n$ (+37 °C).

The T_g s of a few selected polymers in their increasing order are given in Fig. 3.54.

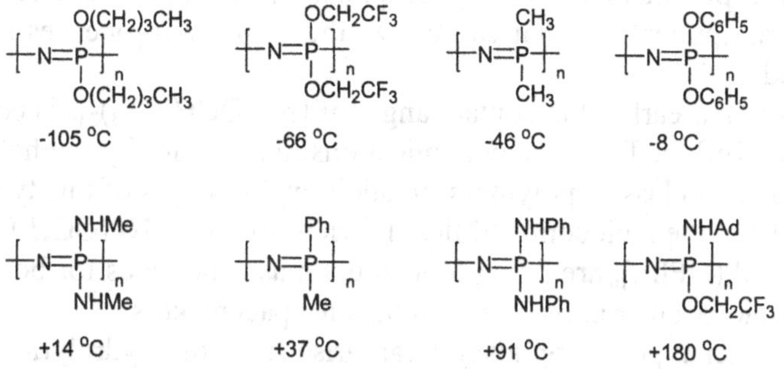

Fig. 3.54 Selected poly(organophosphazene)s in the order of their increasing T_g

What is the relationship of the glass transition temperatures with the structures of polyphosphazenes? Small-sized substituents on phosphorus (such as chlorine or fluorine) aid in low T_g's. Thus, poly(fluorophosphazene)s which contain small-sized fluorine substituents (and because fluorines have very poor van der Waals interactions) have very low T_g's.

Replacing the small-sized groups with larger sized ones always invariably leads to increase in T_g's. Side-chains that themselves are flexible and which create *free-volume* (meaning inefficient packing as a result of a tendency not to crystallize well) also lead to low T_g's.

Groups that can bring about intermolecular alignment will tend to bring more order and result in greater T_g's. Thus, poly(aminophosphazene)s that contain –NHR groups can have large intermolecular hydrogen bonding interactions. This will result in decreased flexibility and increased T_g. Thus, the adamantylamine-containing polymer is the one with the highest T_g measured for any polyphosphazene.

3.3.11 Potential Applications of Polyphosphazenes

Polyphosphazenes are being viewed with interest with regard to various kinds of applications [21]. Their special features such as backbone flexibility and the presence of inorganic elements in the backbone can give new applications. Thus, the fluoroethoxy polymer $[NP(OCH_2CF_3)_2]_n$ has a T_g of -66 °C and is flexible up to this temperature. The copolymer $[NP(OCH_2CF_3)(OCH_2(CF_2)_3CF_2H)]_n$ is flexible up to ~ -70 °C. These polymers can be used in unconventional elastomeric applications where flexibility at low temperatures is critical. The second advantage that polymers such as $[NP(OCH_2CF_3)(OCH_2(CF_2)_3CF_2H)]_n$ possess is that unlike normal organic polymers these are completely resistant to hydrocarbon fluids. Several applications that can utilize these special properties are being explored.

As noted earlier the contact angle of $[NP(OCH_2CF_3)_2]_n$ is comparable to that of Teflon. Thus, many applications that require hydrophobicity can be met by this class of polymers. In addition, polymers of this type have been noted for their biocompatibilty. In view of this, $[NP(OCH_2CF_3)_2]_n$ along with $[NP(OPh)_2]_n$ are being looked upon as candidates for body parts such as cardiovascular parts or as coatings for pacemakers.

Another type of application can arise from the hydrolytic instability of certain poly(aminophosphazene)s such as those that contain amino acid ester side-groups (Fig. 3.55).

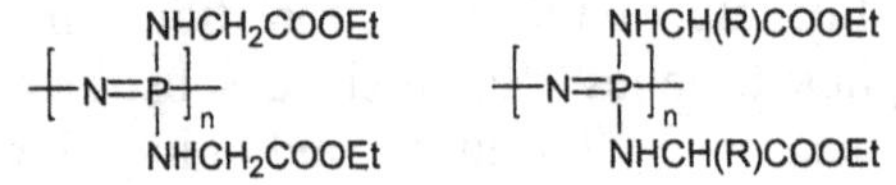

Fig. 3.55. Polyphosphazenes containing amino acid ester side-groups

Polymers $[NP(NHCH_2COOEt)_2]_n$ and $[NP(NHCH(R)COOEt)_2]_n$ degrade slowly in biological media to afford innocuous by-products such as ammonia, phosphate, amino acid and ethanol. These polymers are looked upon for utility in drug-delivery or for soluble sutures. Polyphosphazenes are also being investigated as fire-retarding additives to conventional polymers such as polyurethanes. Many other types of applications can be envisaged for this family of polymers. These can be achieved by appropriate design of the polymeric system. Some of these have already been mentioned in the section on modifications of polyphosphazenes [21]. The important point to note is that unlike other polymeric systems it is easier with polyphosphazenes to visualize an application and tailor the polymer accordingly. Two examples of this approach are shown below.

3.3.11.1 Polyphosphazene Solid Electrolytes

Transport of ions is facilitated through a medium which is known as an electrolyte. This process leads to ionic conduction. Conventionally, the electrolyte is a liquid which mediates ion-transport [36, 37]. However, in recent years there is considerable interest in designing solids in general and polymers in particular that can function as electrolytes. The main idea is that if solids can be used as electrolytes, the weight and volume occupied by the liquid electrolyte can be avoided. Several solids such as β-alumina have been used as solid electrolytes.

It is more advantageous to use a polymer for such a purpose. The polymer has the advantageous features of the solid (reduction in weight and volume) while retaining the benefits of the liquid. Thus, the most important advantage of the liquid electrolyte is that the electrodes (which are almost always solids) are always in contact with the electrolyte. If the electrolyte is also a solid, this is a problem and perfect contact is an issue particularly since charge-discharge cycles lead to change in electrode dimensions. This is to say that the shape of the electrodes is modified while the device is being used and this can cause a problem. This is not a problem if the electrolyte is a liquid; irrespective of the change of shape, the electrodes will be always in contact. But if the electrolyte is also a solid, then retention of contact can be a major problem.

Polymers can be cast into thin films and can be sandwiched between electrodes. This will allow retention of contact since polymers are flexible and can adapt to the dimensional changes that occur at the electrode.

Obviously all polymers cannot be used as electrolytes. They need to be amorphous, have a large free-volume as indicated by low T_g's and should possess donor atoms (such as oxygens) that can bind and transport the metal ions. Poly(ethyleneoxide), $(CH_2CH_2O)_n$, has been the most investigated polymer for the purpose of electrolytes. Through its oxygen atoms in the polymer backbone it can bind to various alkali and alkaline earth metal ions and the complexes formed have been shown to be semiconducting. However, the room temperature conductivity of PEO-metal salt complexes is quite low because PEO has a large crystalline phase.

Polyphosphazenes have an intrinsic skeletal flexibility and therefore can be designed as polymer electrolytes. Reaction of poly(dichlorophosphazene), $[NPCl_2]_n$, with the sodium salt of methoxyethoxyethanol affords the etheroxy-side-chain containing polymer (see Eq. 3.64).

$$\left.\begin{matrix} & Cl \\ | \\ +N\!\!=\!\!P \\ | \\ Cl \end{matrix}\right]_n \xrightarrow{\quad RONa \quad} \left.\begin{matrix} & OR \\ | \\ +N\!\!=\!\!P \\ | \\ OR \end{matrix}\right]_n \qquad (3.64)$$

$$R = -CH_2CH_2-O-CH_2CH_2OCH_3$$

The polymer $[NP(OCH_2CH_2OCH_2CH_2OCH_3)_2]_n$ is a completely amorphous polymer with a T_g of -84 °C. This polymer as well as other related polymers of the above type can be complexed with lithium salts such as $LiCF_3SO_3$, $LiBF_4$, $LiClO_4$ etc., to afford polymer-salt complexes. Essentially the etheroxy side-chain-containing polymers act as solvents and the metal salt dissolves in the polymer to afford a solid solution. The ionic conductivities of such polymers are quite promising (10^{-5}-10^{-4} Scm^{-1}) [21, 38]. Such polymer-salt complexes have been shown to be semi-conducting.

3.3.11.2 Polyphosphazenes with Second-Order Nonlinear Optical Properties

Materials with unusual optical properties such as nonlinear optical properties (NLO) are of great interest from the point of view of developing new optoelectronic materials. The nonlinear optical property is manifested in the material by the conversion of light of frequency ω into light of frequency 2ω. In simple organic molecules this property has been shown to arise from the presence of a donor and an acceptor group that are separated from each other by a conjugated spacer group. Usually such molecules are noncentrosymmetric. The advantages of these polymers such as their film-forming capability and flexibility can be useful in incorporating NLO properties. Thus, one possible approach that can be used for preparing polyphosphazenes that contain NLO properties is outlined in Fig. 3.56. This approach can be summarized as follows:

1. A suitable NLO-exhibiting organic molecule with a terminal functional group such as NH_2 or OH needs to be prepared.
2. Reaction of such functional organic NLO groups with $[NPCl_2]_n$ affords partially or fully substituted derivatives.
3. Such polymer molecules can be aligned by heating them up to their T_g temperatures. This is required so that the polymer molecules are allowed full skeletal mobility, at which point of time application of electrical field is applied so as to preferentially orient the polymer molecules.
4. Examine if such polymers have special NLO properties.

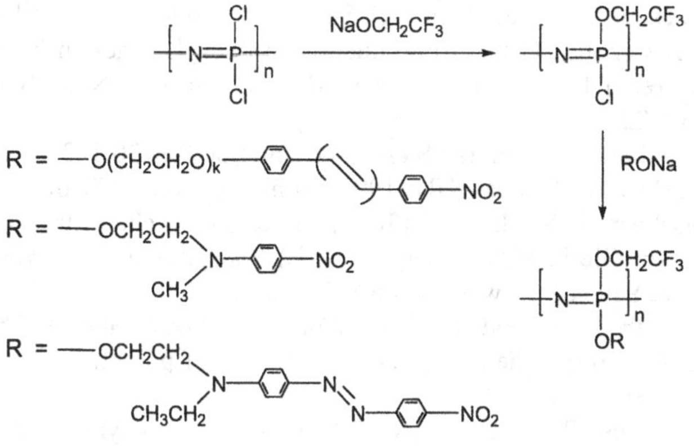

Fig. 3.56. Synthesis of polyphosphazenes containing *NLO-imparting* organic side-groups

The acceptor part of the NLO motif is generally the NO_2 group, while the donor is the amino or the etheroxy unit. These are separated by either a $C_6H_4CH=CHC_6H_4$, C_6H_4 or $C_6H_4N=N-C_6H_4$ spacer group. Polymers prepared in this way have been shown to be amorphous and form optical quality films. Although these polymers showed second-order harmonic generation, this behavior decays after the removal of the electric field [39].

References

1. Gleria M, De Jaeger R (eds) (2004) Phosphazenes - a world-wide insight. Nova Science, New York
2. Chandrasekhar V, Krishnan V (2002) Adv Inorg Chem 53:159
3. Chandrasekhar V, Thomas KRJ (1993) Structure Bonding 81:41
4. Allcock HR (1972) Chem Rev 72:315
5. Allen CW (1994) Coord Chem Rev 120:137
6. Shaw RA, Fitzsimmons BW, Smith BC (1962) Chem Rev 62:248
7. Krishnamurthy SS, Sau AC, Woods M (1978) Adv Inorg Rad Chem 21:41
8. Elias AJ, Shreeve JM (2001) Adv Inorg Chem 52:335
9. Allcock HR (1972) Phosphorus-nitrogen compounds. Academic Press, New York.
10. Heal HG (1980) The inorganic heterocyclic chemistry of sulfur, nitrogen and phosphorus. Academic Press, New York
11. Allcock HR, Nelson JM, Reeves SD, Honeyman CH, Manners I (1997) Macromolecules 30:50
12. Allen CW (1991) Chem Rev 91:119
13. Allcock HR, Desorcie JL, Riding GH (1987) Polyhedron 6:119
14. Neilson RH, Wisian-Neilson P (1988) Chem Rev 88:541

15. Chandrasekhar V, Muralidhara MG, Selvaraj II (1990) Heterocycles 31:2231
16. Allcock HR (2004) General introduction to phosphazenes In: Gleria M, De-Jaeger R (eds) Phosphazenes – a world-wide insight. Nova Science, New York, pp 1-22
17. Chandrasekhar V, Nagendran S (2001) Chem Soc Rev 30:193
18. Chandrasekhar V, Thomas KRJ (1993) Appl Organomet Chem 7:1
19. Chandrasekhar V, Krishnan V (2004) Coordination chemistry of cyclophosphazenes In: Gleria M, De Jaeger R (eds) Phosphazenes – a world-wide insight. Nova Science, New York, pp 827-852
20. Krishnamurthy SS, Woods M (1987) Ann Rep NMR Spectr 19:175
21. Allcock HR (2003) Chemistry and applications of polyphosphazenes. Wiley, Hobolen, New Jersey
22. Mark JE, West R, Allcock HR (1992) Inorganic polymers. Prentice-Hall, Englewood Cliffs
23. De Jaeger R, Gleria M (1998) Prog Polym Sci 23:179
24. De Jaeger R, Potin P (2004) Poly(dichlorophosphazene) from P-trichloro-N-dichlorophosphoryl monophosphazene, $Cl_3PNP(O)Cl_2$ In: Gleria M, De Jaeger R (eds) Phosphazenes – a world-wide insight. Nova Science, New York, pp 25-48
25. Halluin GD, De Jaeger R, Chambrette JP, Potin P (1992) Macromolecules 25:1254
26. Allcock HR (2004) Ambient temperature cationic condensation synthesis of polyphosphazenes In: Gleria M, De Jaeger R (eds) Phosphazenes – a world-wide insight. Nova Science, New York, pp 49-68
27. Allcock HR, Crane CA, Morrissey CT, Nelson JM, Reeves SD, Honeyman CH, Manners I (1996) Macromolecules 29:7740
28. Allcock HR, Ritchie RJ, Harris PJ (1980) Macromolecules 13:1332
29. Allcock HR, McDonnell GS, Desorcie JL (1990) Macromolecules 23:3873
30. Cho Y, Baek H, Soh YS (1999) Macromolecules 32:2167
31. Allcock HR (1979) Acc Chem Res 12:351
32. Allcock HR, Krause WE (1997) Macromolecules 30:5683
33. Wisian-Neilson P (2004) Poly(alkyl/arylphosphazenes) and their derivatives In: Gleria M, De Jaeger R (eds) Phosphazenes – a world-wide insight. Nova Science, New York, pp 109-124
34. Nelson JM, Allcock HR, Manners I (1997) Macromolecules 30:3191
35. Allcock HR, Pucher SR (1991) Macromolecules 24:23
36. Ratner MA, Shriver DF (1988) Chem Rev 88:109
37. Chandrasekhar V (1998) Adv Polym Sci 135:139
38. Allcock HR, Sunderland NJ, Ravikiran R, Nelson JM (1998) Macromolecules 31:8026
39. Allcock HR, Dembek AA, Kim C, Devine RLS, Shi Y, Steier WH, Spangler CW (1991) Macromolecules 24:1000

4 Cyclophosphazene-Containing Polymers

4.1 Introduction

In the previous chapter we have seen inorganic polymers that contained a [P=N] backbone. These polymers can be prepared by two principal methods.

1. Ring-opening polymerization of monomers such as $N_3P_3Cl_6$ or $N_3P_3Cl_5R$ to the corresponding linear polymers followed by replacement of chlorines on the macromolecules by a variety of nucleophiles to afford polymers that contain P-O or P-N linked side-chains.
2. Catalytic or uncatalyzed polymerization of various kinds of N-silylphosphoranimines such as $Me_3SiNPCl_3$, $Me_3SiNPRR'X$ (X=OCH$_2$CF$_3$, OPh, Cl). This method is the most important route for the preparation of poly(alkyl/arylphosphazene)s.

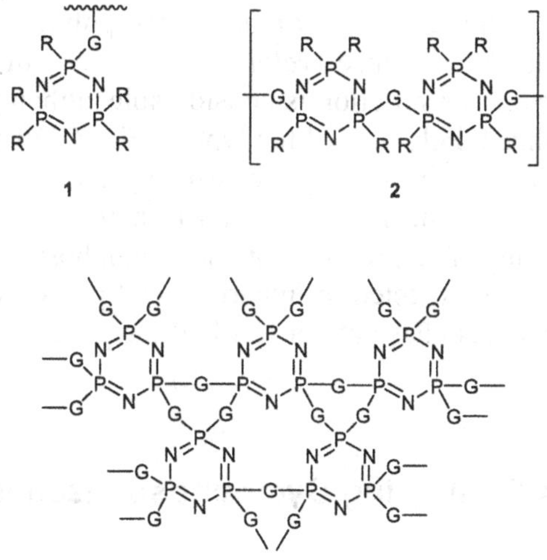

Fig. 4.1. Different kinds of cyclophosphazene-containing polymers

In this chapter we will examine polymers that contain *intact* cyclophosphazene units [1-4]. Examples of these kinds of polymers are shown in Fig. 4.1.

Polymer **1** (Fig. 4.1) is essentially an organic polymer containing a carbon backbone. We have seen in Chap. 2 that several types of organic polymers could be prepared by polymerizing vinyl monomers of the type $CH_2=CHG$ by various types of initiators (see Eq. 4.1). The polymer properties could be varied significantly by change of the group G. Thus, for example, polystyrene prepared by polymerizing styrene (G = Ph) has completely different properties than poly(vinylchloride) prepared by polymerizing vinylchloride (G = Cl). In polymer **1**, the G group is an intact cyclophosphazene unit. These polymers are potentially very promising because the type of cyclophosphazene group that can be appended to the organic backbone can be varied considerably and hence the properties of the resulting polymers can also be accordingly varied.

$$CH_2=\underset{G}{CH} \quad \xrightarrow[\text{Ionic initiators}]{\substack{\text{Free-radical} \\ \text{or}}} \quad \underset{G}{\overset{}{\left[CH_2-CH\right]_n}} \tag{4.1}$$

Polymer **2** (Fig. 4.1) is an example of a condensation polymer where a difunctional cyclophosphazene can react with an organic or an inorganic difunctional reagent to afford a linear polymer. These polymers contain the cyclophosphazene ring as a *repeat motif* in the polymer backbone. These polymers have also been termed *cyclolinear polymers*. In this type of polymers also, the scope for variation is considerable, although in practice this has not been realized. Polymers of the type **3** (Fig. 4.1) are examples of intermolecularly crosslinked cyclophosphazenes. These are reminiscent of thermoset polymers such as phenol-formaldehyde or melamine-formaldehyde resins. The presence of the cyclophosphazene units in the crosslinked matrix is expected to impart special properties. However, this family of polymers also has not yet fulfilled the promise that they seem to hold.

4.2 Polymers Containing Cyclophosphazenes as Pendant Groups

Most of the polymers that contain cyclophosphazenes as pendant groups are prepared by polymerizing cyclophosphazene containing vinyl monomers. The vinyl group is attached to the cyclophosphazene ring either *directly* or through a spacer group. Recently, alternative procedures are also becoming available for the assembly of these types of polymers. These will be considered at the end.

Since cyclophosphazene-containing vinyl monomers are really a special case of $CH_2=CHG$ type of monomers where G is a cyclophosphazene group, in principle all the polymerization methods for polymerizing *conventional vinyl monomers* should also be equally effective for this system. In practice, however, only the free-radical methods have so far been found to be successful. Recently, ring-opening metathesis polymerization has also been successfully implemented in this system. The apparent limitations of the other methods of polymerization with respect to the cyclophosphazene-containing vinyl monomers arise from the following factors:

1. If halogenocyclophosphazenes are the pendant groups, clearly anionic initiators (such as *n*-butyllithium) or reactive organometallic initiators (such as Ziegler-Natta catalysts) cannot be used because they would first react with the more reactive P-X bonds before they can initiate polymerization.
2. Similarly, cationic initiators such as protic or Lewis acids are also not very effective as polymerization catalysts for these systems. The basicity of the skeletal nitrogen atoms of the cyclophosphazene rings overwhelms the reactivity of the cationic initiators towards olefin polymerization. Thus, the cyclophosphazene rings are readily protonated or metalated and thus prevent both protic and Lewis acids from initiation of polymerization.

In view of the above, the most effective method for polymerizing cyclophosphazene-containing vinyl monomers is by using free-radical initiators. Before this aspect is considered let us first examine the strategies for the preparation of such monomers. Clearly the entry to such molecules lies in the application of the nucleophilic substitution of halogenocyclophosphazenes as the synthetic strategy to generate structural units such as cyclophosphazene-R-CH=CH$_2$. This can be achieved by the following means:

1. The vinyl group can be *directly* attached to the cyclophosphazene group through a P-C bond by attack of a suitable carbanion on a halogenocyclophosphazene. Another way this can be done is by having the vinyl group separated from the reactive carbanion part by means of a spacer group.
2. Alternatively, the means of attaching the vinyl group to the cyclophosphazene molecule could be through a P-O or a P-N bond. The simplest example would be to generate a P-O-CH=CH$_2$ linkage. Again, in this case a spacer group could be

used to separate the vinyl moiety from the cyclophosphazene to generate the P-O-R-CH=CH$_2$ unit.

Examples where the vinyl moiety is linked to the cyclophosphazene through an amino group are limited. In the following section we will look at the actual methods of synthesizing cyclophosphazene-containing vinyl monomers. We will first examine monomers where the vinyl group is linked to the cyclophosphazene molecule through a P-C bond. This will be followed by looking at examples where the linkage is through the P-O bond.

4.2.1 Synthesis of Cyclophosphazene Monomers Containing Vinyl Groups

Allen and coworkers have looked at the reactions of N$_3$P$_3$F$_6$ with organo-lithium reagents as a means to achieve the synthesis of cyclophosphazene monomers [5-7]. The choice of N$_3$P$_3$F$_6$ is deliberate. It may be recalled that fluorocylophosphazenes are much better substrates than chlorocyclophos-phazenes towards organolithium reagents. The success would lie in isolat-ing the monosubstituted derivative in preference to other compounds. More than one vinyl group on the cyclophosphazene moiety would not be desirable because polymerization of such monomers would afford crosslinked products. The reaction of N$_3$P$_3$F$_6$ with 2-propenyllithium in an appropriate stoichiometric ratio affords N$_3$P$_3$F$_5$[C(CH$_3$)=CH$_2$] (4) as a dis-tillable liquid [5]. Using a similar strategy monomers 5 and 6 were also synthesized [6, 7] (Fig. 4.2).

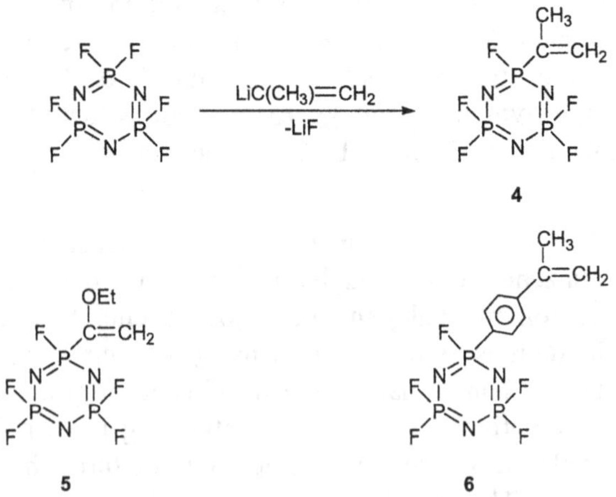

Fig. 4.2. Fluorocyclophosphazene-containing vinyl monomers

In the monomers **4** and **5** the vinyl group is directly attached to the cyclophosphazene unit and these monomers are 1,1-disubstituted ethylenes of the type $RR'C=CH_2$. Monomer **6** also belongs to this type although the cyclophosphazene is separated from the vinyl unit by means of the phenyl group which functions as a *spacer*.

In view of the fact that this synthetic method cannot be applied to chlorocyclophosphazenes, alternative procedures have to be used in this case to prepare monomers where the vinyl group is attached to the cyclophosphazene unit by means of a P-C bond. It may be recalled that the reaction of chlorocyclophosphazenes with Grignard reagents in presence of organocopper reagents affords a phosphazene-magnesio-cuprate intermediate which can lead to the formation of hydridocyclophosphazene by quenching the reaction with isopropanol [8, 9] (Fig. 4.3).

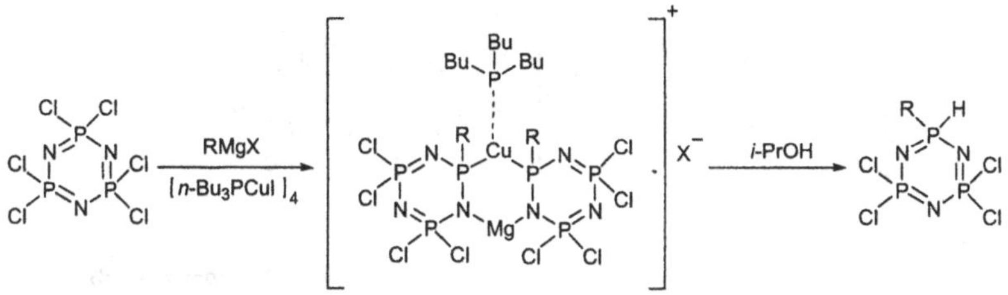

Fig. 4.3. Synthesis of hydridocyclophosphazenes

Reaction of the hydridocyclophosphazene with an excess of acetylchloride in the presence of the base triethylamine results in the formation of the monomer **7** (see Eq. 4.2).

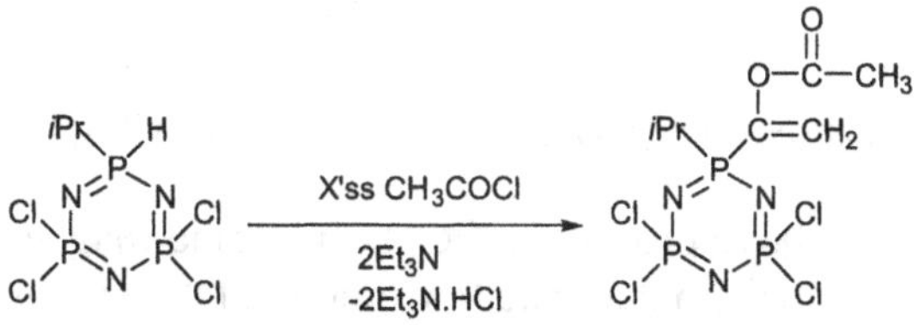

(4.2)

The formation of **7** occurs through a complicated reaction mechanism. It is believed that initial reaction between *gem*-$N_3P_3Cl_4(iPr)(H)$ and acetyl chloride leads to the formation of the *expected* product viz., $N_3P_3Cl_4(iPr)(C(O)CH_3)$. Abstraction of a proton from this molecule by the base leads to the carbanion $[N_3P_3Cl_4(iPr)(C(O)CH_2)]$. Rearrangement of the carbanion occurs so that the enolate anion $[N_3P_3Cl_4(iPr)(C(O)CH_2)]$ is generated. Reaction of the enolate anion with another molecule of acetylchloride leads to the formation of **7** [10]. The by-product in this reaction is triethylammoniumhydrochloride. Compund **7** also is a 1,1-disubstituted ethylene similar to monomers **4-6**. However, unlike the monomers **4, 5** and **6, 7** is a tetrachlorocyclophosphazene where an organic group (*i*Pr) is located geminally on the same phosphorus that contains the vinyl group. Another method of synthesis involves the reaction of the $N_3P_3Cl_6$ with isopropylmagnesium chloride in the presence of the Cu(I) reagent. The resulting phosphazane-magnesio-cuprate intermediate is allowed to react with OHC-C_6H_4-*p*-CH=CH$_2$ to afford **8** (Fig. 4.4). Notice that in this reaction the phosphazene-magnesio-cuprate behaves as a nucleophile and attacks the carbonyl carbon of the aromatic aldehyde. This results in the formation of a P-C bond [11].

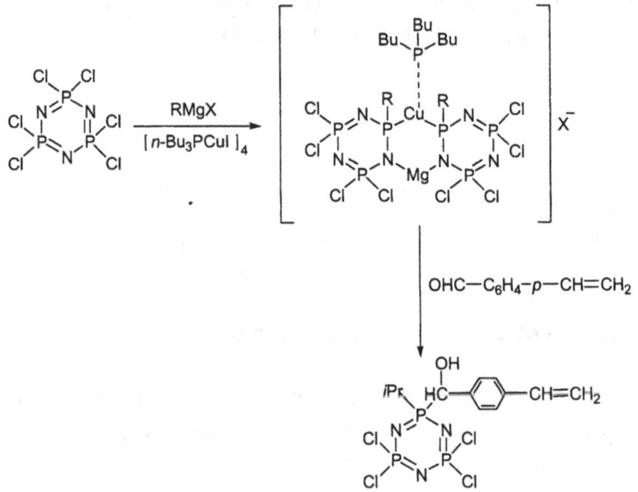

Fig. 4.4. Synthesis of *gem*-$N_3P_3Cl_4$(*i*Pr)[CH(OH)-C_6H_4-*p*-CH=CH$_2$]

In a variation of the above procedure the phosphazane-magnesio-cuprate is reacted with acetaldehyde to generate **9**. Compound **9** contains a reactive hydroxyl group. Consequently, the reaction of **9** with methacryloyl chloride affords **10** (see Eq. 4.3) [11].

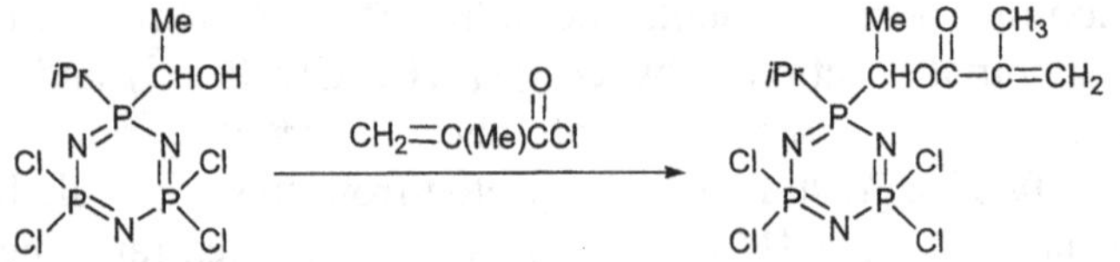

(4.3)

Compound **9** can be dehydrated by an appropriate dehydrating agent to afford a carbon-carbon double bond. However, direct dehydration was not highly successful. A modification of this procedure involving the reaction of **9** with $MeSO_2Cl$ to afford $N_3P_3Cl_4[(iPr)(OSO_2Me)]$. Elimination of the sulfonium group from this compound is assisted by the base DBU to afford the cyclophosphazene-substituted propene, $N_3P_3Cl_4[(iPr)(C(Me)=CH_2]$ (**11**) (Fig. 4.5) [12].

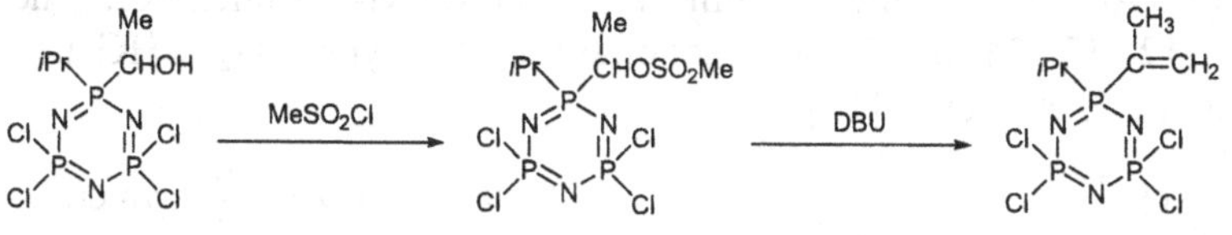

Fig. 4.5. Synthesis of *gem*-$N_3P_3Cl_4(iPr)[C(CH_3)=CH_2]$

In all of the above monomers the vinyl group is attached to the cyclophosphazene ring by means of a P-C bond. A simpler way of preparing the cyclophosphazene containing vinyl monomers is to use oxygen nucleophiles.

The simplest vinyl-containing monomer viz., $N_3P_3Cl_5OCH=CH_2$ (**12**) is prepared by the reaction of $LiOCH=CH_2$ with $N_3P_3Cl_6$ (see Eq. 4.4) [13, 14].

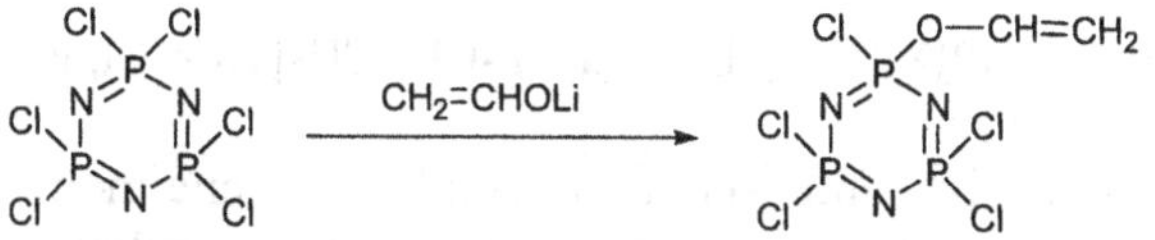

$$(4.4)$$

The anion $LiOCH=CH_2$ is formed in the reaction of tetrahydrofuran with *n*-butyllithium (see Eq. 4.5).

$$C_4H_8O + nC_4H_9Li \longrightarrow LiOCH=CH_2 + C_2H_4 + C_4H_{10} \qquad (4.5)$$

The reaction of $LiOCH=CH_2$ with $N_4P_4Cl_8$ leads to the formation of the monosubstituted derivative $N_4P_4Cl_7(OCH=CH_2)$ (**13**) [15]. This represents a rare example of a cyclotetraphosphazene that contains a vinyl group (see Eq. 4.6).

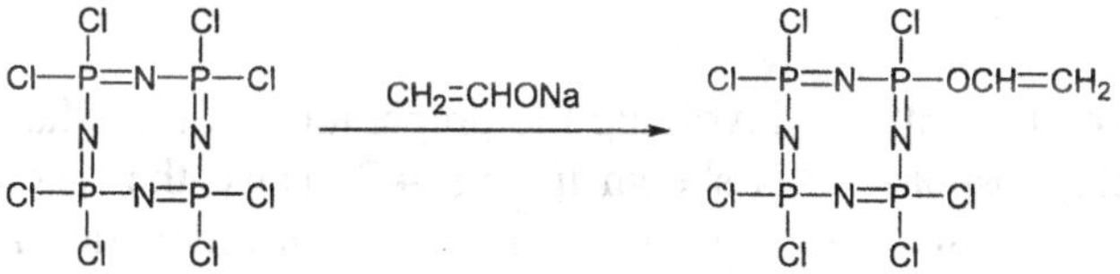

$$(4.6)$$

Inoue and coworkers have used 4'-vinyl-4-biphenylol, $OH-C_6H_4-p-$
$C_6H_4-p-CH=CH_2$ to prepare cyclophosphazenes that contain a long biphenyl spacer group to separate the vinyl group from the cyclophosphazene
ring. Thus, the reaction of $N_3P_3Cl_6$ with $OH-C_6H_4-p-C_6H_4-p-CH=CH_2$ in
the presence of triethylamine as the hydrogen chloride acceptor, affords
the monosubstituted cyclophosphazene $N_3P_3Cl_5[O-C_6H_4-p-C_6H_4-p-$
$CH=CH_2]$ (**14**) (Fig. 4.6). The fluoro-analogue **15** can be prepared by the
reaction of $N_3P_3F_6$ with the sodium salt of 4'-vinyl-4-biphenylol. The
monomers **12**, **13**, **14** and **15** are monomers of the type $CH_2=CHG$ i.e.,
mono substituted ethylenes [16].

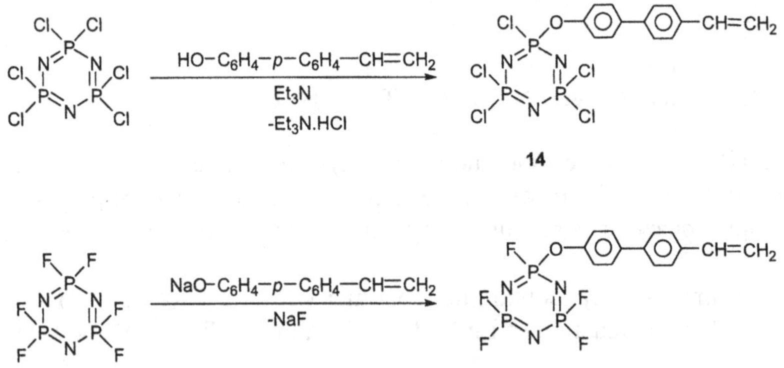

Fig. 4.6. Preparation of $N_3P_3X_5[O-C_6H_4-p-C_6H_4-p-CH=CH_2]$ (X = Cl, F)

Another example of the synthesis of a cyclophosphazene monomer containing the vinyl group relies on an entirely different synthetic strategy.
Thus, partial hydrolysis of hexakis-trifluorethoxycyclotriphosphazene affords $[N_3P_3(OCH_2CF_3)_5(O)]$. This is reacted with methacryloylchloride to
afford $[N_3P_3(OCH_2CF_3)_5(OC(O)C(CH_3)=CH_2)]$ (**16**) (see Eq. 4.7) [17].

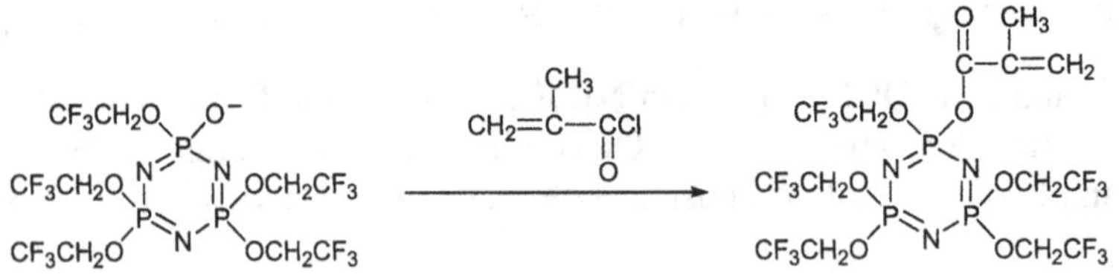

(4.7)

A more elaborate synthesis involving the preparation of a methacryloyl
containing cyclophosphazene is shown in Fig. 4.7. Thus, the reaction of
$N_3P_3Cl_6$ with the sodium salt of *p*-hydroxybenzaldehyde affords
$N_3P_3Cl_5(O-C_6H_4-p-CHO)$. Replacement of all the chlorine atoms from this

compound by trifluoroethoxy substituents leads to the formation of $N_3P_3(OCH_2CF_3)_5(O-C_6H_4-p-CHO)$. At this stage the aldehyde functional group is reduced with the aid of the reducing agent $NaBH_4$. This yields the compound $N_3P_3(OCH_2CF_3)_5(O-C_6H_4-p-CH_2OH)$. The hydroxyl group affords the opportunity of linking the methacryloyl group. Accordingly, reacting $N_3P_3(OCH_2CF_3)_5(O-C_6H_4-p-CH_2OH)$ with methacryloyl chloride affords the monomer $N_3P_3(OCH_2CF_3)_5(O-C_6H_4-p-CH_2OC(O)C(Me)=CH_2)$ (17) (Fig. 4.7) [17].

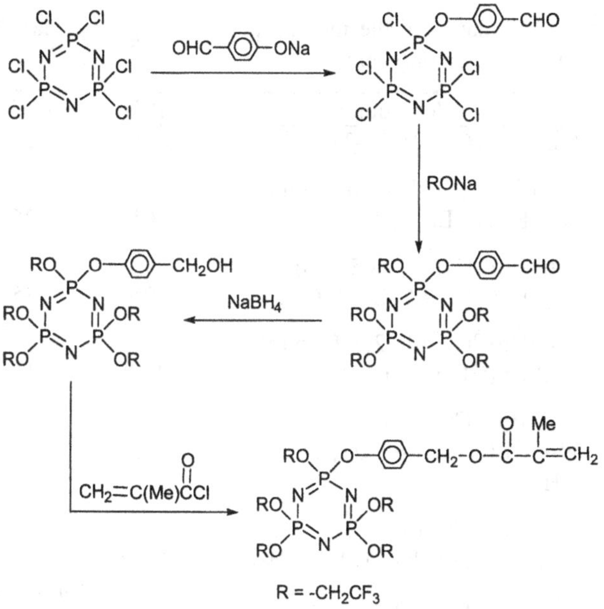

Fig. 4.7. Synthesis of $N_3P_3(OCH_2CF_3)_5[O-C_6H_4-p-CH_2-OC(O)C(Me)=CH_2]$

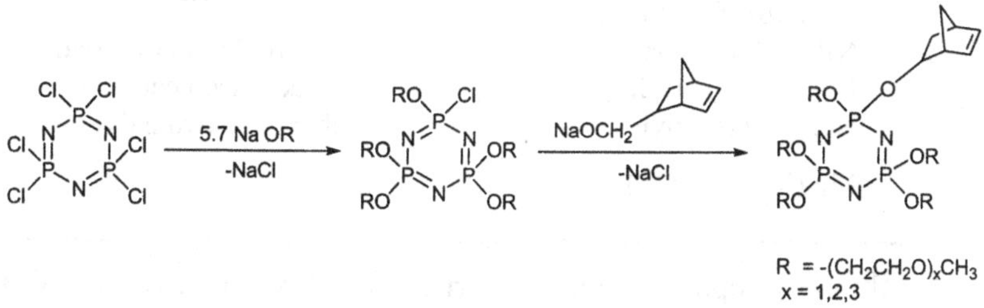

Fig. 4.8. Synthesis of norbornenyl cyclophosphazenes

Recently, norbornenyl cyclophosphazenes, [$N_3P_3(OR)_5(OCH_2Nor)$] (**18**) (Nor = norbornenyl) have been synthesized by the reaction of $N_3P_3(OR)_5Cl$ with Nor-CH_2ONa (Fig. 4.8). Monomers of the type **18** have been converted by ring-opening metathesis polymerization (ROMP) into organic polymers containing cyclophosphazene pendant groups as shown later in this chapter [18].

The ^{31}P-NMR data along with the melting point/boiling point data for a few selected cyclophosphazene monomers are given in Table 4.1.

Table 4.1. ^{31}P-NMR data and the melting point/boiling point data for a few selected cyclophosphazene monomers

S.No	Compound	$\delta\ PX_2$	$\delta\ P(X)(R)$
1	$N_3P_3F_5[C(OMe)=CH_2]$	9.45	25.18
2	Mp (Bp) °C: (25-30, 0.02 mm) $N_3P_3F_5[C(OEt)=CH_2]$	9.05	25.49
3	Mp (Bp) °C: (25, 0.02 mm) $N_3P_3F_5[C_6H_4$-p-$C(CH_3)=CH_2]$	9.25	35.44
4	Mp (Bp) °C: (50-52, 0.04 mm) $N_3P_3Cl_5(OCH=CH_2)$	23.40	13.20
5	Mp (Bp) °C: 50-52 $N_3P_3Cl_5(OC_6H_4$-p-C_6H_4-p-$CH=CH_2)$	21.46	11.25
6	Mp (Bp) °C: 113 *gem*-$N_3P_3Cl_4(iPr)[C(Me)=CH_2]$ Mp (Bp) °C: 48.5-50.5	18.20	40.20 ($\delta P(iPr)(C(Me)=CH_2)$)
7	*gem*-$N_3P_3Cl_4(iPr)[C(OCOMe)=CH_2]$ Mp (Bp) °C: 34.0-36.0	19.40	30.5 ($\delta P(iPr)(C(OCOMe)=CH_2)$)
8	$N_3P_3(OR)_5(OCH_2Nor)$ R = CH_2CH_2-OCH_3 Nor = Norbornyl Mp (Bp) °C: (Oil)	14.8-16.7 ppm (a broad peak is seen centered above the two halves)	

The fluorocyclophosphazene derivatives $N_3P_3F_5[C(OMe)=CH_2]$, $N_3P_3F_5[C(OEt)=CH_2]$ and $N_3P_3F_5[C_6H_4$-p-$C(CH_3)=CH_2]$ as well as the pentachlorovinyloxycyclotriphosphazene, $N_3P_3Cl_5(OCH=CH_2)$ are distillable liquids. In contrast the norbornenyl compounds $N_3P_3(OR)_5(OCH_2Nor)$ are oils. The geminal cyclophosphazenes *gem-*

$N_3P_3Cl_4(iPr)[C(OCOMe)=CH_2]$ and gem-$N_3P_3Cl_4(iPr)[C(Me)=CH_2]$ are low-melting solids while the biphenyloxy derivative $N_3P_3Cl_5(OC_6H_4$-p-C_6H_4-p-$CH=CH_2)$ is a solid which melts at 113 °C. Except of the norbornenyl derivatives the rest of these show an AB_2 or AX_2 type of ^{31}P-NMR spectra.

4.2.2 Polymerization of Cyclophosphazene Monomers

4.2.2.1 Polymerization of Fluorocyclophosphazene Monomers

Attempts to polymerize the propenylfluorophosphazene monomer $N_3P_3F_5[C(Me)=CH_2]$ by free-radical methods have not been successful (see Eq. 4.8) [19].

(4.8)

This was attributed to two features: (1) The $N_3P_3F_5$ group exerts a σ-electron withdrawing effect on the vinyl group. (2) The vinyl monomer is sterically hindered as it has two substituents on the same carbon. One of these is the $N_3P_3F_5$ group which is sterically encumbered.

The monomers, $N_3P_3F_5[C(OMe)=CH_2]$ and $N_3P_3F_5[C(OEt)=CH_2]$ have been designed with a view to alleviate the electronic factors by having the electron-releasing -OR groups directly placed on the vinyl group. The monomer $N_3P_3F_5[C_6H_4$-p-$C(CH_3)=CH_2]$ contains a phenyl group that separates the vinyl moiety from the cyclophosphazene ring. This structural feature is aimed at reducing the steric effects at the vinyl group. In spite of modifying the original monomer design by these innovations it was observed that none of these monomers could be homopolymerized by free-radical initiators [20, 21].

However, copolymerization of the monomers **4-6** with organic monomers such as styrene or methyl methacrylate was successful (Fig. 4.9) [19-21].

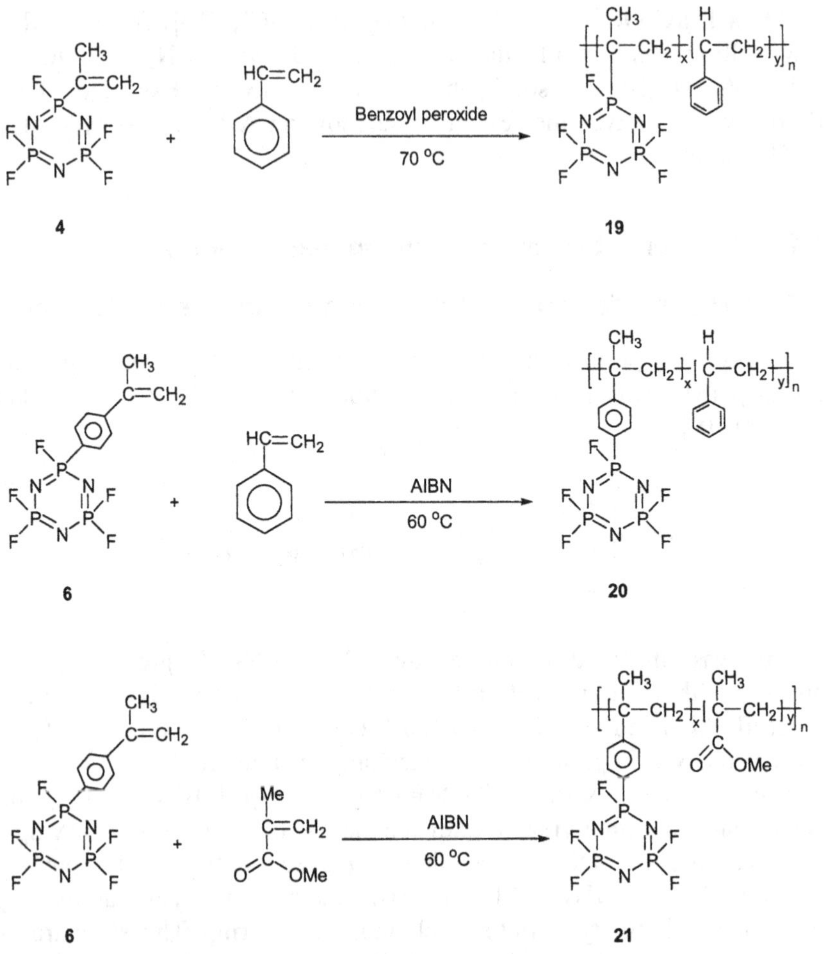

Fig. 4.9. Copolymerization of fluorocyclophosphazene-containing vinyl monomers with organic monomers

The following is the summary of the copolymerization behavior of the fluorocyclophosphazene monomers.

1. The copolymers obtained are *true copolymers* and *not* a *mixture* of two homopolymers. This is indicated by the presence of a *single* glass transition temperature. For example, glass transition temperatures of polystyrene and poly(methylmethacrylate) are 100 and 114 °C. Single glass transition temperatures at 112 °C for polymer **19** (about 43% of cyclophosphazene) and 151 °C for polymer **20** (about 40% of cyclophosphazene) were detected.

2. The cyclophosphazene group does not undergo degradation during the process of polymerization. The ^{31}P-NMR chemical shifts of the co-polymers are similar to those observed for the monomers.

3. The molecular weights of the co-polymers vary considerably. For example, polymer **19** has M_w's up to 90-100,000 with a PDI of 1.1 to 1.7. The general trend observed is that the molecular weights of the polymers *decrease* with an *increase* in the cyclophosphazene content.

4. The maximum incorporation of cyclophosphazene in the co-polymer occurs in the case of the least sterically hindered monomer **6** (in the copolymer **20**). The maximum amount of cyclophosphazene incorporation in **20** has been estimated to be about 67%. In contrast, the incorporation of the sterically hindered monomer $N_3P_3F_5[C(CH_3)=CH_2]$ **4** in its polystyrene copolymers **19** is considerably less. Up to a maximum of 38% incorporation has been observed.

5. All the copolymers are self-extinguishing polymers. Incorporation of the cyclophosphazene units imparts the organic polymer with excellent flame-retardant properties. Other thermal properties for these copolymers are not drastically altered in comparison to the homopolymers.

4.2.2.2 Polymerization of Chlorocyclophosphazene Monomers

The 1,1'-disubstituted olefins such as *gem*-$N_3P_3Cl_4(iPr)[C(OC(O)CH_3)$ $=CH_2)]$ (**7**) and *gem*-$N_3P_3Cl_4(iPr)[C(CH_3)=CH_2)]$ (**11**), also do not undergo homopolymerization. This is similar to what was observed with the fluoro-cyclophosphazene monomers **4-6** [17]. Copolymerization of **7** and **11** with monomers such as styrene and methyl methacrylate does, however, occur. Other chlorocyclophosphazene monomers that have a spacer unit that separates the vinyl group from the cyclophosphazene ring have been successfully homopolymerized [17, 22]. The *spacer* can just be an oxygen atom as it is for the monomers $N_3P_3Cl_5(OCH=CH_2)$ (**12**) or $N_4P_4Cl_7(OCH=CH_2)$ (**13**). Examples of homopolymers prepared from chlorocyclophosphazene monomers are shown in Fig. 4.10.

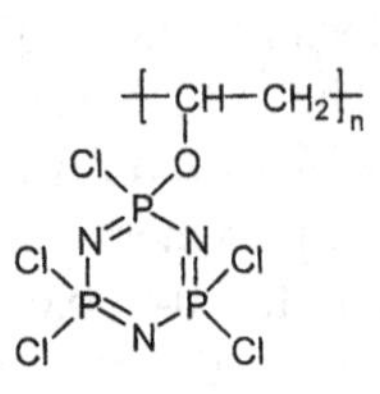

Fig. 4.10. Homopolymers containing cyclophosphazene pendant groups

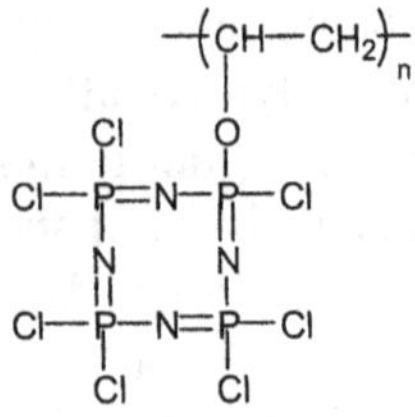

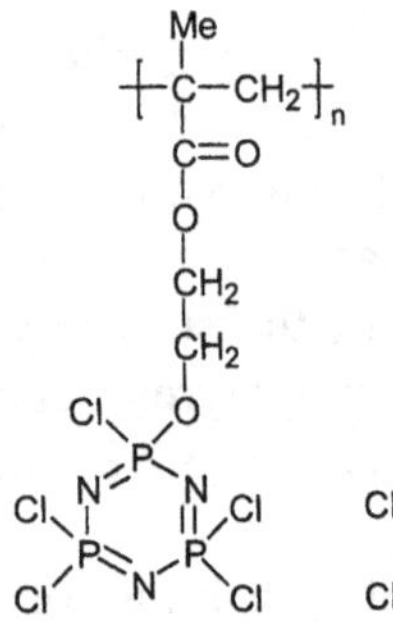

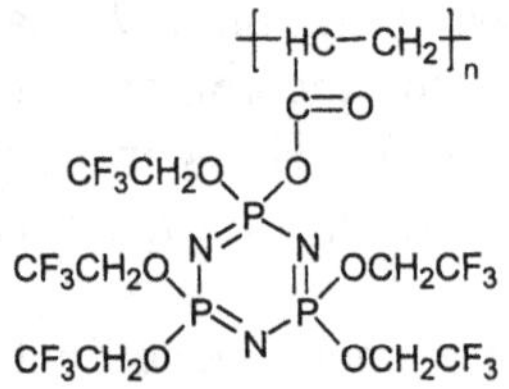

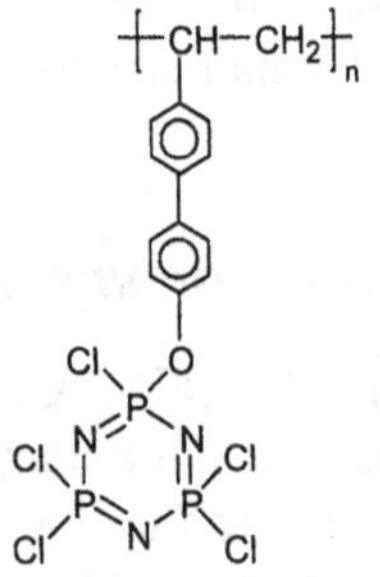

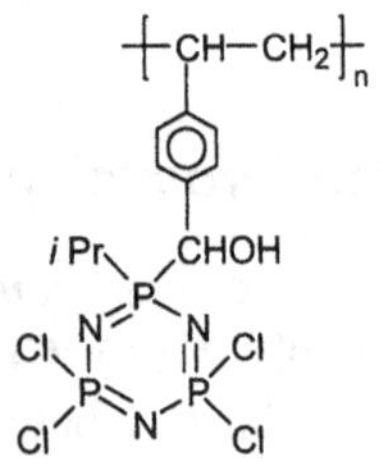

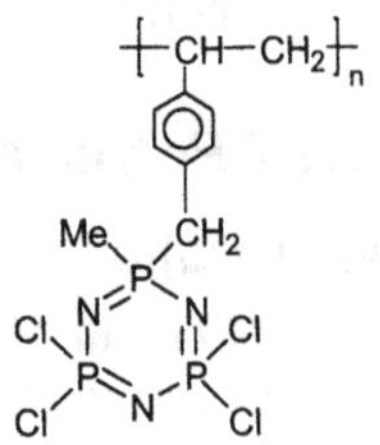

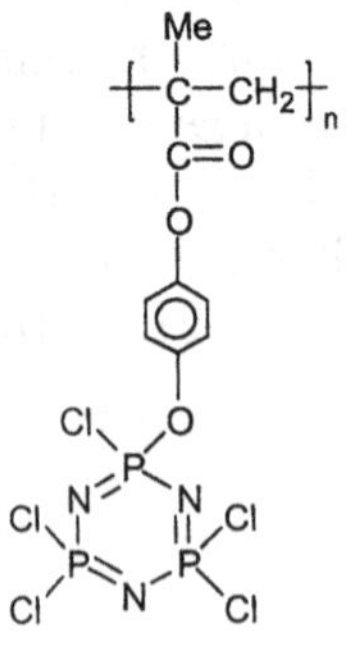

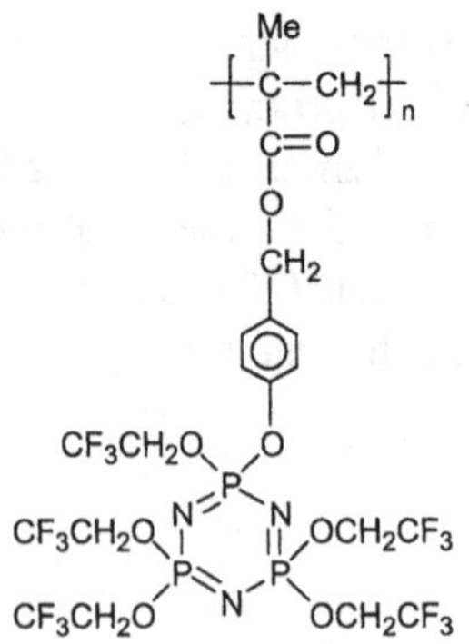

The homopolymerization of $N_3P_3Cl_5(OCH=CH_2)$ (12) and $N_4P_4Cl_7(OCH=CH_2)$ (13) can be accomplished successfully in the bulk to afford high-molecular-weight polymers (Fig. 4.11) [22]. The polymerization of the vinyloxy cyclophosphazenes $N_3P_3Cl_5(OCH=CH_2)$ (12) and $N_4P_4Cl_7(OCH=CH_2)$ (13) occurs in the bulk and is catalyzed by AIBN at 65 °C. This process affords the homopolymers 22 and 23. However, the attempted polymerization reactions of $N_3P_3X_5(OCH=CH_2)$ [(X = F,OCH_3,OCH_2CF_3,N(CH_3)_2] have not been successful. The M_w of the polymer 22 was found to be around 500,000 with a PDI of 1.29. The M_w of 23 was determined as 210,000, while the M_n was found to be 8,000. Determination of the molecular weight by membrane osmometry revealed it to be 136,000. The molecular weights of both 22 and 23 are sufficiently high to allow them to form films. Both 22 and 23 are soluble in a large variety of organic solvents and are film-forming polymers. The glass transition temperatures of 22 and 23 are 72 and 76 °C. These polymers decompose by a two-step decomposition process, the first one occurring at 110 °C accompanied by the removal of two molecules of HCl. The second decomposition occurs at 400 °C. The char yield of these polymers at 700 °C is around 30%. These homopolymers are inflammable which is surprising in view of the self-extinguishing nature of the copolymers of styrene containing fluorocyclophosphazenes such as 19.

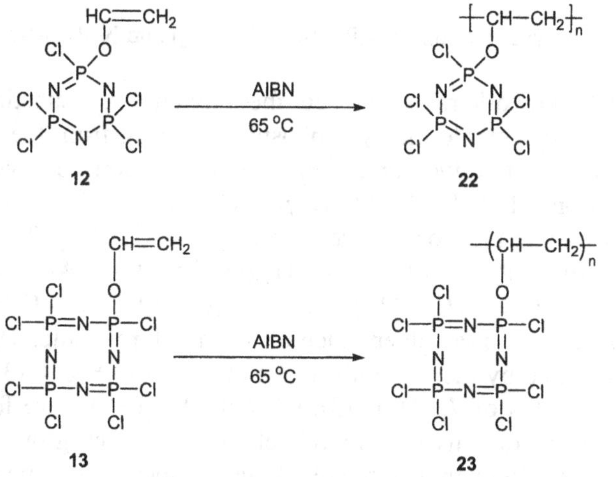

Fig. 4.11. Homopolymerization of $N_3P_3Cl_5(OCH=CH_2)$ and $N_4P_4Cl_7(OCH=CH_2)$

The monomers $N_3P_3Cl_5(OCH=CH_2)$ (12) and $N_4P_4Cl_7(OCH=CH_2)$ (13) are reminiscent of vinylacetate in their reluctance to copolymerize with other organic monomers such as styrene. On the other hand, the monomers 12 and 13 could be copolymerized with each other to afford the interesting copolymer **31** which contains an almost equal proportion of the two monomers (Fig. 4.12). This copolymer has a M_w of about 1,054,000 and a M_n of 14,700.

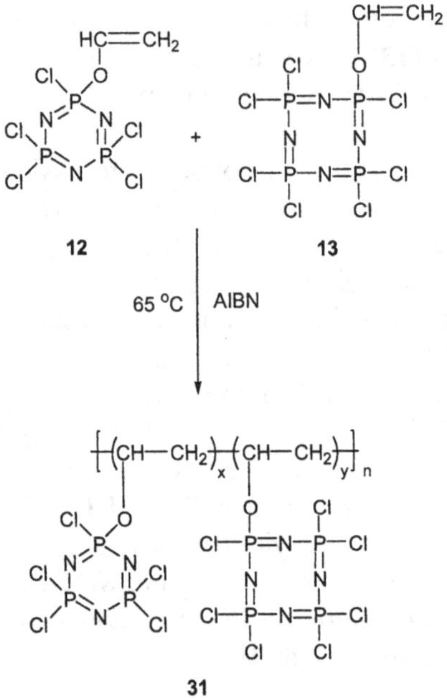

Fig. 4.12. Co-polymerization of $N_3P_3Cl_5(OCH=CH_2)$ and $N_4P_4Cl_7(OCH=CH_2)$

Inoue and coworkers have used the biphenyloxy substituent as the spacer group to separate the cyclophosphazene ring from the vinyl group. This is very effective and homopolymerization as well as co-polymerization of $N_3P_3Cl_5(O-C_6H_4-p-C_6H_4-p-CH=CH_2)$ **(14)** and other monomers of this type occurs very readily [16, 23, 24]. The homopoly-merization of $N_3P_3Cl_5(O-C_6H_4-p-C_6H_4-p-CH=CH_2)$ **(14)** and $N_3P_3F_5(O-C_6H_4-p-C_6H_4-p-CH=CH_2)$ **(15)** or $N_3P_3(OCH_2CF_3)_5(O-C_6H_4-p-C_6H_4-p-CH=CH_2)$ occurs in a solvent such as 1,2-dichloroethane. The polymeri-zation is initiated by the free-radical initiator AIBN (Fig. 4.13). Maximum molecular weights of 7,30,000 (M_w) and 98,000 (M_n) were found for the polymer which contains the chlorocyclophosphazenes group as the pen-dant. These polymers are soluble in a large number of organic solvents and can be cast into films. Bulk polymerization of $N_3P_3Cl_5(O-C_6H_4-p-C_6H_4-p-CH=CH_2)$ **(14)** occurs between 170-250 °C to afford high-molecular-weight polymers. Interestingly, heating at 250 °C does not lead to the ring-opening of the cyclophosphazene ring.

The thermal behavior of the polymer **26** (X = Cl) is quite remarkable. It is a self-extinguishing polymer as shown by simple flame-tests. Secondly, the polymers are thermally stable up to 300-400 °C. The char yield of the polymer **26** (X = Cl) at 800 °C is greater than 60% even under air.

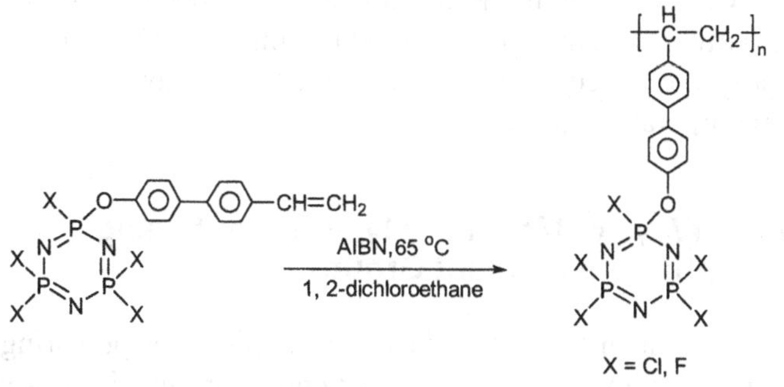

Fig. 4.13. Homopolymerization of $N_3P_3X_5(O\text{-}C_6H_4\text{-}p\text{-}C_6H_4\text{-}p\text{-}CH=CH_2)$ (X = Cl, F)

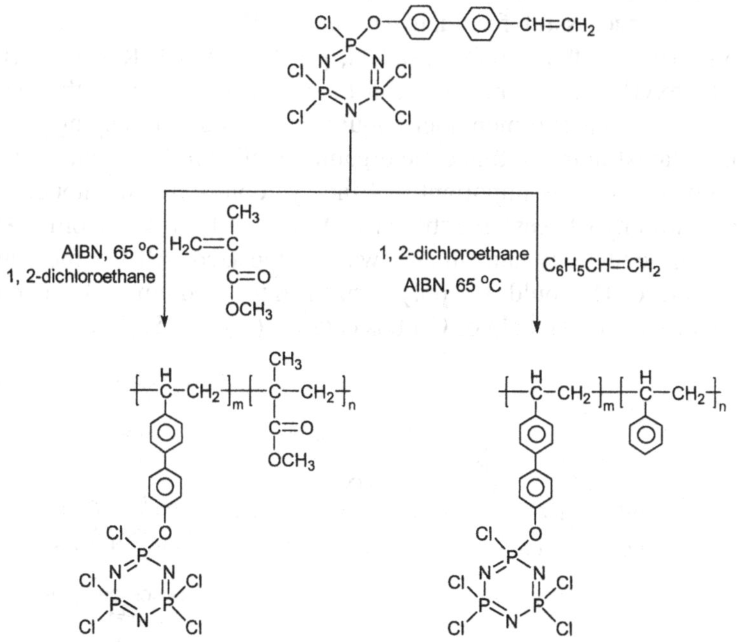

Fig. 4.14. Copolymerization of $N_3P_3Cl_5(O\text{-}C_6H_4\text{-}p\text{-}C_6H_4\text{-}p\text{-}CH=CH_2)$ with styrene and acrylate monomers

$N_3P_3Cl_5(O\text{-}C_6H_4\text{-}p\text{-}C_6H_4\text{-}p\text{-}CH=CH_2)$ (**14**) can be copolymerized with a variety of monomers such as styrene or acrylates such as methyl methacrylate, methyl acrylate or ethyl acrylate (Fig. 4.14) [25]. The T_g's of **26** (R = Cl) and its copolymer with styrene (**32**) are around 76-88 °C. This may be compared with the value found for polystyrene (100 °C). Copolymers such as **32** are flame-resistant and have good thermal stability. A 64% char yield for the polystyrene copolymer is achieved with about 50% cyclophosphazene incorporation in the copolymer.

4.2.3 Other Ways of Making Polymers Containing Cyclophosphazene Pendant Groups

One of the recent methods that has been applied for preparing polymers containing cyclophosphazene rings as pendant groups is by ring-opening metathesis polymerization (ROMP). The principles involved in the application of this strategy to cyclophosphazene monomers are as follows.

Monomers that are amenable for this type of polymerization are cyclic compounds that contain a double bond. Normal methods of polymerization will not be successful for such monomers. But by using metal carbene catalysts such as the Grubb's catalyst, [{P(Cyc)$_3$}$_2$(Cl)$_2$Ru(=CHPh)] [Cyc = cyclohexyl], these monomers can be polymerized. Thus, the design of the cyclophosphazene monomer should be carried out keeping the nature of this catalyst in mind. Since the organometallic Grubb's catalyst is sensitive to halogens, it is important that the cyclophosphazene monomer does not contain any of these substituents. Accordingly, the monomer **18** containing oligoetheroxy substituents was synthesized. All the monomers of the structure **18** could be polymerized into medium-molecular-weight polymers **34** with the help of Grubbs catalyst (Fig. 4.15) [18].

R = (-CH$_2$CH$_2$O)$_x$CH$_3$
x = 1, 2, 3

34

Fig. 4.15. ROMP of norbornenyl cyclophosphazenes

The polymers **34** have very low glass transition temperatures. Thus, the T_g's of the polymers **34** vary as a function of x in the oligoetheroxy side-chains from -50 (x = 1) to -66.5 °C (x = 3). The presence of the double bonds in the backbone allows further reactions to be carried out. Thus, epoxidation of the polymer **34** (x = 3) was carried out which converts the gum-like polymer in to a solid albeit still with a lower T_g (-51.1 °C)

An interesting way of anchoring a cyclophosphazene to an organic polymer consists of choosing a preformed organic polymer on which the cyclophosphazene can be attached. This method differs from the previous methods seen so far. Thus, this can be construed as a polymer modification reaction involving an organic polymer. For example, a polymer obtained from a modified styrene viz., Ph_2P-p-C_6H_4-CH=CH_2 contains phosphine units as side-groups. Reaction of such a polymer can occur with azides by the formation of P=N bonds. This reaction known as the Staudinger reaction converts phosphorus in the oxidation state of +III to +V. Utilization of this strategy on a copolymer of styrene and Ph_2P-p-C_6H_4-CH=CH_2 is shown in Fig. 4.16 [26].

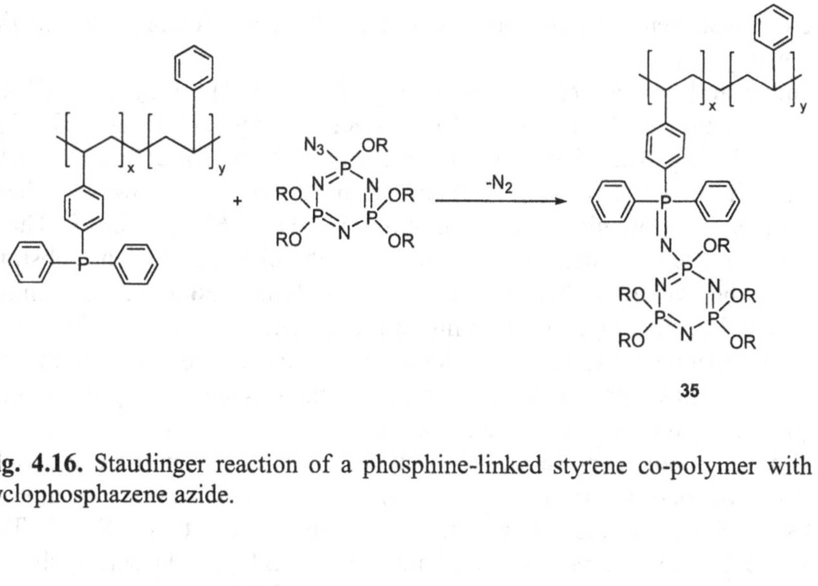

Fig. 4.16. Staudinger reaction of a phosphine-linked styrene co-polymer with a cyclophosphazene azide.

4.2.4 Applications of Polymers Containing Cyclophosphazene Pendant Groups

One of the important applications of cyclophosphazene pendant polymers appears to be in the field of thermally stable and flame-retardant materials. Thus, polymers such as **19**, **20** and **26** have been shown to be self-extinguishing [19-21]. Polymer **26** has shown exceptional thermal stabilities and has char yields as high as 60-65% even at high temperatures such as 800 °C [23, 24]. Also it has been shown that incorporation of $N_3P_3Cl_5(O$-C_6H_4-p-C_6H_4-p-CH=$CH_2)$ in styrene and acrylate polymers

(polymers **32** and **33**) accords them with favorable flame-retardant and high-temperature stability properties [25].

Many other applications of these polymers depend upon their design. Analogous to the polyphosphazene family discussed in Chap. 3 these polymers also can be modified considerably by careful design of the monomers. Two types of applications that emanate from such designs are illustrated below.

4.2.4.1 Polymer-Solid Electrolytes

We have seen in Chap. 3 that polyphosphazenes containing oligoethylene-oxy side-groups, such as $[NP(OCH_2CH_2OCH_2CH_2OCH_3)_2]_n$ were very good solvents for lithium salts such as $LiClO_4$, $LiCF_3SO_3$ etc. The polymer-salt complexes, thus formed, showed good ionic conductivity of the order of 10^{-5} Scm^{-1}. Similar applications have also been found for suitable cyclophosphazene pendant polymers. This has been demonstrated in the following way.

Inoue and coworkers have polymerized $N_3P_3(OR)_5(O-C_6H_4-p-C_6H_4-p-CH=CH_2)$ to afford medium-molecular-weight polymers **36** and **37** (Fig. 4.17) [27, 28]. Note that these polymers do not contain a flexible back-bone. These are essentially substituted poly(styrene)s. However, these polymers have low glass transition temperatures of -62 and -65 °C. These low values are attributed to the flexibility of the oligooxoethylene substitu-ents on the cyclophosphazene side-groups. Polymers **36** and **37** are amor-phous and form complexes with lithium salts. These are semiconductors at room temperature. Although the ionic conductivities are not high enough for any practical applications to be realized, there is sufficient promise that improved design can lead to better properties.

Similar to polymers **36** and **37**, polynorbornenes containing cyclophos-phazene pendant groups (x = 2, 3) (**34**) also form complexes with lithium salts. These show ionic conductivities of the order of 10^{-5} Scm^{-1}. The polymer-Li salt complex formed with polymer **34** (x = 1) with a shorter oligoetheroxy substituent does not show appreciable ambient temperature ionic conductivity.

The ionic conductivity studies carried out on the lithium salt complexes of **34**, **36** and **37** reveals that the backbone flexibility is not the only crite-rion for mediating ion transport. Facile pathways for allowing ion-transport can exist in stiff backbone polymers also provided the side-chains are suf-ficiently flexible.

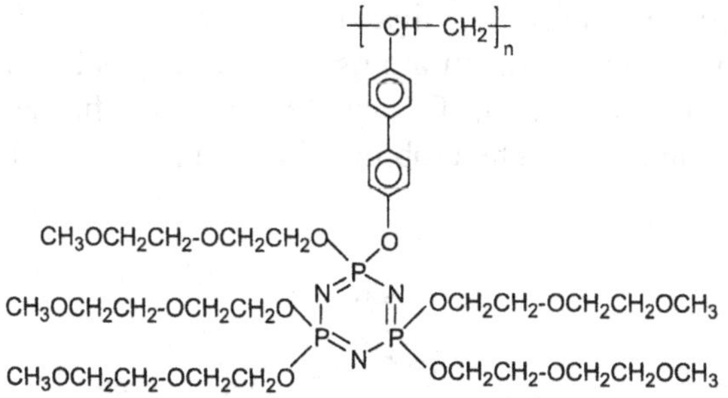

36

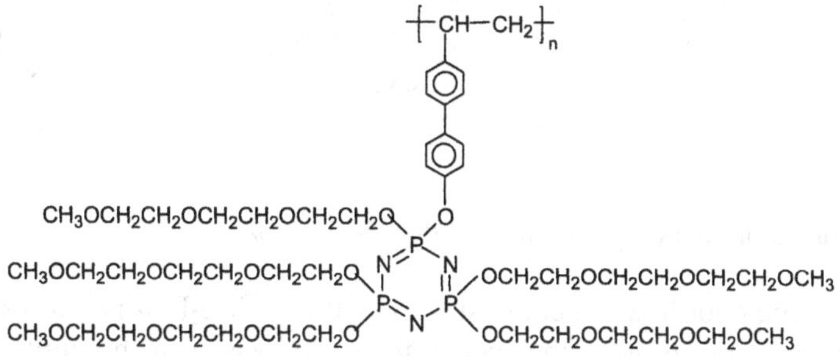

37

Fig. 4.17. Polymers containing oligoethyleneoxycyclophosphazenes as pendant groups

4.2.4.2 Polymeric Ligands

We have seen in Chap. 3 that cyclophosphazenes can be utilized to build multi-site coordination ligands. Thus, for example, the reaction of $N_3P_3Cl_6$ with 3,5-dimethylpyrazole affords the multi-site coordination ligand, $N_3P_3(3,5\text{-}Me_2Pz)_6$ (see Eq. 4.9) [29, 30].

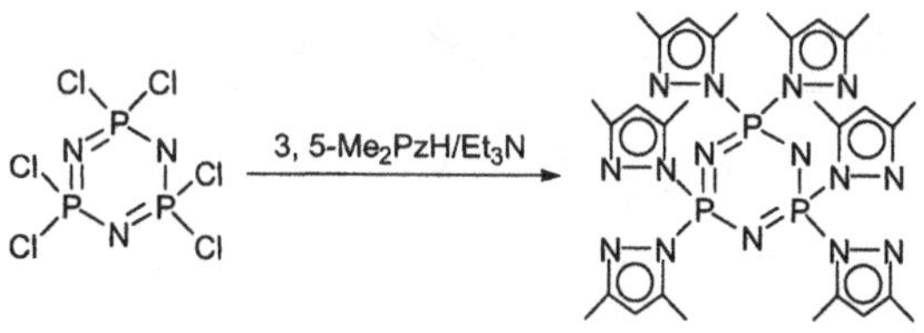

(4.9)

$N_3P_3(3,5\text{-}Me_2Pz)_6$

The compound $N_3P_3(3,5\text{-}Me_2Pz)_6$ has been shown to be an excellent ligand towards transition metal ions. For example, this ligand interacts with $CuCl_2$ to afford transition metal complexes $N_3P_3(3,5\text{-}Me_2Pz)_6{\cdot}CuCl_2$

where the cyclophosphazene ligand coordinates to the copper atom through two pyrazolyl nitrogen atoms and one cyclophosphazene ring nitrogen atom. This complex in fact can also be used further to form a homobimetallic complex or a heterobimetallic complex (Fig. 4.18) [29-36].

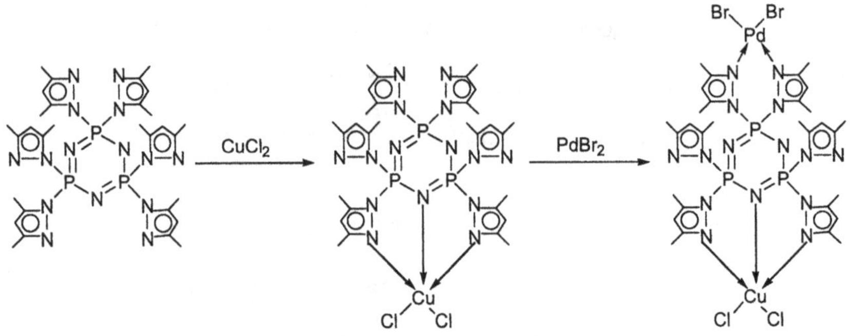

Fig. 4.18. Coordination behavior of $N_3P_3(3,5-Me_2Pz)_6$

While the coordination chemistry of $N_3P_3(3,5-Me_2Pz)_6$ is interesting, the more important question that needs to be addressed in the context of the subject matter being discussed in this chapter, is whether (and how) this chemistry can be translated to the polymeric system.

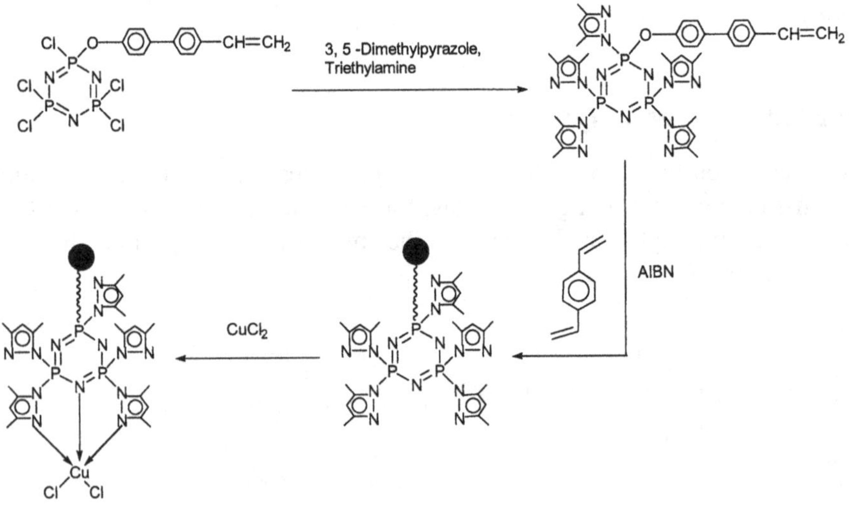

Fig. 4.19. Preparation of polymeric ligands containing pyrazolyl cyclophosphazenes as pendant groups

The monomer $N_3P_3Cl_5(O-C_6H_4-p-C_6H_4-p-CH=CH_2)$ (14) has five reactive P-Cl groups. In principle it is possible to substitute the chlorines with a variety of nucleophiles. We have seen a few examples involving oxygen-based nucleophiles previously. It is also possible to perform the reactions of 14 with amines to generate coordination ligands with nitrogen donor sites. Thus, the reaction of $N_3P_3Cl_5(O-C_6H_4-p-C_6H_4-p-CH=CH_2)$ (14) with 3,5-dimethylpyrazole affords the compound $N_3P_3(3,5-Me_2Pz)_5(O-C_6H_4-p-C_6H_4-p-CH=CH_2)$ [37]. This can be polymerized in the presence of divinylbenzene to afford a crosslinked polymer 38 that contains the multi-site coordination ligand. Reaction of this crosslinked polymer with $CuCl_2$ generates a copper-containing cross-linked polymeric system 39 (Fig. 4.19) [38]. The crosslinked metalated polymer has been used as a heterogeneous catalyst for the hydrolysis of phosphate esters [38, 39]. This catalyst is robust and can be recycled several times.

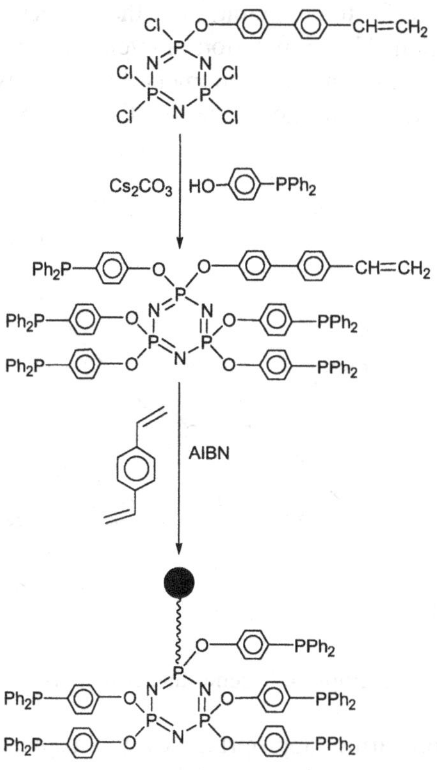

40

Fig. 4.20. Preparation of polymeric phosphine-containing ligands

Another design of a heterogeneous polymeric ligand based on the cyclophosphazene ligand is given in Fig. 4.20. The reaction of $N_3P_3Cl_5$(O-C_6H_4-p-C_6H_4-p-CH=CH$_2$) with OH-C_6H_4-p-PPh$_2$ affords a multi-phosphine derivative which has been converted to the crosslinked polymeric ligand **40**. This has been used to bind palladium and the metalated polymer has been found to be useful as a catalyst for the Heck arylation reactions [40].

4.3 Cyclolinear and Cyclomatrix Polymers

Cyclolinear polymers contain cyclophosphazenes as a repeat unit of a polymer chain. Cyclomatrix polymers contain crosslinked cyclophosphazene units. These polymers have not been studied as well as the other members of the phosphazene polymer family. Some of the strategies that have been employed for the preparation of these types of polymers are discussed in this section. The preparation of cyclolinear polymers depends on the design of cyclophosphazene monomers that retain two reactive sites. A few strategies for achieving the synthesis of such monomers are shown in Fig. 4.21 [17].

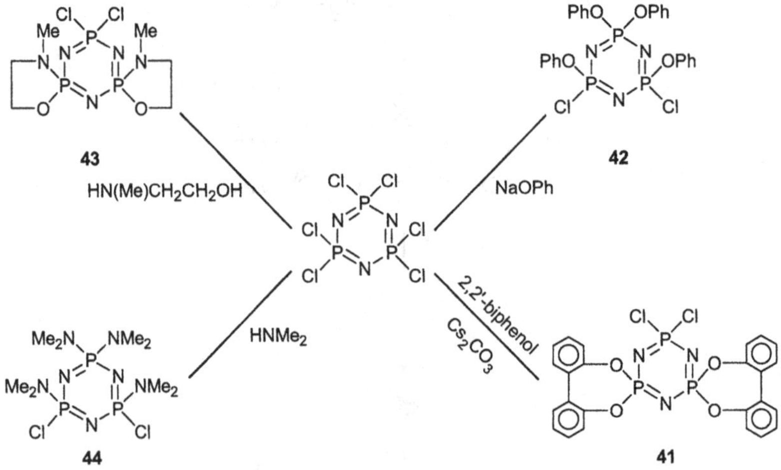

Fig. 4.21. Synthesis of cyclophosphazenes that contain two reactive sites

The task of preparing appropriate cyclophosphazene monomers for cyclolinear polymers consists of the following. $N_3P_3Cl_6$ consists of six reactive P-Cl units. Clearly this compound as such would not be a suitable monomer for preparing soluble cyclolinear polymers. Recall that condensation polymers involving organic monomers relied on difunctional monomers such as diols, diacids, diamines etc. Replacing four chlorine atoms on $N_3P_3Cl_6$ by other substituents would lead to compounds of the type $N_3P_3Cl_2R_4$. The substituents (R) themselves should not be reactive towards reagents that would lead to polymerization. The remaining two chlorines in $N_3P_3Cl_2R_4$ could be in a *geminal* or *nongeminal* orientation. Thus, the reaction of $N_3P_3Cl_6$ with 2,2'-dihydroxybiphenyl or N-methylethanolamine leads to compounds **41** and **43,** respectively, where the two remaining chlorines are *geminal* to each other. On the other hand, reaction of $N_3P_3Cl_6$ with sodium phenoxide or dimethylamine leads to the *nongeminal* compounds **42** and **44,** respectively [8]. Further replacement of the remaining chlorine atoms can be accomplished in a variety of ways to afford a variety of difunctional cyclophosphazenes which can be used for further polymerization reactions. Thus, compounds containing *two* terminal amino (**45, 48**), hydroxyl (**47**), or vinyl groups (**46**) groups can be assembled (Fig. 4.22).

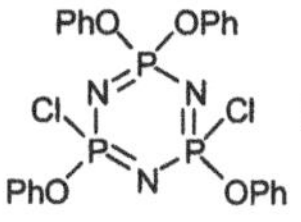

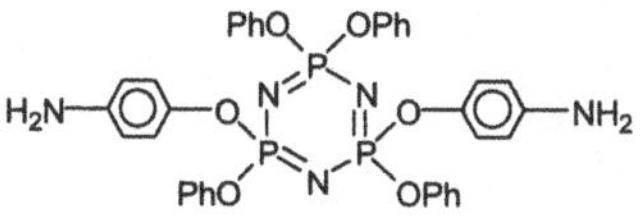

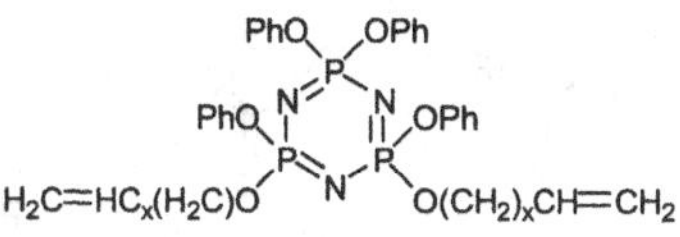

Fig. 4.22. Difunctional cyclophosphazenes

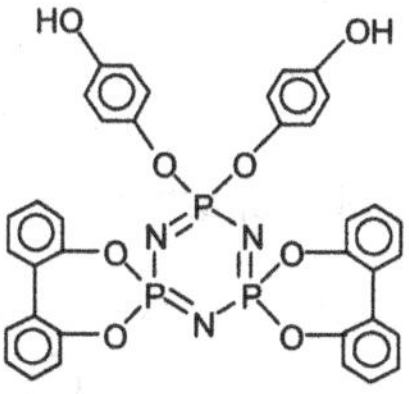

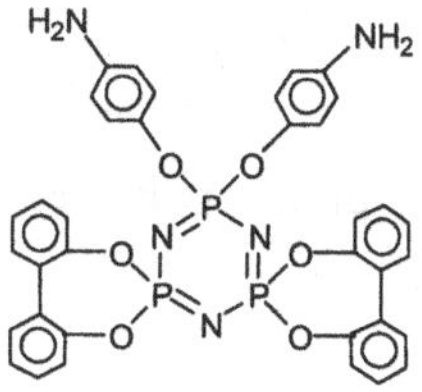

Using the above monomers several polymers have been prepared. Thus, monomer **46** has been polymerized by the acyclic diene metathesis polymerization method (ADMET) by using Grubb's catalyst to afford medium-molecular-weight polymers [41]. On the other hand, the reaction of the dichloro derivative **42** with NaO- C_6H_4-p-SO_2-C_6H_4-p-ONa yields the polymer **49** (Fig. 4.23). Similarly the monomer **47** has been used in condensation polymerization with hexamethylenediisocyanate to afford the polyurethane **50**. Polyimides such as **51** have been prepared by using monomer **48**.

Cyclomatrix polymers containing cyclophosphazenes have received some interest. Although $N_3P_3Cl_6$ itself would be an ideal monomer for this purpose as it contains six reactive groups, this has not had much success.

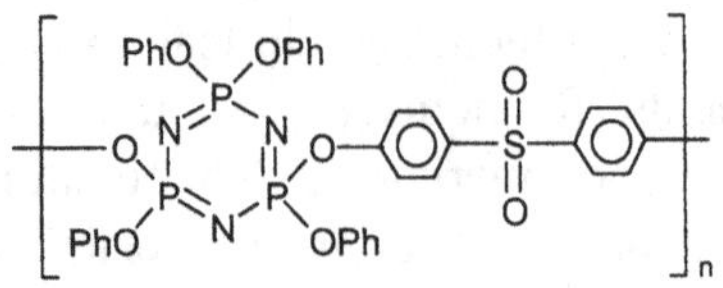

49

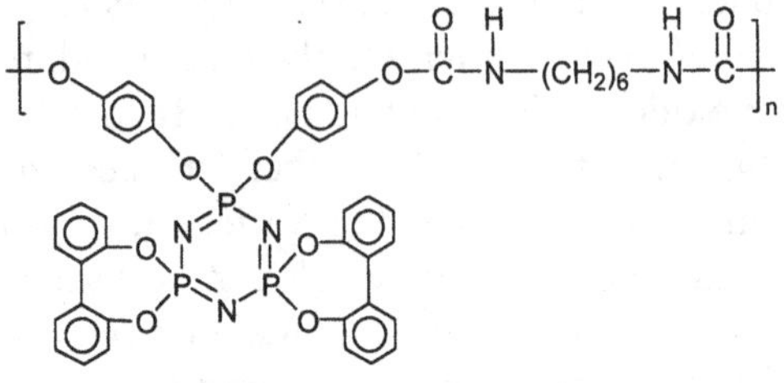

50

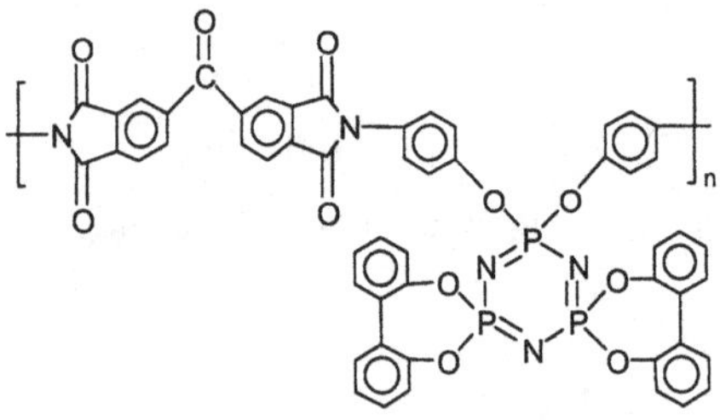

51

Fig. 4.23. Some cyclolinear polymers containing cyclophosphazenes

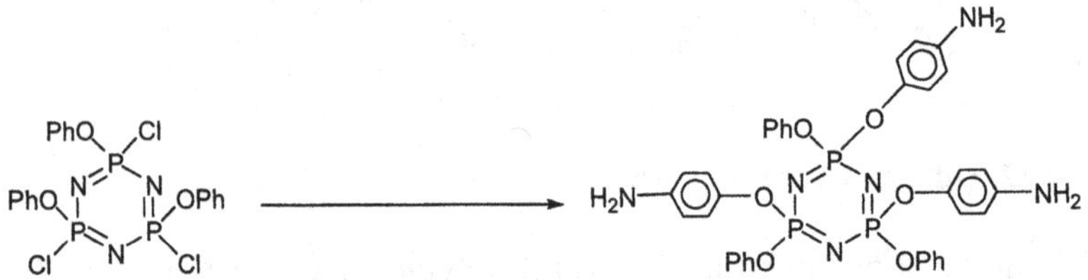

52

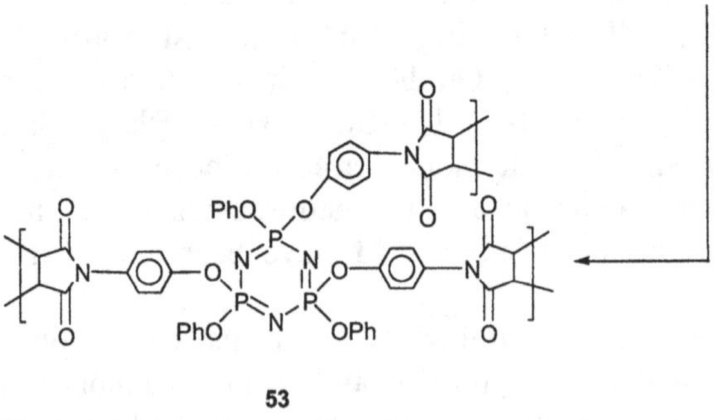

53

Fig. 4.24. Cyclomatrix polymers containing cyclophosphazenes

Monomers such as $N_3P_3(NHCH_2CH=CH_2)_6$ or $N_3P_3(OCH_2CH=CH_2)_6$ have been used as multifunctional monomers for preparing crosslinked polymers. A crosslinked polyimide polymer containing cyclophosphazenes has been prepared by using the trifunctional derivative *nongem-*$N_3P_3(OPh)_3(O\text{-}C_6H_4\text{-}p\text{-}NH_2)_3$ (Fig. 4.24).

The utility and practical applications of the cyclolinear and the cyclomatrix polymers have not yet been established and perhaps this would depend on improved design of the cyclophosphazene monomers.

References

1. Allcock HR (2003) Chemistry and applications of polyphosphazenes. Wiley, Hoboken, New Jersey
2. Gleria M, De Jaeger R (eds) (2004) Phosphazenes - a world-wide insight. Nova Science, New York
3. De Jaeger R, Gleria M (1998) Prog Polym Sci 23:179
4. Allen CW (1994) Coord Chem Rev 130:137
5. Dupont JG, Allen CW (1978) Inorg Chem 17:3093
6. Allen CW, Bright RP (1983) Inorg Chem 22: 1291
7. Shaw JC, Allen CW (1986) Inorg Chem 25:4632
8. Chandrasekhar V, Krishnan V (2002) Adv Inorg Chem 53:159
9. Chandrasekhar V, Thomas KRJ (1993) Struct Bond 81:41
10. Bosscher G, Meetsma A, Van de Grampel JC (1996) Inorg Chem 35:6646
11. Van de Grampel JC, Alberda van Ekenstein GOR, Baas J, Buwalda PL, Jekel AP, Oosting GE (1992) Phosphorus Sulfur Silicon 64:91
12. Bosscher C, Meetsma A, Van de Grampel JC (1997) Dalton Trans 1667
13. Ramachandran K, Allen CW (1983) Inorg Chem 22:1445
14. Allen CW, Ramachandran K, Brown DE (1987) Inorg Synth 25:74
15. Brown DE, Allen CW (1987) Inorg Chem 26:934
16. Inoue K, Takagi M, Nakano M, Nakamura H, Tanigaki T (1988) Makromol Chem Rapid Commun 9:345
17. Van de Grampel JC (2004) Hybrid inorganic-organic phosphazene polymers In: Gleria M, De Jaeger R (eds) Phosphazenes – a world-wide insight. Nova Science, New York, pp 143-170
18. Allcock HR, Laredo WR, Kellam EC, Morford RV (2001) Macromolecules 34:787
19. DuPont JG, Allen CW (1979) Macromolecules 12:169
20. Allen CW, Bright RP (1986) Macromolecules 19:571
21. Allen CW, Shaw JC, Brown DE (1988) Macromolecules 21:2653
22. Brown DE, Ramachandran K, Carter K, Allen CW (2001) Macromolecules 34:2870
23. Inoue K, Nakano M, Takagi M, Tanigaki T (1989) Macromolecules 22:1530

24. Inoue K, Nakamura H, Ariyoshi S, Takagi M, Tanigaki T (1989) Macromolecules 22:4466
25. Selvaraj II, Chandrasekhar V (1997) Polymer 38:3617
26. Allcock HR, Hartle TJ, Taylor JP, Sunderland NJ (2001) Macromolecules 34:3896
27. Inoue K, Nishikawa Y, Tanigaki T (1991) Macromolecules 24:3464
28. Inoue K, Kinoshita K, Nakahara H, Tanigaki T (1990) Macromolecules 23:1227
29. Chandrasekhar V, Thomas KRJ (1993) Appl Organomet Chem 7:1
30. Chandrasekhar V, Nagendran S (2001) Chem Soc Rev 30:193
31. Chandrasekhar V, Krishnan V (2004) Coordination chemistry of cyclophosphazenes In: Gleria M, De Jaeger R (eds) Phosphazenes – a world-wide insight. Nova Science, New York, pp 827-852
32. Thomas KRJ, Chandrasekhar V, Pal PS, Scott R, Hallford R, Cordes AW (1993) Inorg Chem 32:606
33. Thomas KRJ, Tharmaraj P, Chandrasekhar V, Bryan CD, Cordes AW (1994) Inorg Chem 33:5382
34. Thomas KRJ, Chandrasekhar V, Pal PS, Scott R, Hallford R, Cordes AW (1993) Dalton Trans 2589
35. Thomas KRJ, Tharmaraj P, Chandrasekhar V, Tiekink ERT (1994) Dalton Trans 1301
36. Thomas KRJ, Chandrasekhar V, Bryan CD, Cordes AW (1995) J Coord Chem 35:337
37. Chandrasekhar V, Athimoolam AP, Vivekanandan K, Nagendran S (1999) Tetrahed Lett 40:1185
38. Chandrasekhar V, Athimoolam AP, Srivatsan SG, Sundaram PS, Varma S, Steiner A, Zacchini S, Butcher R (2002) Inorg Chem 41:5162
39. Chandrasekhar V, Deria P, Krishnan V, Athimoolam AP, Singh S, Madhavaiah C, Srivatsan SG, Varma S (2004) Bio Org Med Chem Lett 14:1559
40. Chandrasekhar V, Athimoolam AP (2002) Org Lett 4:2113
41. Allcock HR, Kellam EC, Hoffmann MA (2001) Macromolecules 34:5140

5 Other Inorganic Polymers that Contain Phosphorus, Boron and Sulfur

There are many other inorganic polymers that contain main group elements in their backbone [1-3]. While some of these polymers were discovered long ago, some others are quite recent. Representative examples of these polymers that contain P-N-C, P-N-S, P-N-M, P-B, P-C, S-N and B-N linkages are shown in Fig. 5.1.

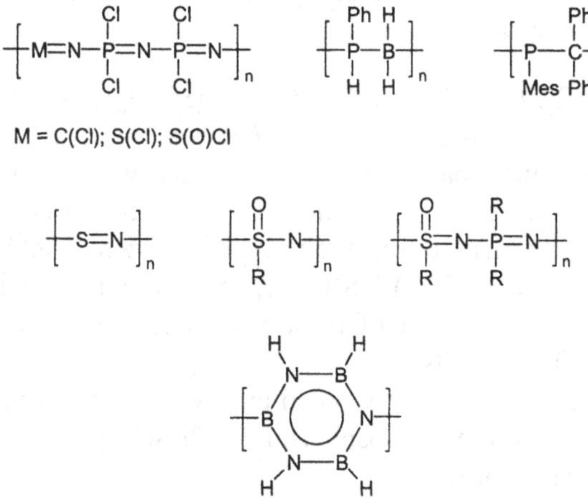

M = C(Cl); S(Cl); S(O)Cl

Fig. 5.1. Representative examples of polymers containing P-N-C, P-N-S, P-B, P-C, S-N and B-N linkages

We will first examine polymers that contain P-N-C, P-N-S and P-N-M linkages. These are also called poly(heterophosphazene)s since they are prepared from the corresponding heterocyclophosphazenes.

5.1 Poly(heterophosphazene)s

Poly(heterophosphazene)s are a recently emerging polymer family that contain a heteroatom such as carbon or sulfur or a transition metal as part of the polymer backbone. Examples of these types of polymers are shown in Fig. 5.2.

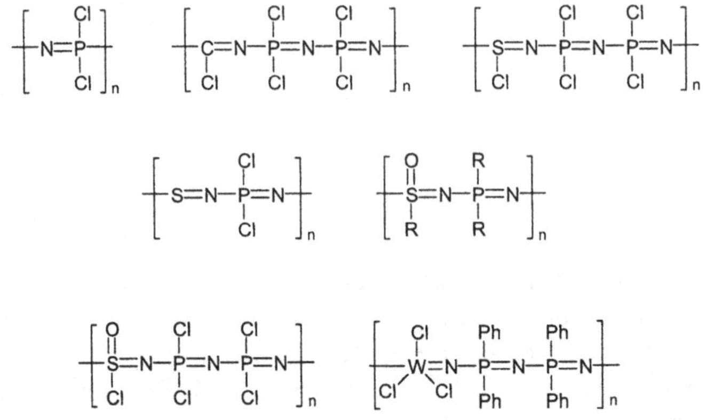

Fig. 5.2. Examples of poly(heterophosphazene)s. Poly(dichlorophosphazene) is also shown for comparison

Polymers $[(NPCl_2)_2(NCCl)]_n$, $[(NPCl_2)_2(NSCl)]_n$, $[(NPCl_2)_2(NS(O)Cl)]_n$, $[(NPCl_2)_2(NWCl_3)]_n$ contain one *extra heteroatom* viz., C, S or W in the six-atom repeat unit of backbone of the polymer which otherwise contains phosphorus and nitrogen atoms. Thus, each of these hetero atoms replaces *every third* phosphorus atom in the polyphosphazene backbone. Polymers $[(SN)(NPCl_2)_2]_n$ and $[(NPR_2)(NS(O)R)]_n$ can be considered as alternating copolymers. The repeat unit of these polymers consist of alternate P=N and S=N (or S(O)=N) units.

How are these polymers assembled and what are their properties? How are they different from polyphosphazenes? These aspects will be addressed in the following sections.

5.2 Poly(carbophosphazene)s

The six-membered heterocyclic ring, cyclocarbophosphazene, $[(NPCl_2)_2CCl]$ serves as the precursor for poly(carbophosphazene)s. The pentachlorocyclocarbophosphazene $N_3P_2CCl_5$ can be considered as a hybrid of cyanuric chloride, $N_3C_3Cl_3$ and hexachlorocyclotriphosphazene, $N_3P_3Cl_6$. Thus, while cyanuric chloride has three carbons and the hexachlorocyclotriphosphazene has three phosphorus atoms, the cyclocarbophosphazene has one carbon and two phosphorus atoms (Fig. 5.3) [4, 5].

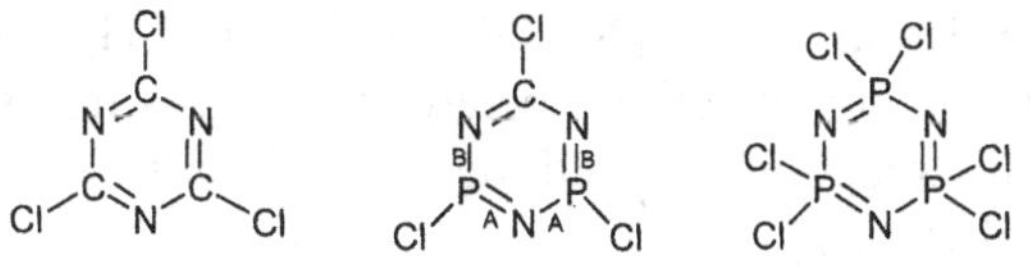

Fig. 5.3. Relationship between [(NPCl$_2$)$_2$CCl], N$_3$C$_3$Cl$_3$ and N$_3$P$_3$Cl$_6$

The X-ray crystal structure of [(NPCl$_2$)$_2$CCl] has been determined. The inorganic ring is nearly planar in spite of the mismatch of sizes between phosphorus and carbon. The C-N distance in [(NPCl$_2$)$_2$CCl] is 1.33 Å, compared to 1.33 Å found in N$_3$C$_3$Cl$_3$. The other bond-parameters are also not very different. Thus, the P-N distances A (av. 1.58 Å) and B (av.1.61 Å) in [(NPCl$_2$)$_2$CCl] (Fig. 5.3) are comparable to the P-N distance of 1.58 Å found in N$_3$P$_3$Cl$_6$. The N-P-N (117.4°) and P-N-P (115.8°) angles in [(NPCl$_2$)$_2$CCl] are *normal* while the N-C-N angle is quite wide (134°).

The preparation of [(NPCl$_2$)$_2$CCl] is achieved by the condensation reaction of [Cl$_3$P=N=PCl$_3$]$^+$[Cl]$^-$ with cyanamide, NC-NH$_2$, in methylene chloride. Cyanamide serves as the source of carbon and nitrogen. Although, intermediate products have not been isolated, the formation of [(NPCl$_2$)$_2$CCl] is believed to proceed through a condensation reaction to afford the intermediate Cl$_3$PNPCl$_2$NCN. The latter adds hydrogen chloride across the C-N triple bond leading to the possible precursor Cl$_3$PNPCl$_2$NC(Cl)NH. Loss of hydrogen chloride followed by ring closure leads to the carbophosphazenes, [(NPCl$_2$)$_2$CCl] (Fig. 5. 4) [4, 5].

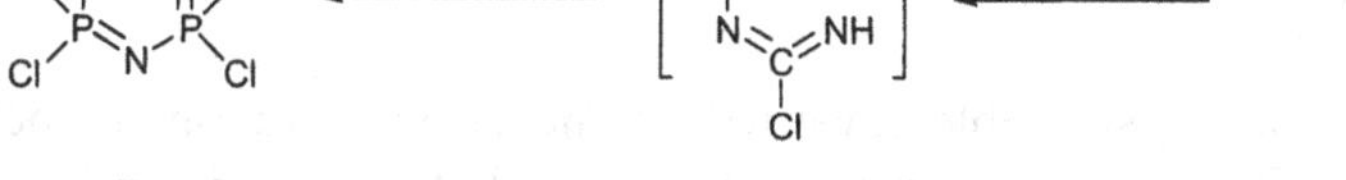

Fig. 5.4. The mechanism of formation of [(NPCl$_2$)$_2$CCl]

The carbophosphazene [(NPCl$_2$)$_2$CCl] undergoes nucleophilic substitution reactions analogous to that found for N$_3$P$_3$Cl$_6$. Even though the chemistry of [(NPCl$_2$)$_2$CCl] has not been studied to the same extent as that of N$_3$P$_3$Cl$_6$ the reactions shown in Fig. 5.5 indicate that many nucleophiles such as primary and secondary amines, aniline, alkoxides, trifluoroethoxide as well as phenoxide react to afford fully substituted products [NP(NHR)$_2$C(NHR)],   [NP(NR$_2$)$_2$C(NR$_2$)],   [NP(NHPh)$_2$C(NHPh)],

[NP(OCH$_3$)$_2$C(OCH$_3$)],   [NP(OCH$_2$CF$_3$)$_2$C(OCH$_2$CF$_3$)]   and [NP(OPh)$_2$C(OPh)]. This reactivity behavior suggests that similar nucleophilic substitution reactions should be possible for the linear polymer [(NPCl$_2$)$_2$CCl]$_n$ also [6, 7].

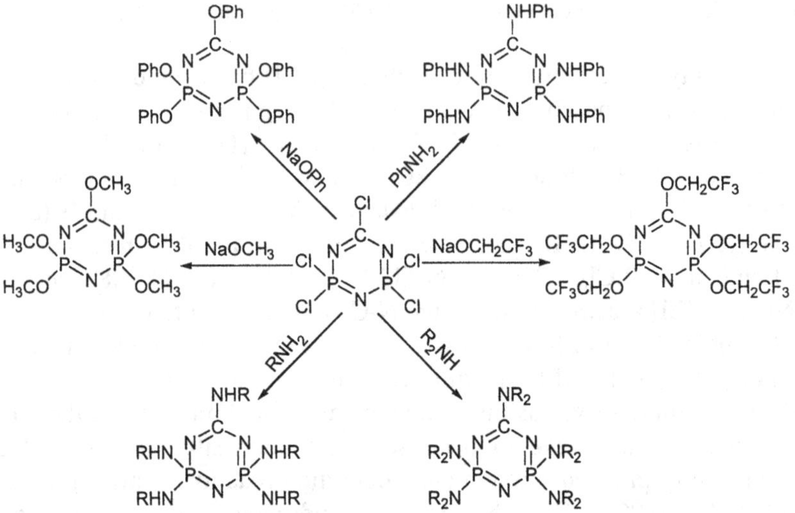

Fig. 5.5. Reactions of [(NPCl$_2$)$_2$CCl] with various nucleophiles such as amines, alkoxides and aryloxides to afford fully substituted products

The pentachlorocyclocarbophosphazene, [(NPCl$_2$)$_2$CCl], has been found to undergo ROP at 120 °C to afford poly(chlorocarbophosphazene), [(NPCl$_2$)$_2$CCl]$_n$ (see Eq. 5.1) [4, 5].

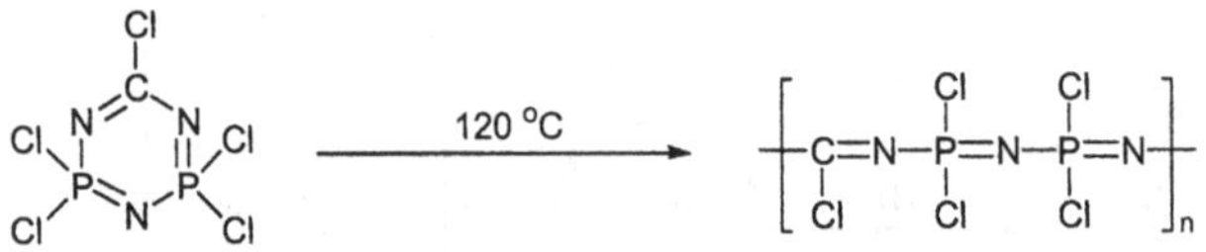

(5.1)

Note the considerable lowering of the polymerization temperature of ROP for [(NPCl$_2$)$_2$CCl] in comparison to that observed for N$_3$P$_3$Cl$_6$ (250 °C). Heating [(NPCl$_2$)$_2$CCl] at higher temperatures (150-190 °C) affords a crosslinked polymer. The lower temperature of ROP has been attributed to an increase in the ring strain suffered by [(NPCl$_2$)$_2$CCl] as a result of replacing the larger skeletal phosphorus of N$_3$P$_3$Cl$_6$ by the smaller carbon atom.

The polymer [(NPCl$_2$)$_2$CCl]$_n$ is hydrolytically sensitive and rapidly reacts with moisture with concomitant chain degradation. Its ^{31}P-NMR spectrum consists of a single resonance observed -3.7 ppm. This is upfield shifted in comparison to the cyclic precursor (+36.5 ppm). A similar shift is observed for $[NPCl_2]_n$ (δ = -18.4 ppm) in comparison to $N_3P_3Cl_6$ (δ = +19.3 ppm). The observation of a single resonance in the ^{31}P-NMR spectrum suggests a head-tail arrangement in the polymer backbone (Fig. 5.6). Alternative arrangements such as head-head would have resulted in more number of ^{31}P resonances.

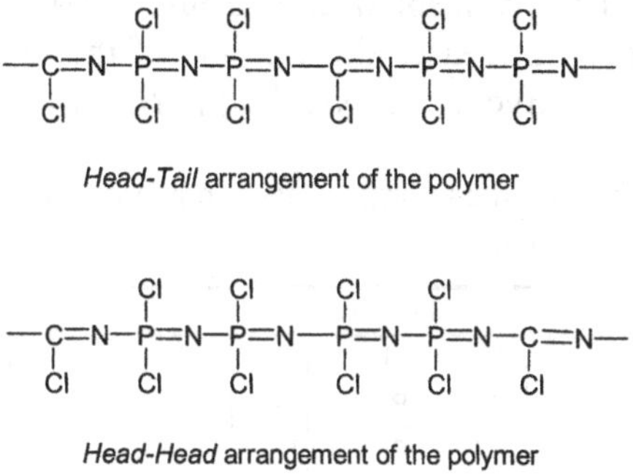

Head-Tail arrangement of the polymer

Head-Head arrangement of the polymer

Fig. 5.6. Head-tail and head-head arrangements of $[(NPCl_2)_2CCl]_n$

The glass transition temperature of $[(NPCl_2)_2CCl]_n$ is -21 °C. This is considerably higher than that observed for $[NPCl_2]_n$ (T_g = -63 °C).This suggests that skeletal rigidity increases upon replacing phosphorus by carbon. One possible reason for this is the more efficient π-bonding formed between C and N in comparison to P and N.

The chlorine substituents of poly(chlorocarbophosphazene) can be replaced with other nucleophiles to afford poly(organocarbophosphazene)s which are hydrolytically more stable in comparison to the parent polymer (Fig. 5.7). There are no examples of poly(alkyl/arylcarbophosphazene)s. The molecular weights of poly(organocarbophosphazene)s are moderately high (M_w = 50,000-150,000).

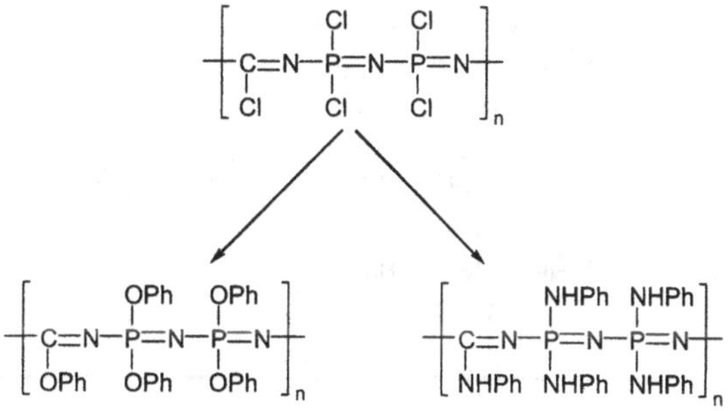

Fig. 5.7. Formation of poly(organocarbophosphazene)s

The ^{31}P-NMR data and glass transition temperatures of some poly(organocarbophosphazene)s are given in Table 5.1. Notice the near invariance of the ^{31}P NMR chemical shifts for a given class of polymer. Thus, for example, the ^{31}P-NMR chemical shifts for most aryloxy polymers are seen at about -10 ppm.

The trends in the glass transition temperatures are similar to what have been observed in polyphosphazenes. Thus, replacing the chlorines with aryloxy substituents increases the T_g. One of the highest T_g's is observed for the anilino derivative, [{NP(NHPh)$_2$}{CNHPh}]$_n$ (+112 °C).

Table 5.1. ^{31}P-NMR data and glass transition temperatures of selected poly(organocarbophosphazene)s

S.No	Polymer	^{31}P	T_g (°C)
1		-3.7	-21
2	Ar = Ph	-10.4	+18
3	Ar = -C$_6$H$_4$-p-CF$_3$	-10.8	+31
4		-9.1	+112
5	R = nPr	6.7	+74
6	R = CH$_3$	0.5	+28

5.3 Poly(thiophosphazene)s

Poly(thiophosphazene)s contain S(IV) along with phosphorus and nitrogen atoms in their backbone [8, 9]. Similar to poly(dichlorophosphazene) and poly(chlorocarbophosphazene), poly(chlorothiophosphazene) is also prepared by the ring-opening polymerization method. The monomer for the preparation of this polymer is the pentachlorocyclothiophosphazene, $N_3P_2SCl_5$. The latter can be considered as a hybrid of the two heterocyclic rings $N_3P_3Cl_6$ and $N_3S_3Cl_3$. The compound $N_3P_2SCl_5$ contains a tricoordinate sulfur(IV) and two tetracoordinate phosphorus atoms (Fig. 5.8).

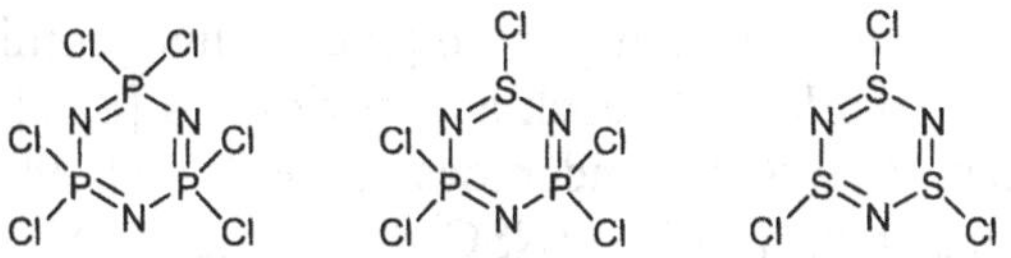

Fig. 5.8. The heterocyclic ring $N_3P_2SCl_5$ and its relationship to $N_3P_3Cl_6$ and $N_3S_3Cl_3$

Pentachlorocyclothiophosphazene, $N_3P_2SCl_5$, can be synthesized in two closely related ways. The key precursor is the bis-trimethylsilylsulfurdiimide, $S[N(SiMe_3)]_2$, which is itself prepared by the reaction of the lithium salt of hexamethyldisilazane with thionyl chloride. The synthon $S[N(SiMe_3)]_2$ provides the NSN fragment for the heterocyclic ring. The reaction of $S[N(SiMe_3)]_2$ with either PCl_5 or $[Cl_3P=N-PCl_3]^+Cl^-$ affords $N_3P_2SCl_5$ in moderate yields as a colorless distillable liquid which shows a single ^{31}P NMR resonance at $\delta = +24.5$ ppm.

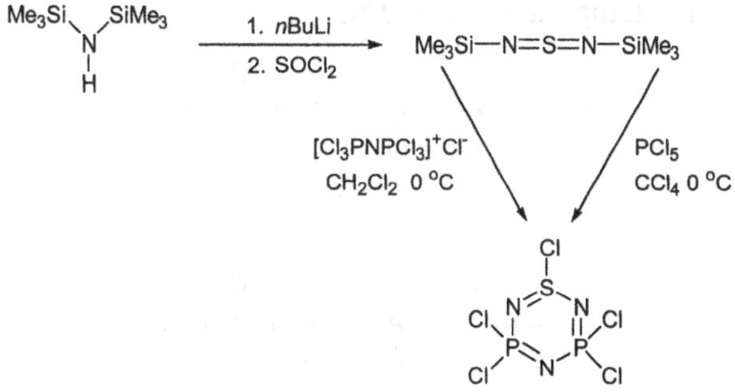

Fig. 5.9. Synthesis of $N_3P_2SCl_5$

An important reaction that revealed that the S-Cl bond in $N_3P_2SCl_5$ can be heterolytically cleaved has been carried out by Roesky [10]. Thus, the reaction of $N_3P_2SCl_5$ with $SbCl_5$ affords $[N_3P_2SCl_4]^+[SbCl_6]^-$ (see Eq. 5.2). The crystal structure of $[N_3P_2SCl_4]^+[SbCl_6]^-$ revealed that it has a planar ring where the sulfur is dicoordinate as a result of the heterolytic cleavage of the S-Cl bond. It may be recalled that in the ROP of $N_3P_3Cl_6$ a phosphonium cation intermediate has been proposed; however, such an intermediate has not actually been isolated in that instance.

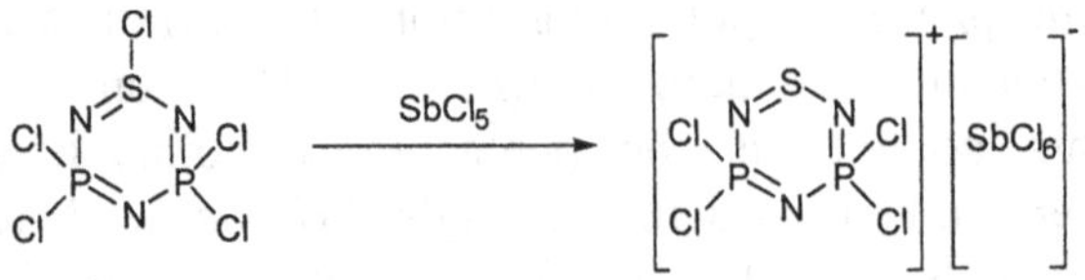

$$(5.2)$$

The conversion of $N_3P_2SCl_5$ to its cation implied that ROP can be utilized for its conversion to a linear polymer. This promise was realized much later by the work of Allcock and coworkers [8, 9]. Accordingly, it was found that $N_3P_2SCl_5$ undergoes a ROP at 90 °C to afford poly(thiophosphazene), $[(NPCl_2)_2(NSCl)]_n$ (see Eq. 5.3).

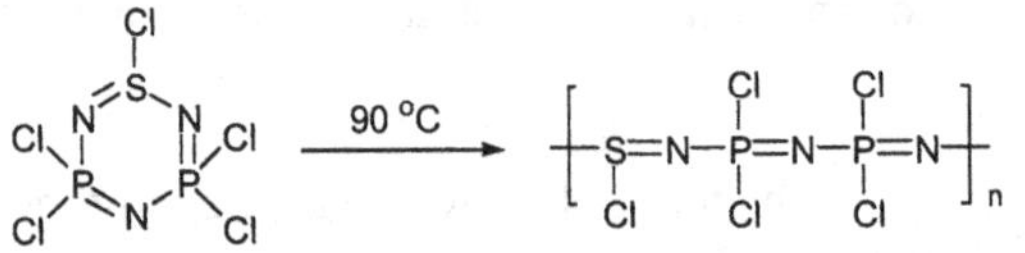

$$(5.3)$$

As in other instances, crosslinked products are formed if the polymerization is not carried out carefully. In fact it has been found that heating is not required for polymerization. *Aged* $N_3P_2SCl_5$ (not freshly distilled) undergoes ROP at ambient temperature merely by mechanical shaking. Heating is required for polymerizing a freshly distilled sample of $N_3P_2SCl_5$. The polymer $[(NPCl_2)_2(NSCl)]_n$ is *extremely* sensitive to hydrolysis. It has a glass transition temperature of -40 °C.

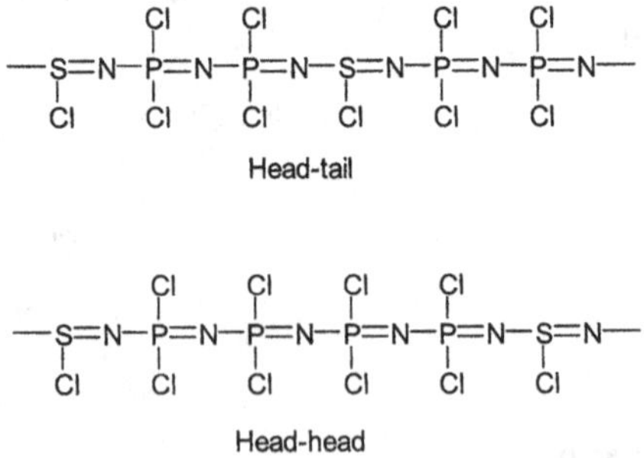

Head-tail

Head-head

Fig. 5.10. Head-tail and head-head arrangements in $[S(Cl)=NPCl_2=NPCl_2]_n$

The ^{31}P-NMR spectrum of the polythiophosphazene, $[(NPCl_2)_2(NSCl)]_n$, shows a single resonance at $\delta = -4.6$. The chemical shift of the polymer is about 30 ppm upfield shifted in comparison to the six-membered ring.

Also the presence of a single resonance indicates a predominantly head-tail arrangement as in the case of the poly(carbophosphazene) (Fig. 5.10).

The chlorine atoms in the polymer $[(NPCl_2)_2(NSCl)]_n$ can be readily replaced by nucleophiles such as phenoxide or p-phenyl phenoxide. However, unlike $[NP(OPh)_2]_n$, $[\{NP(OPh)_2\}_2S(OPh)]_n$ is not hydrolytically as stable. This has been attributed to the inherent reactivity of S(IV) centers. Further, reaction of $[(NPCl_2)_2(NSCl)]_n$ with substituted phenols such as 2-phenylphenol or 3-phenylphenol do not afford fully substituted polymers. However, these polymers $[N_3P_2S(OC_6H_4\text{-}m\text{-}C_6H_5)_{4.4}(Cl)_{0.6}]$ and $[N_3P_2S(OC_6H_4\text{-}o\text{-}C_6H_5)_{3.25}(Cl)_{1.75}]$, which contain P-Cl bonds, are hydrolytically stable.

Although the molecular weight of the parent polymer $[(NPCl_2)_2(NSCl)]_n$ could not be determined, measurements on $[N_3P_2S(OC_6H_4\text{-}p\text{-}C_6H_5)_{4.7}(Cl)_{0.3}]$ found the latter to have high molecular weight ($M_w = 3.1 \times 10^5$ and $M_n = 1.0 \times 10^5$).

An interesting aspect of the chemistry of $[(NPCl_2)_2(NSCl)]_n$ is the difference of reactivity of the S-Cl and P-Cl bonds. Thus, substitution of S-Cl bonds occurs much faster than the P-Cl bonds. Consequently, mixed substituent polymers can be readily prepared.

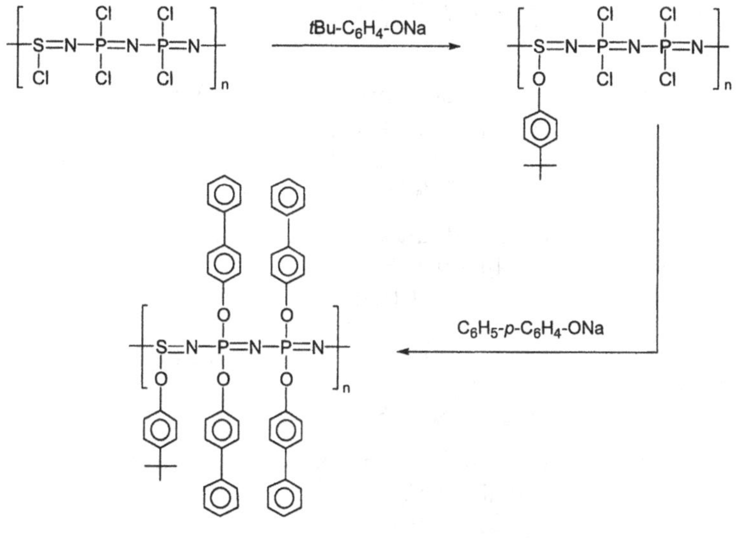

Fig. 5.11. Regioselective nucleophilic substitution reactions of $[S(Cl)=NPCl_2=NPCl_2]_n$

^{31}P-NMR data and T_g data for some selected poly(thiophosphazene)s are summarized in Table 5.2. The polymer $[(NPCl_2)_2(NSCl)]_n$ has a T_g of -40 °C. Thus, in comparison with $[NPCl_2]_n$ the glass transition temperature of

$[(NPCl_2)_2(NSCl)]_n$ is higher. This is similar to what has been observed for the poly(carbophosphazene) $[(NPCl_2)_2(NCCl)]_n$. However, the reasons for this increase in T_g of $[(NPCl_2)_2(NSCl)]_n$ are not entirely clear. While in the case of $[(NPCl_2)_2(NCCl)]_n$ it could be reasoned that the improved π-bonding in the C-N segment leads to a decrease in the skeletal flexibility such reasoning is not applicable for the sulfur-containing polymer. Substitution of the S-Cl and P-Cl bonds in $[(NPCl_2)_2(NSCl)]_n$ by aryloxy substituents increases the T_g of the polymers (Table 5.2). This trend is similar to what has been found in the polyphosphazene family. However, the number of poly(thiophosphazene)s are very limited and therefore a proper structure-property relationship cannot be drawn for this class of polymers.

The ^{31}P-NMR data of poly(thiophosphazene)s indicate that the trends observed are similar to those found in polyphosphazenes. Thus, while $[(NPCl_2)_2(NSCl)]_n$ shows a resonance at -4.6 ppm, replacing the chlorines by biphenyloxy substituents shifts the ^{31}P resonance upfield (Table 5.2).

Table 5.2. ^{31}P-NMR data and T_g data for poly(thiophosphazene)s

S.No	Polymer	[31]P	T_g(°C)
1	$\begin{array}{c} \text{Cl} \quad \text{Cl} \\ \mid \quad \mid \\ -\text{S}=\text{N}-\text{P}=\text{N}-\text{P}=\text{N}- \\ \mid \quad \mid \quad \mid \\ \text{Cl} \quad \text{Cl} \quad \text{Cl} \end{array}_n$	-4.6	-40
2	$\begin{array}{c} \text{OAr} \quad \text{OAr} \\ \mid \quad \mid \\ -\text{S}=\text{N}-\text{P}=\text{N}-\text{P}=\text{N}- \\ \mid \quad \mid \quad \mid \\ \text{OAr} \quad \text{OAr} \quad \text{OAr} \end{array}_n$ Ar = $-C_6H_4$-4-C_6H_5 (Idealized structure for $[N_3P_2S(OC_6H_4\text{-}p\text{-}C_6H_5)_{4.7}(Cl)_{0.3}]$	-11.0	69
3	$\begin{array}{c} \text{OAr} \quad \text{OAr} \\ \mid \quad \mid \\ -\text{S}=\text{N}-\text{P}=\text{N}-\text{P}=\text{N}- \\ \mid \quad \mid \quad \mid \\ \text{OAr} \quad \text{OAr} \quad \text{Cl} \end{array}_n$ (Idealized structure for $[N_3P_2S(OC_6H_4\text{-}m\text{-}C_6H_5)_{4.4}(Cl)_{0.6}]$	-12.1	32

5.4 Poly(thionylphosphazene)s

The unfavorable properties of poly(chlorothiophosphazene)s and poly(organothiophosphazene)s can be traced to the presence of the reactive S(IV) in the backbone of the polymer. Replacement of S(IV) by the oxidatively and hydrolytically more stable S(VI) would overcome this difficulty. This involves incorporation of a S(O)Cl moiety in place of the S(Cl)unit. Cyclic rings containing the S(O)Cl group are known. Thus, analogous to the P-N heterocyclic ring, $N_3P_3Cl_6$, the corresponding S-N ring $(NSOCl)_3$ is also known (Fig. 5.12). The hybrid P-N-S ring $[\{NPCl_2\}_2\{NS(O)Cl\}]$ is also known and can serve as the precursor for the preparation of the linear polymer [11, 12].

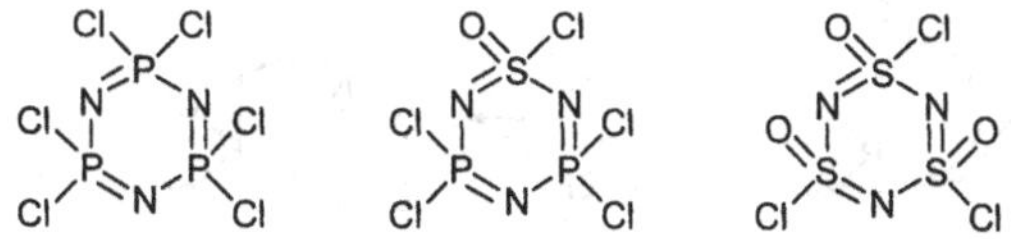

Fig. 5.12. The inorganic heterocyclic ring $[\{NPCl_2\}_2\{NS(O)Cl\}]$ and its relation with the other related rings $N_3P_3Cl_6$ and $[NS(O)Cl]_3$

The monomer $[\{NPCl_2\}_2\{NS(O)Cl\}]$ was originally prepared by a vacuum thermolysis of the linear derivative $Cl_3PNPCl_2NSO_2Cl$ along with Cl_3PNPSO_2Cl (see Eq. 5.4).

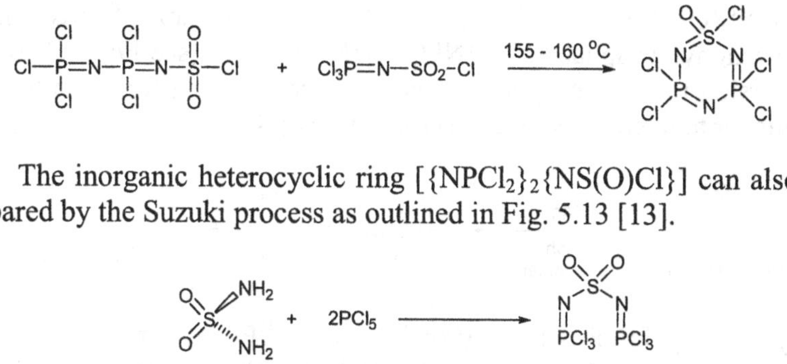

$$(5.4)$$

The inorganic heterocyclic ring $[\{NPCl_2\}_2\{NS(O)Cl\}]$ can also be prepared by the Suzuki process as outlined in Fig. 5.13 [13].

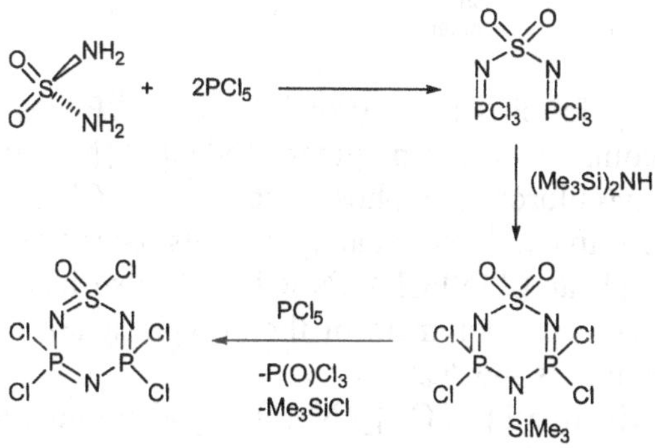

Fig. 5.13. Preparation of $[\{NPCl_2\}_2\{NS(O)Cl\}]$ by the Suzuki process

The Suzuki process involves the reaction of $SO_2(NH_2)_2$ with PCl_5 to afford $(Cl_3PN)_2SO_2$. The latter reacts with hexamethyldisilazane to afford the cyclic ring which already contains the N_3P_2S framework. This can further react with PCl_5 to afford $[\{NPCl_2\}_2\{NS(O)Cl\}]$ over 70% yield. The latter has a ^{31}P-NMR resonance at $\delta = 26.5$. Many nucleophilic substitution reactions of $[\{NPCl_2\}_2\{NS(O)Cl\}]$ are known. The important observation is that in the aminolysis reactions of $[\{NPCl_2\}_2\{NS(O)Cl\}]$, the P-Cl bond is more reactive than the S(O)-Cl bond. In contrast, reaction of $[\{NPCl_2\}_2\{NS(O)Cl\}]$ with AgF_2 affords the mono-fluoro product $[\{NPCl_2\}_2\{NS(O)F\}]$ (see Eq. 5.5). Notice that fluorination occurs exclusively at the sulfur center. The fluoro compound $[\{NPCl_2\}_2\{NS(O)F\}]$ has a ^{31}P-NMR resonance at +26.1 ppm.

$$(5.5)$$

5.4.1 Polymerization of $[\{NPCl_2\}_2\{NS(O)Cl\}]$ and $[\{NPCl_2\}_2\{NS(O)F\}]$

Thermal treatment of $[\{NPCl_2\}_2\{NS(O)Cl\}]$ and $[\{NPCl_2\}_2\{NS(O)F\}]$ affords the corresponding linear polymers. Thus, heating $[\{NPCl_2\}_2\{NS(O)Cl\}]$ at 165 °C for 4 h in vacuum affords the poly(chlorothionylphosphazene), $[\{NPCl_2\}_2\{NS(O)Cl\}]_n$ (see Eq. 5.6). An 80% conversion to the high polymer was noticed. Heating for more than 4 h or at higher temperatures leads to crosslinking [14, 15].

$$(5.6)$$

The cyclic ring $[\{NPCl_2\}_2\{NS(O)Cl\}]$ has a ^{31}P-NMR resonance at +26.1 ppm. In comparison the polymer $[\{NPCl_2\}_2\{NS(O)Cl\}]_n$ resonates at -10.0 ppm. Poly(chlorothionylphosphazene), $[\{NPCl_2\}_2\{NS(O)Cl\}]_n$ is a hydrolytically sensitive polymer analogous to its other congeners $[NPCl_2]_n$, $[\{NPCl_2\}_2\{NCCl\}]_n$ and $[\{NPCl_2\}_2\{NSCl\}]_n$. It also has a flexible backbone analogous to other polymers of the polyphosphazene family. Thus, the T_g of this polymer is -46 °C.

The fluoro derivative $[\{NPCl_2\}_2\{NS(O)F\}]$ also undergoes a ROP, albeit at a higher temperature, to afford the linear polymer $[\{NPCl_2\}_2\{NS(O)F\}]_n$ (see Eq. 5.7). The T_g of the latter is -56 °C.

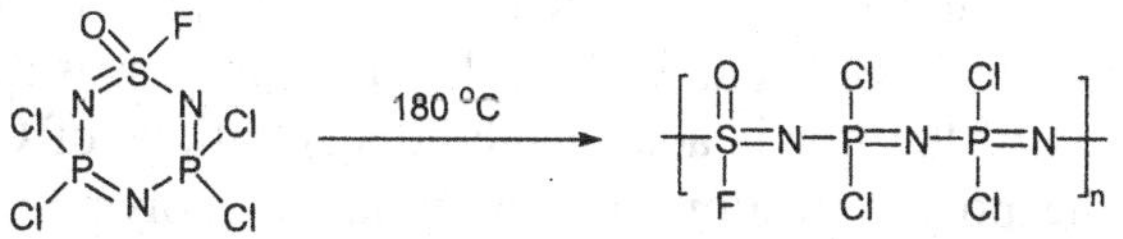

(5.7)

An interesting aspect of the thermal treatment of [{NPCl$_2$}$_2${NS(O)Cl}] is the formation of the cyclic rings [{NPCl$_2$}$_2${NS(O)Cl}]$_n$ (n = 2, 4) (Fig. 5.14). However, these cyclic rings are formed in small quantities.

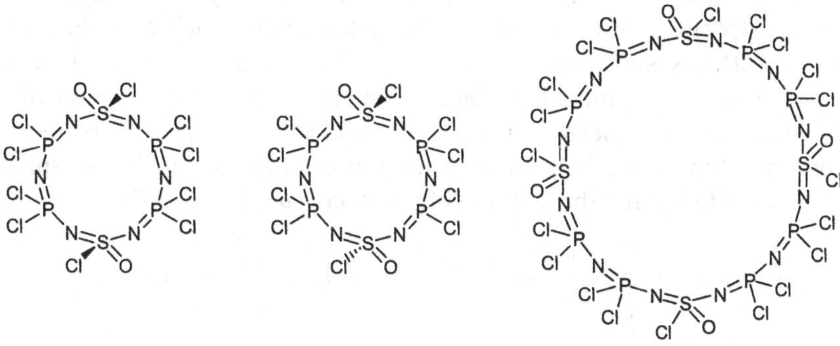

Fig. 5.14. Cyclic 12- and 24-membered rings made up of N-P-S framework

Both *cis*- and *trans*-isomers of the 12-membered rings [{NPCl$_2$}$_2${NS(O)Cl}]$_2$ have been isolated. A 24-membered macrocyclic ring [{NPCl$_2$}$_2${NS(O)Cl}]$_4$ has also been isolated in the thermolysis experiment. All of these macrocyclic rings are nonplanar [1, 2].

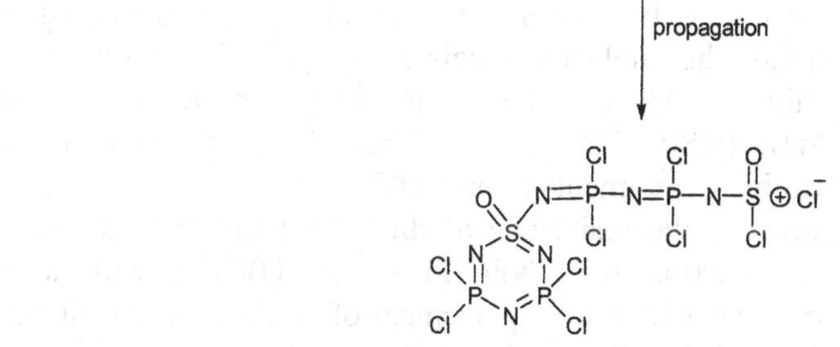

Fig. 5.15. Mechanism of polymerization of [{NPCl$_2$}$_2${NS(O)Cl}]

An ambient temperature ROP of $[\{NPCl_2\}_2\{NS(O)Cl\}]$ is mediated by Lewis acids such as $GaCl_3$, $SbCl_5$ and $AlCl_3$ [16]. This process is dependant on the concentration of the Lewis acid. High concentrations of $GaCl_3$ are needed to effect the polymerization. The initiation of polymerization of $[\{NPCl_2\}_2\{NS(O)Cl\}]$ by Lewis acids supports the mechanism of polymerization as occurring through a S-Cl heterolytic cleavage. This is analogous to that proposed for the polymerization of $N_3P_3Cl_6$ and is outlined in Fig. 5.15.

The polymer $[\{NPCl_2\}_2\{NS(O)Cl\}]_n$ is not hydrolytically stable. However, replacement of chlorines by other nucleophiles such as aryloxides or amines affords stable polymers. Interestingly, reactions with alkoxides seem to cause polymer degradation. On the other hand, reaction of the polymer with aryloxides occurs in a regioselective manner. Thus, its reaction with NaOPh leads to the substitution of chlorines at the phosphorus centers while leaving the sulfur center *untouched* (see Eq. 5.8)

$$\begin{array}{c}\text{(structure: } {-}[S(=O)(Cl){=}N{-}P(Cl)_2{=}N{-}P(Cl)_2{=}N]{-}_n \xrightarrow{\text{NaOPh}} {-}[S(=O)(Cl){=}N{-}P(OPh)_2{=}N{-}P(OPh)_2{=}N]{-}_n \text{)} \end{array} \qquad (5.8)$$

The reaction of the heterocyclic ring $[\{NPCl_2\}_2\{NS(O)Cl\}]$ with NaOPh also does not effect the S(O)Cl bond (see Eq. 5.9). Thus, in this instance also the reactivity of the cyclic ring system can be a good model for the corresponding polymer reactivity.

$$\text{(cyclic structure with S(=O)Cl, P Cl}_2\text{ groups)} \xrightarrow{\text{NaOPh}} \text{(cyclic structure with S(=O)Cl, P(OPh)}_2\text{ groups)} \qquad (5.9)$$

Poly(thionylphosphazenes) are amorphous, high-molecular-weight polymers [3, 11, 12]. The phenoxy polymer $[\{NP(OPh)_2\}_2\{NS(O)Cl\}]_n$ has a M_w of 30,500 and a M_n of 16,030. It has a T_g of -21.5 °C. Interestingly in spite of the presence of the S-Cl bond this polymer is stable not only in atmosphere but also in a mixture of dioxane containing 10% water.

In general, the molecular weights of poly(thionylphosphazene)s are quite high. Thus, the absolute molecular weight of $[\{NP(OAr)_2\}_2\{NS(O)Cl\}]_n$ $(Ar = C_6H_4\text{-}p\text{-}C_6H_5)$ has been determined by light-scattering experiments to be 64,000.

In contrast to the resistance of the S-Cl bond towards reactivity with aryloxides, reaction of $[\{NP(OPh)_2\}_2\{NS(O)Cl\}]_n$ with a number of amines proceeds with total replacement of all the chlorine atoms. Thus, the reactions of $[\{NP(OPh)_2\}_2\{NS(O)Cl\}]_n$ with primary amines such as n-butylamine leads to $[\{NP(NHnBu)_2\}_2\{NS(O)NHnBu\}]_n$ [17]. Similarly, reactions of $[\{NP(OPh)_2\}_2\{NS(O)Cl\}]_n$ with secondary amines such as piperidine or pyrrolidine also leads to the fully substituted polymers (Fig. 5.16) [18].

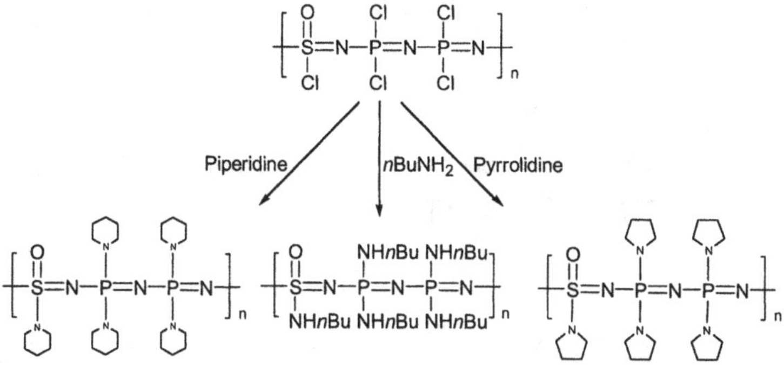

Fig. 5.16. Aminolysis reactions of $[\{NPCl_2\}_2\{NS(O)Cl\}]_n$

As mentioned earlier the glass transition temperatures of poly(thionylphosphazene)s indicate that these are polymers with a flexible backbone. These polymers have five side-groups per repeat unit which is similar to poly(carbophosphazene)s and poly(thiophosphazene)s. On the other hand, a comparable polyphosphazene unit has six side-chains. This has interesting consequences on the glass transition temperatures of poly(thionylphosphazene)s (Table 5.3).

Table 5.3. ^{31}P NMR and T_g data of poly(thionylphosphazene)s

S.No	Polymer	^{31}P	T_g (°C)
1	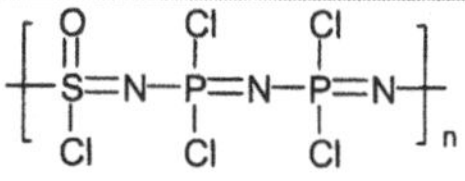	-10.0	-46
2	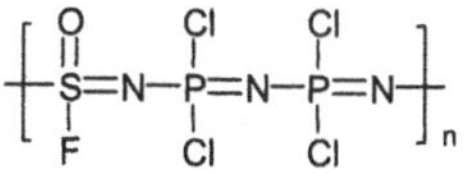	-8.6	-56
3	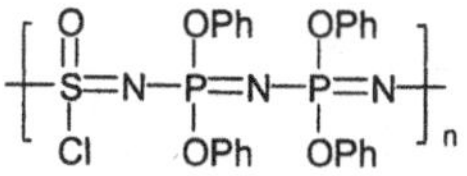	-21.5	+10
4	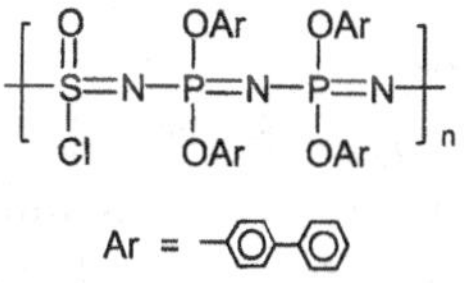	-20.9	+55

Table 5.3. (contd.)

5	$\begin{bmatrix} & \overset{O}{\underset{Cl}{\overset{\|}{S}}}=N-\overset{OAr}{\underset{OAr}{P}}=N-\overset{OAr}{\underset{OAr}{P}}=N \end{bmatrix}_n$ Ar = (phenyl with F₃C)	-21.0	-25
6	$\begin{bmatrix} & \overset{O}{\underset{F}{\overset{\|}{S}}}=N-\overset{OAr}{\underset{OAr}{P}}=N-\overset{OAr}{\underset{OAr}{P}}=N \end{bmatrix}_n$ Ar = Ph	-20.3	-15
7	$\begin{bmatrix} & \overset{O}{\underset{F}{\overset{\|}{S}}}=N-\overset{OAr}{\underset{OAr}{P}}=N-\overset{OAr}{\underset{OAr}{P}}=N \end{bmatrix}_n$ Ar = (biphenyl)	-19.5	+48
8	$\begin{bmatrix} & \overset{O}{\underset{NHR}{\overset{\|}{S}}}=N-\overset{NHR}{\underset{NHR}{P}}=N-\overset{NHR}{\underset{NHR}{P}}=N \end{bmatrix}_n$ R = Me	-	+22
9	$\begin{bmatrix} & \overset{O}{\underset{NHR}{\overset{\|}{S}}}=N-\overset{NHR}{\underset{NHR}{P}}=N-\overset{NHR}{\underset{NHR}{P}}=N \end{bmatrix}_n$ R = Et	-	+4
10	$\begin{bmatrix} & \overset{O}{\underset{NHR}{\overset{\|}{S}}}=N-\overset{NHR}{\underset{NHR}{P}}=N-\overset{NHR}{\underset{NHR}{P}}=N \end{bmatrix}_n$ R = nhex	-	-18
11	$\begin{bmatrix} & \overset{O}{\underset{NHR}{\overset{\|}{S}}}=N-\overset{NHR}{\underset{NHR}{P}}=N-\overset{NHR}{\underset{NHR}{P}}=N \end{bmatrix}_n$ R = C_6H_5	-	82

Poly(chlorothionylphosphazene)[$\{NPCl_2\}_2\{NS(O)Cl\}$]$_n$ has a higher T_g (-46 °C) in comparison to [$NPCl_2$]$_n$ (-63 °C). This indicates a slight stiffening of the backbone as a result of the presence of the sulfonyl group in the backbone. Replacing the chlorines with aryloxy substituents increases the T_g (entries 3 and 4, Table 5.3). Among the amino derivatives, polymers containing *small* primary amino substituents [{NP(NHMe)$_2$}$_2$}{NS(O)NHMe}]$_n$ have a higher T_g (+22 °C) while those that contain long-chain alkylamino substituents such as [{NP(NHnHex)$_2$}$_2${S(O)(NHnHex)}]$_n$ have a lower glass transition temperature (-18 °C). The absence of poly(alkyl/arylthionylphosphazene)s does not allow a full structure-property relationship to be drawn. However, in general the thionyl group in the backbone of the poly(thionylphosphazene)s seems to prevent the polymer chains from having sufficient intermolecular interactions so that no genuine melting transition is seen for any poly(thionylphosphazene) [3, 11, 12].

Although no major applications of poly(heterophosphazene)s have yet been indicated, many of these polymers seem to have good gas-transport properties. The *n*-butylamino polymer [{NP(NHnBu)$_2$}$_2${S(O)(NHnBu)}]$_n$ has shown interesting applications as matrices for oxygen-sensors. The principle of this application emerges from the following considerations. The excited electronic (phosphorescent) states of transition-metal derivatives such as ruthenium-phenanthroline complexes [Ru(phenPh$_2$)$_3$]$^{2+}$ are quenched by oxygen. Such complexes are dispersed in polymer films and exposed to various oxygen pressures. The intensity of phosphorescence emission is directly related to the concentration (pressure) of oxygen that the polymer film experiences. It is believed that these applications will be particularly useful in aircraft design [3, 11, 12].

5.5 Other P-N-S Polymers

In an interesting experiment Turner and coworkers have shown that heating a 1:1 mixture of the N-silylphosphoranimine, Me$_3$SiNPMe$_2$(OCH$_2$CF$_3$), and the N-silylsulfonimide, Me$_3$SiNS(O)Me(OCH$_2$CF$_3$), affords a regio-selective coupling to afford Me$_3$SiNSMe(O)NPMe$_2$(OCH$_2$CF$_3$), which contains the trifluoroethoxy group on the phosphorus and the NSiMe$_3$ group on sulfur. Heating Me$_3$SiNSMe(O)NPMe$_2$(OCH$_2$CF$_3$) affords a true alternating poly(thionylphosphazene) copolymer [MeS(O)NPMe$_2$N]$_n$ where the MeS(O)=N and Me$_2$P=N units alternate with each other [19]. Interestingly, treating Me$_3$SiNSMe(O)NPMe$_2$(OCH$_2$CF$_3$) with alcohol converts it into HNSMe(O)NPMe$_2$(OCH$_2$CF$_3$) which contains a S=NH end-group. Heating this at 100 °C also leads to the formation of the polymer [RS(O)NPR$_2$N]$_n$ by the expulsion of trifluoroethanol. The polymer [RS(O)NPR$_2$N]$_n$ is also probably the first example of a poly(thionylphosphazene) that contains alkyl groups attached to sulfur and phosphorus (Fig. 5.17). The molecular weight of this polymer has been found to be moderate (M_n = 8,000). Further examples of these interesting polymers will cast more light on the structure-property relationship of poly(thionylphosphazene)s.

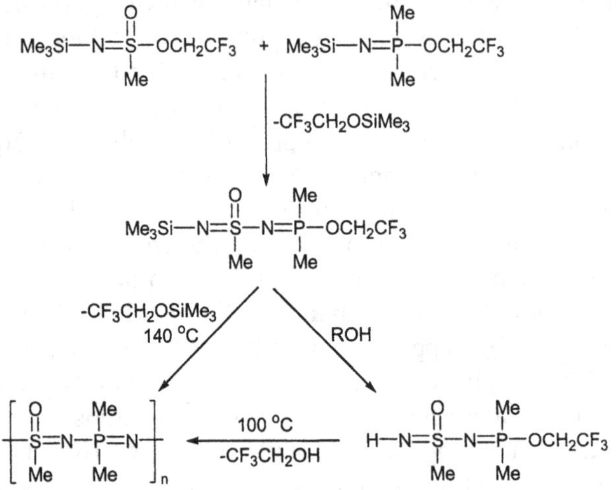

Fig. 5.17. Synthesis of poly(thionylphosphazene)s containing alkyl side-groups

Chivers and coworkers have reported that the reaction of sulfur monochloride, S_2Cl_2 with $Cl_2P(NSiMe_3)(NSiMe_3)_2$ affords a material with a composition that is consistent with the polymer $[SNPCl_2N]_n$ (see Eq. 5.10) [3]. However, the discouraging solubility properties of this polymer did not allow its full characterization.

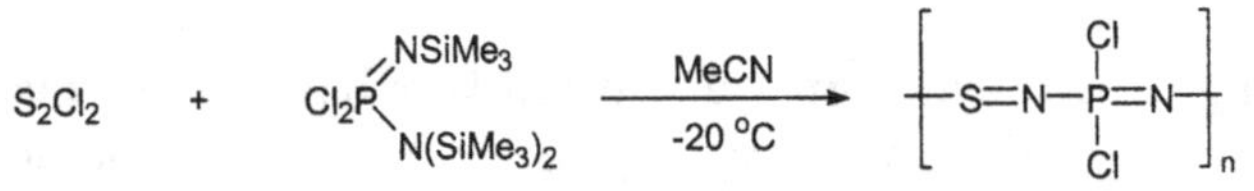

$$(5.10)$$

5.6 Metallacyclophosphazenes

Metallacyclophosphazenes are a new class of heterocyclophosphazenes that contain a metal atom in the phosphazene ring [20-23]. Apart from their novelty there have been preliminary experiments which seem to suggest that a few representative metallacyclophosphazenes can function as monomers for ROP to the corresponding poly(metallaphosphazene)s [24].

The Bezman's salt $[\{Ph_2P(NH_2)\}_2N]^+[Cl]^-$ has served as a good synthon for the preparation of metallacyclophosphazenes providing, a five-atom, NPNPN, fragment. Ring-closure can occur if an appropriate reagent is allowed to react with this synthon. Thus, the reaction of Bezman's salt with WCl_6 in refluxing chloroform afforded the first metallacyclophosphazene, $[\{NPPh_2\}_2\{WCl_3\}]$ as a crystalline solid. The latter can be fluorinated by AgF or NaF to afford $[\{NPPh_2\}_2\{WF_3\}]$. The fluoro derivative can also be obtained by the reaction of Bezman's salt with WF_6 (Fig. 5.18). The molybdenum derivative $[\{NPPh_2\}_2\{MoCl_3\}]$ can be prepared by the reaction of Bezman's salt with molybdenum pentachloride or molybdenum trichloride nitride (Fig. 5.18).

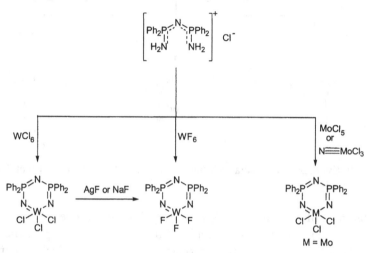

Fig. 5.18. Synthesis of metallacyclophosphazenes

The metallacyclophosphazenes $[\{NPPh_2\}_2\{VCl_2\}]$, $[\{NPPh_2\}_2\{WBr_3\}]$, or $[\{NPPh_2\}_2\{ReCl_4\}]$ can be prepared by the reaction of Bezman's salt with the corresponding metal salt in the presence of Br_2 or Cl_2 (Fig. 5.19).

The characterization of these metallacyclophosphazenes has been carried out by NMR as well as single-crystal X-ray structural studies. Thus, the compound $[\{NPPh_2\}_2\{WCl_3\}]$ has a ^{31}P-NMR signal at 39.2 ppm. This may be compared with the ^{31}P-NMR signal of *gem*-$N_3P_3Cl_2Ph_4$ (δ PPh_2 = 17.2) and $N_3P_3Ph_6$ (δ = 14.3). The molybdenum analogue $[\{NPPh_2\}_2\{MoCl_3\}]$ has a similar chemical shift (δ = 42.4). The crystal structures of $[\{NPPh_2\}_2\{WCl_3\}]$ and the $[\{NPPh_2\}_2\{MoCl_3\}]$ show that the six-membered rings in these metallacyclophosphazenes are nearly planar. The metal atoms deviate from the plane of the rest of the ring. Thus, in the molybdenum derivative, the molybdenum is about 0.13 Å away from the mean-plane of the ring. The P-N distances in the molybdenum derivative are of two types. Short distances (av.1.583 Å) are associated with the nitrogen that is farther away from the metal. The P-N distances involving the nitrogens that are adjacent to the metal are longer (1.650 Å). Similarly, the bond angles involving the nitrogens in the Mo-N-P bonds are larger (av. 134. 5°) than the nitrogen in the P-N-P segment (127.9 °).

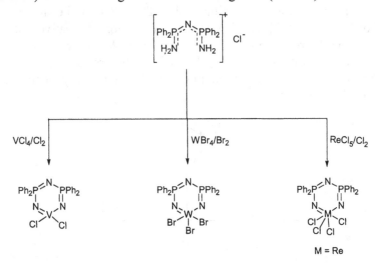

Fig. 5.19. Synthesis of metallacyclophosphazenes in the presence of metal halides and Br$_2$ or Cl$_2$

Examples of eight-membered metallacyclophosphazene rings also exist. (see Eq. 5.11). There is no example of a metallacyclophosphazene of the type [{NPCl$_2$}$_2${MCl$_x$}]. This is probably because of the lack of appropriate synthons. However, the recent utility of the chlorophosphoranimine Me$_3$SiNPCl$_3$ suggests that this compound could be a useful synthon for the synthesis of chlorometallacyclophosphazenes.

$$(5.11)$$

$$(5.12)$$

It has been briefly reported that the metallacyclophosphazenes [{NPPh$_2$}$_2${WCl$_3$}] and [{NPPh$_2$}$_2${MoCl$_3$}] can be polymerized in boiling xylene to afford the corresponding poly(metallaphosphazene)s (see Eq. 5.12) [24]. However, no further studies on these polymers have appeared.

5.7 Other Phosphorus-Containing Polymers

5.7.1 Polyphosphinoboranes

Polyphosphinoboranes, $[RPH-BH_2]_n$, are a recent class of polymers that contain a P-B repeat unit in the polymer backbone [25, 26]. These polymers are prepared by a rhodium-catalyzed dehydropolymerization procedure. The monomers that have been used for these polymers are the phosphine-borane adducts, $RPH_2.BH_3$. Dehydropolymerization of these adducts occurs by the use of rhodium complexes such as $[\{Rh(\mu-Cl)(1,5-cod)\}_2]$ or even by simple rhodium salts such as $RhCl_3$ or $RhCl_3.xH_2O$. The polymerization is carried out in neat conditions (no solvent) (see Eq.5.13).

$$R-PH_2 \cdot BH_3 \xrightarrow[\substack{\text{Rh catalyst} \\ -H_2}]{100\text{-}120\,^{\circ}C} \left[\begin{array}{cc} R & H \\ | & | \\ P-B \\ | & | \\ H & H \end{array} \right]_n \qquad (5.13)$$

$$R = Ph,\ iBu,\ p\text{-}nBuC_6H_4,\ p\text{-}C_{12}H_{25}\text{-}C_6H_4$$

The molecular weights of these polymers have been determined by various methods. For example, the molecular weights (M_w) of $[PhPHBH_2]_n$ and $[iBuPHBH_2]_n$ as determined by static light scattering methods are 20,000 and 13,100, respectively. The phosphorus chemical shifts of the phosphine-borane adducts and the linear polymers are nearly similar. Thus, while the ^{31}P-NMR signal of $PhPH_2.BH_3$ is at -47 ppm, the linear polymer resonates at -48.9 ppm.

Polyphosphinoboranes are reasonably flexible at room temperature. This is reflected in the T_g's of these polymers: $[iBuPHBH_2]_n$ (5 °C); $[p\text{-}nBuC_6H_4)PHBH_2]_n$ (8 °C); $[(p\text{-}C_{12}H_{25}\text{-}C_6H_4)PHBH_2]_n$ (-1 °C). The few examples of polyphosphinoboranes that are known are all thermally quite stable. Interestingly, $[PhPHBH_2]_n$ shows a ceramic yield of 75-80 % at 1000 °C indicating a possible utility for these polymers as polymer precursors for boron phosphide-based solid-state materials [26].

5.7.2 Polymers Containing P=C bonds

Recently, an addition polymerization of the phosphaalkene, $(Mes)_2P=CPh_2$ (Mes = $2,4,6$-Me_3-C_6H_2) has been reported [27]. Thus, after the vacuum distillation of the crude monomer (150 °C; 0.1 mmHg) a residue was left behind. Analysis of this revealed it to be a high polymer. It was suspected that radical impurities might have caused the polymerization. Subsequently it was shown that the phosphaalkene can be polymerized by free-radical or anionic initiators to afford a low molecular weight (M_n = 5,000-10,000) polymer (see Eq. 5.14) [27]. The ^{31}P-NMR chemical shift of the polymer (-10 ppm) is *considerably* upfield shifted vis-à-vis the monomer (+233 ppm).

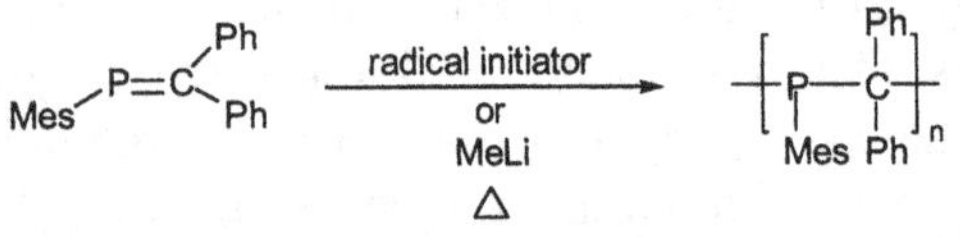

$$(5.14)$$

The polymer $[MesP\text{-}CPh_2]_n$ is reasonably stable towards moisture and air. However, the trivalent phosphorus center in this polymer can be oxidized by treatment with H_2O_2 to afford $[MesP(O)\text{-}CPh_2]_n$. Similarly, treating $[MesP\text{-}CPh_2]_n$ with sulfur converts it into $[MesP(S)\text{-}CPh_2]_n$.

An interesting condensation reaction to afford a polymer with P=C bonds has been recently reported [28]. The reaction involves a [1,3]-silatropic rearrangement (Fig. 5.20).

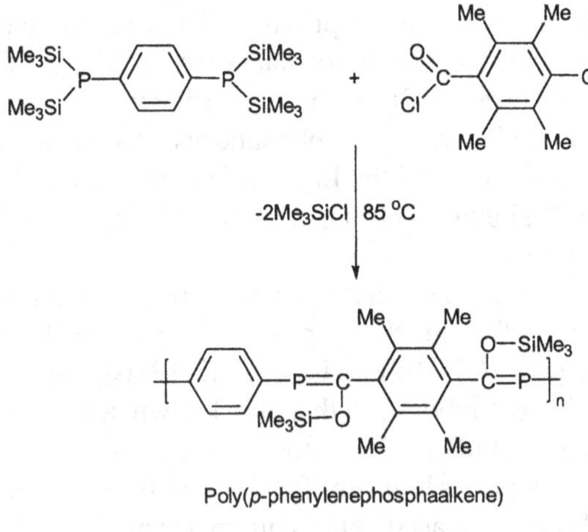

Poly(*p*-phenylenephosphaalkene)

Fig. 5.20. Preparation of poly(*p*-phenylenephosphalakene)

The reaction of the difunctional reagent $ClC(O)\text{-}Ar\text{-}C(O)Cl$ [Ar = 2, 3, 5, 6-Me_4C_6] with $(SiMe_3)_2P\text{-}C_6H_4$-$p$-$P(SiMe_3)_2$ leads not only to the elimination of Me₃SiCl, but also of the migration of the remaining SiMe₃ group on phosphorus to the carbonyl oxygen. This overall reaction generates poly(*p*-phenylenephosphaalkene). This polymer is isolated as a mixture of various regioisomers. Poly(*p*-phenylenephosphaalkene) is a heavier congener of the well-studied organic luminescent polymer poly(*p*-phenylene vinylene) (Fig. 5. 21).

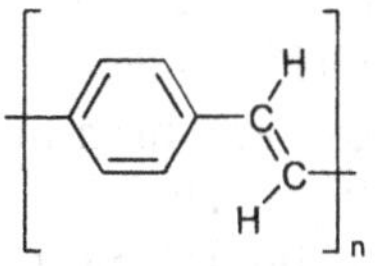

Fig. 5.21. Poly(*p*-phenylenevinylene)

These new phosphorus containing, polymers reveal the potential utility of incorporating the main-group linkages in polymer chains.

5.8 Poly(alkyl/aryloxothiazenes)

The polymers $[S]_n$ and $[SN]_n$ have been mentioned in Chapter 1. Other sulfur-containing polymers that have already been dealt with so far include those that have P-N-S and P-N-S(O) units as well as those that contain alternate S(O)N and PN units in the polymer backbone. In this section we will briefly look at polymers containing alternate sulfur and nitrogen atoms in the polymer backbone. In these polymers sulfur is hexavalent. An oxygen atom and an alky or an aryl group is present as the side-chain substituents on sulfur.

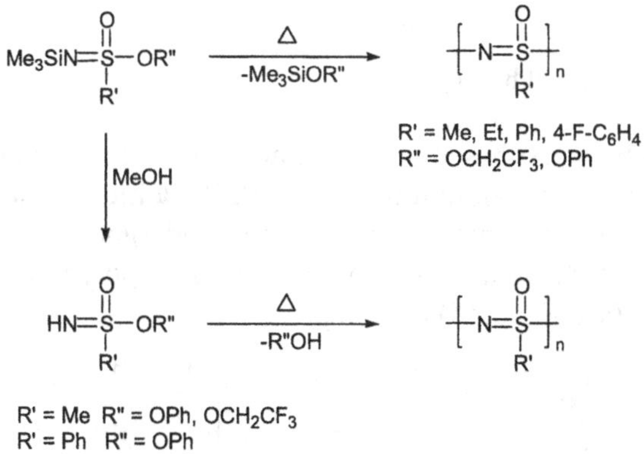

Fig. 5.22. Synthesis of poly(alkyl/aryloxothiazene)s

Poly(alkyl/aryloxothiazenes), [N=S(O)R]$_n$, have been prepared by the thermolysis of N-silylsulfonimidates, [Me$_3$SiN=S(O)R'(OR'')]. The thermal condensation is catalyzed by Lewis acids such as BF$_3$.Et$_2$O, AlCl$_3$ or by bases such as F$^-$ and the phenoxide ion (Fig. 5.22) [29]. These polymers can also be prepared by using an alternative pathway. Thus, the N-silylsulfonimidates can be desilylated to the corresponding free sulfonimidates [HN=S(O)R'(OR'')]. The latter can be converted to the polymers, also by a thermolysis reaction (Fig. 5.22).

Poly(alkyl/aryloxothiazenes) are high-molecular-weight polymers and have good solubility properties. Their glass transition temperatures are quite high. For example, while [N=S(O)Me]$_n$ has a T_g of 55-65 °C, [N=S(O)Ph]$_n$ has a T_g of 85 °C. Although experimental structural studies have been limited, based on theoretical calculations, a *cis-trans* helical conformation has been proposed for poly(methyloxothiazene), [N=S(O)Me]$_n$ [29].

5.9 Borazine-based Polymers

Borazine is an important and well-studied inorganic heterocyclic ring. It is isoelectronic with benzene and has been known as *inorganic benzene* (5.23).

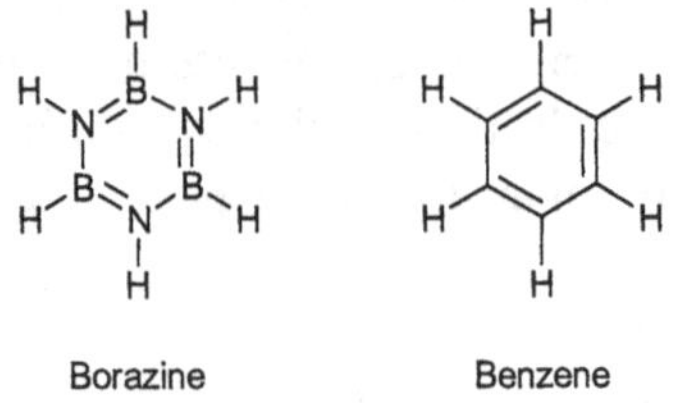

Borazine   Benzene

Fig. 5.23. Borazine and benzene

Interest in borazine-based polymers arises out of their possible utility as precursors for the preparation of boron nitride, a high value ceramic material. It was discovered that borazine readily undergoes a dehydropolymerization at 70-110 °C to afford a soluble polymer, polyborazylene, in about 80-90% yield (see Eq. 5.15) [30, 31].

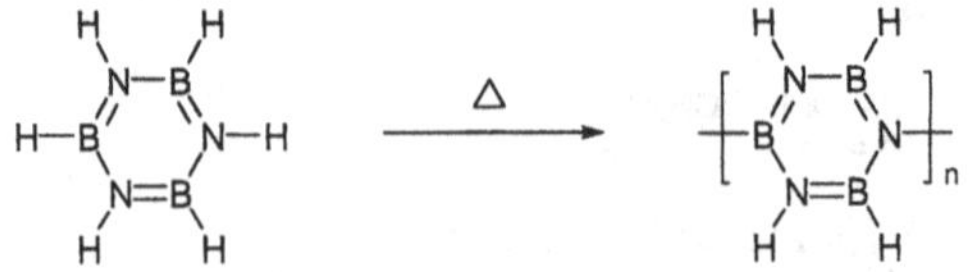

(5.15)

Based on extensive spectroscopic studies it has been suggested that the six-membered B-N rings are retained in the polymer. The structure of the polymer is not entirely linear but also contains branched-chain and fused-cyclic segments [31]. Polyborazylene is soluble in solvents such as THF. Measurements by low-angle laser light scattering techniques reveal that the molecular weights vary from 3,000-8,000. Most interestingly, it has been shown that polyborazylenes are excellent polymer precursors for BN ceramics. High ceramic yields (84-93 %) have been obtained.

Recently, modified polyborazylenes have been prepared by reactions with secondary amines containing alkyl chains such as diethylamine, dipentylamine as well as silyl amines such as hexamethyldisilazane (see Eq. 5.16) [32].

$$(5.16)$$

Modified-polyborazylenes

Modified polyborazylenes have better solubility properties than the parent polyborazylene. Also, these new versions of polyborazylenes are melt-spinnable and have been shown to afford boron nitride ceramic fibers on thermal treatment [33].

References

1. Gates DP, Manners I (1997) Dalton Trans 2525
2. Manners I (1996) Angew Chem Int Ed 35:1602
3. McWilliams AR, Dorn H, Manners I (2002) Top Curr Chem 220:141
4. Manners I, Allcock HR (1989) J Am Chem Soc 111:5478
5. Allcock HR, Coley SM, Manners I, Visscher KB Parvez M, Nuyken O, Renner G (1993) Inorg Chem 32:5088
6. Allcock HR, Coley SM, Manners I, Renner G (1991) Macromolecules 24:2024
7. Allcock HR, Coley SM, Morrissey CT (1994) Macromolecules 27:2904
8. Dodge JA, Manners I, Allcock HR, Renner G, Nuyken O, (1990) J Am Chem Soc 112:1268
9. Allcock HR, Dodge JA, Manners I (1993) Macromolecules 26:11
10. Roesky HW (1972) Angew Chem Int Ed 11:642
11. Manners I (1994) Coord Chem Rev 137:109

12. McWilliams AR, Manners I (2004) Polythionylphosphazenes: inorganic polymers with a main chain of phosphorus, nitrogen and sulfur(VI) atoms. In: Gleria M, De Jaeger R (eds) (2004) Phosphazenes - a world-wide insight. Nova Science, New York pp 191-208
13. Suzuki D, Akagi H, Matsumura K (1983) Synthesis Commun 369
14. Liang M, Manners I (1991) J Am Chem Soc 113:4044
15. Ni Y, Stammer A, Liang M, Massey J, Vancso GJ, Manners I (1992) Macromolecules 25:7119
16. McWilliams AR, Gates DP, Edwards M, Liable-Sands LM, Guzei I, Rheingold AL, Manners I (2000) J Am Chem Soc 122:8848
17. Ni Y, Park P, Liang M, Massey J, Waddling C, Manners I (1996) Macromolecules 29:3401
18. Nobius MN, McWilliams AR, Nuyken O, Manners I (2000) Macromolecules 33:7707
19. Chunechom V, Vidal TE, Adams H, Turner ML (1998) Angew Chem Int Ed 37:1928
20. Witt M, Roesky HW (1994) Chem Rev 94:1163
21. Roesky HW, Katti KV, Seseke U, Witt M, Egert E, Herbst R, Sheldrick GM (1986) Angew Chem Int Ed 25:477
22. Roesky HW, Katti KV, Seseke U, Schmidt HG, Egert E, Herbst R, Sheldrick GM (1987) Dalton Trans 847
23. Katti KV, Roesky HW, Rietzel M (1987) Inorg Chem 26:4032
24. Roesky HW, Lücke M (1989) Angew Chem Int Ed 28:493
25. Dorn H, Singh RA, Massey JA, Nelson JM, Jska CA, Lough AJ, Manners I (2000) J Am Chem Soc 122:6669
26. Dorn H, Rodezno JM, Brunnhöfer B, Rivard E, Massey JA, Manners I (2003) Macromolecules 36:291
27. Tsang CW, Yam M, Gates DP (2003) J Am Chem Soc 125:1480
28. Wright VA, Gates DP (2002) Angew Chem Int Ed 41:2389
29. Roy AK, Burns GT, Lie GC, Grigoras S (1993) J Am Chem Soc 115:2604
30. Fazen PJ, Beck JS, Lynch AT, Remsen EE, Sneddon LG (1990) Chem Mater 2:96
31. Fazen PJ, Remsen EE, Beck JS, Carroll PJ, McGhie AR, Sneddon LG (1995) Chem Mater 7:1942
32. Wideman T, Sneddon LG (1996) Chem Mater 8:3
33. Wideman T, Remsen EE, Cortez E, Chlanda VL, Sneddon LG (1998) Chem Mater 10:412

6 Polysiloxanes

6.1 Introduction

Polysiloxanes are inorganic polymers that contain alternate silicon and oxygen atoms in their backbone [1-6]. While the silicon atom has two side-groups the oxygen atom carries none. In this regard they are similar to polyphosphazenes that we have seen in Chap. 3. The most widely studied polymer of this family is poly(dimethylsiloxane) (PDMS) which contains methyl groups as the substituents on silicon. PDMS is isoelectronic with poly(isobutylene) as well as poly(dimethylphosphazene) although its properties are dramatically different (Fig. 6.1).

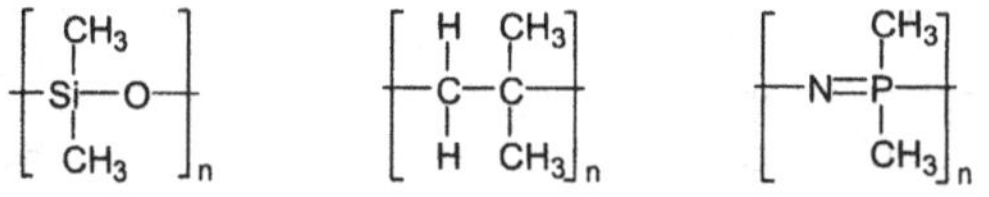

Fig. 6.1. PDMS, poly(isobutylene) and poly(dimethylphosphazene)

PDMS has several unusual properties [1, 5]. These are listed as follows.

1. Extremely low T_g (-123 °C)
2. Very high hydrophobicity
3. High thermal stability
4. Low surface tension
5. Low variation of viscosity with temperature
6. High gas permeability
7. Non-toxicity

These properties have made PDMS and related polymers the most important inorganic polymers from a commercial point of view. In fact, they are the only inorganic polymers that have achieved a remarkably high level of utility and consequently are produced in large quantities throughout the world. Thus, it is estimated that nearly 800,000 metric tons of this family of polymers is manufactured every year with a commercial value of over 4 billion U.S. dollars [7]. The applications of these polymers are in extremely diverse areas. These include areas such as *high-temperature insu-lators, antifoam applications, biotransplants, contact lenses, drug-delivery systems, flexible elastomers, personal care* products and many more [1-7]. The greatest advantage that these polymers seem to possess is the ease with which subtle modifications can be brought about in the polymer structure which has a profound influence on their properties and hence on the applications. Thus, although PDMS as written above is depicted with a linear backbone all the applications mentioned above are not achieved by such a linear polymer alone but by its many modifications. Thus, for many *elastomeric* applications PDMS is used in a crosslinked form after being reinforced with a material such as fumed silica [5]. The idealized structures of such crosslinked polymers are depicted in Fig. 6.2. Crosslinks are formed between different polysiloxane chains as a result of *intermolecular* Si-O-Si linkages or Si-CH$_2$CH$_2$-Si links. Thus, the cross-linking principle is similar to that of organic rubbers except that the method of achieving such crosslinks is different.

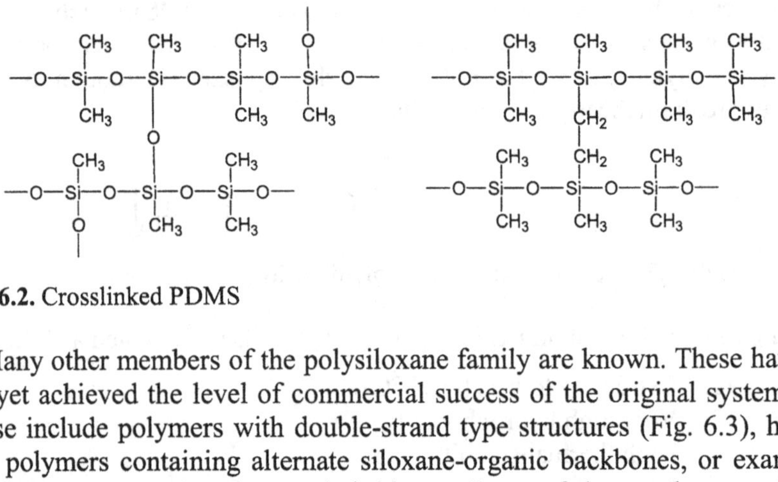

Fig. 6.2. Crosslinked PDMS

Many other members of the polysiloxane family are known. These have not yet achieved the level of commercial success of the original systems. These include polymers with double-strand type structures (Fig. 6.3), hybrid polymers containing alternate siloxane-organic backbones, or examples with siloxane-phosphazene hybrids etc. Some of these polymers will also be described in this chapter.

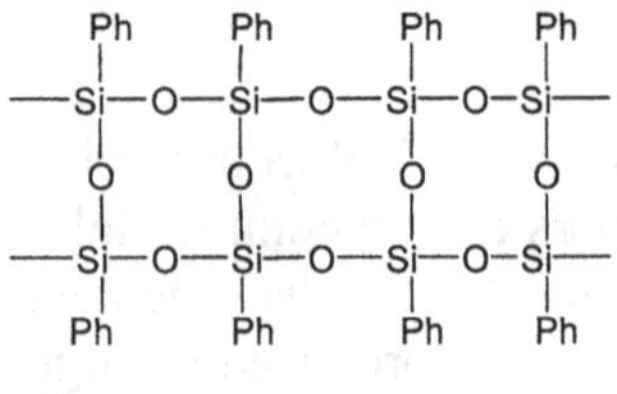

Fig. 6.3. Double-strand polysiloxanes

Even before we consider polysiloxanes we should be aware that there are other silicon-oxygen compounds and polymers that are all around us. In fact, compounds containing silicon-oxygen bonds are ubiquitous [8]. Oxygen (49.5 %) and silicon (27.3%) are the two most abundant elements in the earth's crust. Thus, almost inevitably most of the earth's crust - its rocks, clays, soils and sands - are made up of entirely silica (SiO_2) or silicates. In fact, silicon does not exist in the free form in the earth's crust. Silicon-oxygen compounds found in the earth's crust (silica or mineral silicates) have an enormous diversity of structures. Thus, silica itself is present in over 22 crystalline phases with different types of structures. For example, the structure of β-cristobalite is analogous to the structure of diamond. Thus, analogous to diamond structure in β-cristobalite also the structure is built from interconnected tetrahedral units and it belongs to the tetragonal system. In contrast, although the structure of the most common form of silica quartz although built from corner shared [SiO_4] tetrahedral units has a helical architecture. Quartz itself exists in two different forms; thus, while the thermodynamically more stable α-quartz has a trigonal symmetry, the other crystalline modification β-quartz possesses an hexagonal symmetry. The Si-O bond distances in α-quartz are 1.597 and 1.617 Å, while the Si-O-Si bond angle is 144° [8].

In contrast to silica which is neutral and has a chemical composition of SiO_2, the mineral silicates are composed of the basic tetrahedral structural unit [SiO_4]$^{4-}$. The variation in structural types arises from a sharing of oxygens between silicon centers. This essentially gives rise to structures where the tetrahedral [SiO_4] units are joined together. The corners of the tetrahedra are shared between such fused structures. Because of the various ways in which [SiO_4] tetrahedra can be linked with each other, the silicate structures are quite diverse. For a detailed description of these silicate structures the reader is advised to see well-known textbooks on Inorganic Chemistry [8-9]. A brief summary of the structural information of silicates is as follows. Interestingly, some of the mineral silicate structures are mimicked by the corresponding organosilicon compounds.

1. Orthosilicates (*nesosilicates*) such as Mg_2SiO_4 or $ZrSiO_4$. These are also called *nesosilicates*. These silicate minerals contain discrete [SiO_4]$^{4-}$ units. Silicon is surrounded by a tetrahedral coordination of oxygen atoms. The oxygen centers are arranged in a close-packed or nearly close-packed manner. The interstices of this close-packed layer are occupied by the cations present in the mineral silicate. The other examples of this mineral class include Be_2SiO_4, Mn_2SiO_4 or Fe_2SiO_4. Garnets such as [$M^{II}_3M^{III}_2(SiO_4)_3$] ($M^{II}$ = Ca, Mg, Fe; M^{III}= Al, Cr, Fe) also be-

long to the orthosilicate family. The simplest organosilicon analogue of this family is tetramethylsilane $Si(CH_3)_4$. Notice that
$[SiO_4]^{4-}$ and $Si(CH_3)_4$ are isoelectronic (Fig. 6.4).

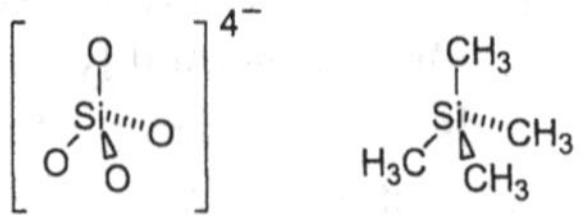

Fig. 6.4. The simplest structural unit of orthosilicates, $[SiO_4]^{4-}$ and its organosilicon analogue $Si(CH_3)_4$

2. Disilicates (*sorosilicates*) contain $[Si_2O_7]^{6-}$ units. In this type of
    silicates two silicons share one oxygen atom. This results in two
    $[SiO_4]$ tetrahedra sharing one common corner having a Si-O-Si
    bond. The bond angle at this junction oxygen is quite variable
    and ranges from 130 to 180°. The examples of these mineral
    silicates are, $Sc_2Si_2O_7$, $Lu_2Si_2O_7$, $Nd_2Si_2O_7$, $[Zn_4(OH)_2Si_2O_7]$.
    The organosilicon analogue of this type of disilicates is the disi-
    loxane, $Me_3Si\text{-}O\text{-}SiMe_3$ (Fig. 6.5)

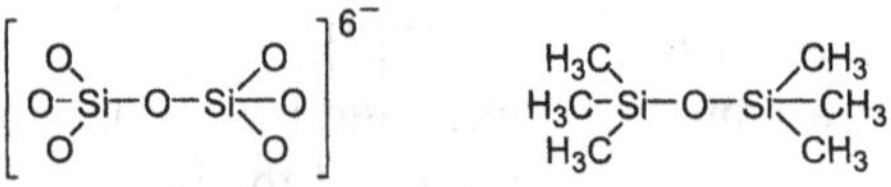

Fig. 6.5. $[Si_2O_7]^{6-}$ and its organosilicon analogue $Me_3Si\text{-}O\text{-}SiMe_3$

3. Cyclic silicates (*cyclosilicates*). This type of silicates, as the
    name suggests has a cyclic structure with alternate silicon and
    oxygen atoms in the cyclic ring. There is no example of a four-
    membered ring because at least three tetrahedral $[SiO_4]$ units
    have to join together to form a cyclic ring. Common ring motifs
    are the six-membered $[Si_3O_9]^{6-}$ and the twelve-membered
    $[Si_6O_{18}]^{12-}$ structures. The examples of minerals that possess
    these structures are $[Ba,Ti\{Si_3O_9\}]$ and $[Be_3Al_2\{Si_6O_{18}\}]$. In
    these types of silicates two oxygen atoms per each $[SiO_4]$ are
    shared. Thus, the empirical formula of these structural types is
    $[SiO_3]_n^{2n-}$. Several organocyclosiloxanes are known that are
    structurally analogous to the cyclosilicates. For example, hexa-
    methylcyclotrisiloxane, $[Me_2SiO]_3$ is a structural analogue of
    the minerals possessing $[Si_3O_9]^{6-}$ structural units (Fig. 6.6).

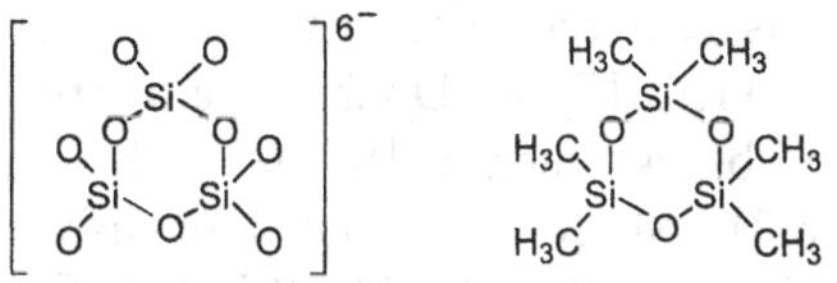

Fig. 6.6. The cyclic structural unit $[Si_3O_9]^{6-}$ and the organosiloxane analogue $[Me_2SiO]_6$

4. Polymeric linear single-chain silicates (single-chain *inosilicates*). Chain silicates with successive repeat units of $[SiO_3]^{2-}$ are among the most common type of silicate minerals. In this type of minerals the fused $[SiO_4]$ tetrahedra is organized linearly instead of in a cyclic manner (as was the case with the cyclosilicates). In this structural type also every repeat unit has two common oxygen centers that are shared. Several variations of this structural type are found in nature. Many minerals that belong to the pyroxene minerals are of this type. Examples of this family include $[Mg_2Si_2O_6]$, $[Na,Al\{Si_2O_6\}]$. The organosilicon example that is the structural analogue of these linear polymeric silicates is polydimethylsiloxane $[Me_2SiO]_n$ (Fig. 6.7). The structural relationship between the cyclic and polymeric silicates is similar to that between cyclo-$[Me_2SiO]_n$ and poly-$[Me_2SiO]_n$.

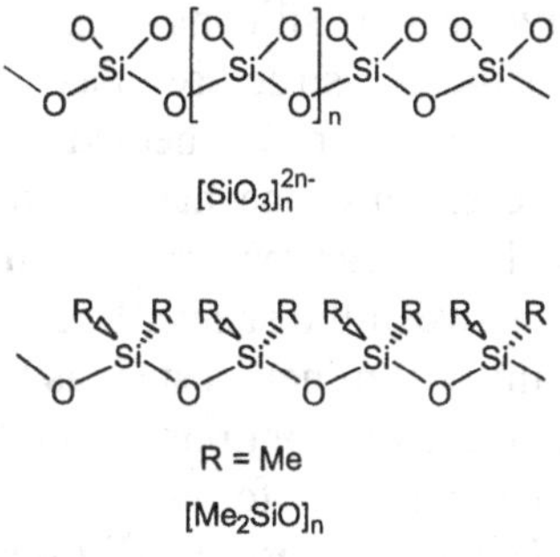

$[SiO_3]_n^{2n-}$

R = Me

$[Me_2SiO]_n$

Fig. 6.7. Linear single-chain silicates $[SiO_3]_n^{2n-}$ and their organosilicon analogues $[Me_2SiO]_n$

5. Polymeric linear double chains (double-chain *inosilicates*). If two linear silicates are linked with each other it results in double chains. We have seen that in order to generate cyclic or linear silicate structures every silicon has to share two oxygens with its neighbors. When such linear chains are linked to each other obviously additional sharing of oxygen atoms between silicon

centers is required. This gives rise to silicon:oxygen ratios of 2: 5, 4:11, 6:17 etc. Double-chain structures having structural building blocks such as $[Si_2O_5^{2-}]_n$, $[Si_4O_{11}^{6-}]_n$ are quite common (Fig. 6.8). Many amphiboles or asbestos minerals such as $[Mg_3(Si_2O_5)(OH)_4]$, $[Ca_2Mg_5(Si_4O_{11})_2(OH)_2]$ belong to this class of silicates.

$[Si_4O_{11}^{6-}]_n$ $[Si_2O_5^{2-}]_n$

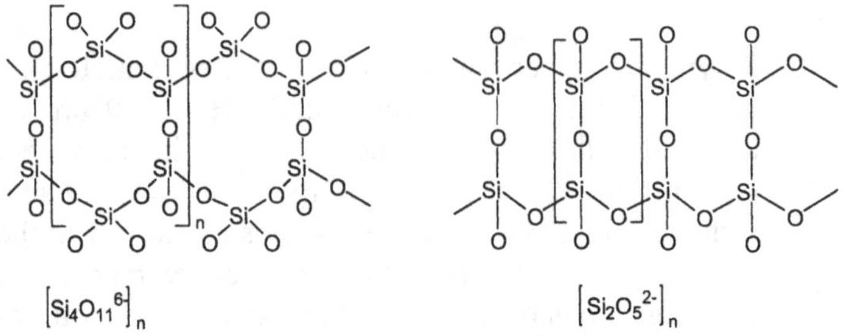

Fig. 6.8. Examples of linear double-chain silicate structures

6. A further extension of the silicate mineral structures is possible by forming sheet-type layered structures (*phyllosilicates*). In this class of silicate structures cyclic rings of $[SiO_4]$ tetrahedra are linked by shared oxygens to other cyclic rings to generate two-dimensional sheet-like structures. One oxygen is exclusively bonded to every silicon. The remaining oxygens are shared between two silicons (Fig. 6.9). Thus, every silicon in the phyllosilicate structure has 2.5 oxygens. Such two-dimensional sheets are connected with each other by layers of cations. The cations also have a coordination sphere of water molecules. It is clear that many neutral molecules also can be trapped in between the sheets. Because of these structural peculiarities many soft minerals such as talc and many types of clays such as kaolinite or montmorillonite belong to this class of minerals. The general formula of the kaolinite clays is $[Al_2Si_2O_5(OH)_4]$. In these the Si_2O_5 sheets are bound to the aluminum hydroxide $[Al_2(OH)_4]$ layers. Montmorillonite clays have a general formula of $[(Ca,Na,H)(Al,Mg,Fe,Zn)_2(SiAl)_4O_{10}(OH)_2].xH_2O$. In these structures aluminum hydroxide is sandwiched between two silicate layers. The interlayer space is filled with the cations and variable amounts of water.

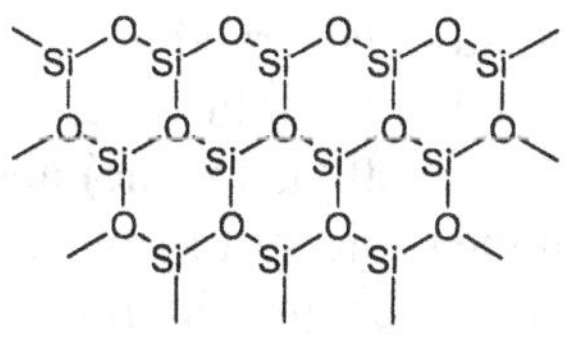

Fig. 6.9. Layered silicate structure

7. The most complex elaboration of silicates leads to the genera-
    tion of three-dimensional structures (tectosilicates). These
    classes of minerals are also called *framework silicates* because
    their structures are composed of interconnected tetrahedral net-
    works forming intricate frameworks similar to those found in
    houses. In these framework silicates *all the oxygens* are shared
    giving rise to a silicon to oxygen ratio of 1:2. The framework
    silicates can have a bewildering array of compositions because
    many cations such as the aluminum ion can substitute for the
    silicon ion up to 50%. Replacing each silicon by one aluminum
    increases the negative charge by one unit. To compensate this
    and maintain the charge balance additional cations are needed.
    Minerals such as feldspars and zeolites belong to the framework
    silicate family.

6.2 Historical Aspects of Poly(dimethysiloxane)

Although many of the silicates discussed above are polymers they do not
possess the properties that we associate with organic polymers. Thus, these
minerals are neither thermoplastic nor elastomeric. This is primarily due to
their three-dimensional structures and/or their ionic nature. Thus, neutral
substances such as silica are three-dimensional structures analogous to
highly crosslinked organic thermoset materials. The thermal stability of
silica and related substances is, of course, much higher. On the other hand,
even two-dimensional silicates are not tractable and cannot be processed
by methods that are commonly employed for organic polymers. This is
probably due to the fact that the countercations present in mineral silicates
function as effective crosslinking agents.

The search for organosilicon polymers in general and
poly(dimethylsiloxane) in particular has its origins in attempts to bridge
the gap between mineral silicates, on the one hand, and organic polymers
on the other. The need for flexible substances that can serve as high-

temperature insulation materials served as the impetus for this search [5]. These aspects are discussed in the following.

The discovery of poly(dimethylsiloxane) and the foundation of the silicones industry has been the subject of many reviews and for the full details of this interesting historical account the reader is directed to these appropriate sources [5, 7]. A brief summary of this is presented here.

The development of the silicones industry depended very critically on the ability to make compounds containing Si-C bonds. From the point of view of commercial viability it was also important that organosilicon compounds should be prepared by efficient synthetic methodologies which would contain a minimum number of synthetic steps and which avoid expensive reagents. The achievement of these goals culminated in the current silicone technology.

The first efforts of making organosilicon compounds are due to Friedel and Crafts in the nineteenth century. They alkylated silicon tetrachloride with the help of zinc alkyls or mercury alkyls. Silicon tetrachloride is prepared by the chlorination of silicon and the latter itself is produced by the Moissan method of high temperature reduction of SiO_2 with carbon which involves reduction in an electric furnace by the use of graphite electrodes. The whole process can be summarized as follows (see Eq. 6.1) [5, 7].

$$Si + 2Cl_2 \longrightarrow SiCl_4 \tag{6.1}$$

$$SiO_2 + 2C \longrightarrow Si + 2CO$$

$$2\,Zn(C_2H_5)_2 + SiCl_4 \longrightarrow Si(C_2H_5)_4 + 2\,ZnCl_2$$

$$2\,Hg(C_6H_5)_2 + SiCl_4 \longrightarrow Si(C_6H_5)_4 + 2\,HgCl_2$$

From these preparative procedures tetraalkyl and tetraarysilanes could be synthesized. It was observed that compounds such as tetraphenylsilane, $SiPh_4$, have a remarkable thermal stability (boiling point 530 °C). Seyferth, in a recent review on dimethyldichlorosilane, points out that Friedel and Crafts by oxidation of tetraethylsilane, $SiEt_4$, were able to isolate $[Et_2SiO]_n$- probably the first example of a polysiloxane [7]. Ladenburg in Germany was able to extend the methodology of Friedel and Crafts to the alkylation of tetraethoxysilane to prepare $EtSi(OEt)_3$, $Et_2Si(OEt)_2$, $Et_3Si(OEt)$ and $MeSi(OEt)_3$. He also isolated products of the type $[Et_2SiO]_n$, $[EtSiO_{1.5}]$ as well as $[MeSiO_{1.5}]_n$ [7].

The next leap in the development of organosilicon compounds was due to the efforts of F. S. Kipping in England and A. Stock in Germany. Kipping applied the then newly discovered Grignard reagents for alkylating and arylating silicon (see Eq. 6.2).

$$C_2H_5MgBr \; + \; SiCl_4 \; \longrightarrow \; (C_2H_5)SiCl_3 \qquad (6.2)$$

$$2\,C_2H_5MgBr \; + \; SiCl_4 \; \longrightarrow \; (C_2H_5)_2SiCl_2$$

Thus, the sequential reaction of the alkyl/aryl Grignard reagents with silicon tetrachloride allowed Kipping to isolate products such as R_3SiCl, R_2SiCl_2 or $RSiCl_3$. Kipping further found out that the organochlorosilanes hydrolyze to afford the corresponding hydroxy derivatives which further condense to give other products (see Eq. 6.3). Thus, Kipping found out that hydrolysis of diorganodichlorosilanes results in final products with the composition $[R_2SiO]_n$. These products were called *silicones* in analogy with *ketones* [5, 7]. However, these compounds do not contain a silicon-oxygen double bond (Si=O). Indeed, there has been no authentic or-ganosilicon compound to date that has been shown to contain a Si=O double bond.

$$R_3SiCl \; + \; H_2O \; \longrightarrow \; R_3SiOH \; + \; HCl \qquad (6.3)$$

$$R_3SiOH \; + \; R_3SiOH \; \longrightarrow \; R_3Si\text{-}O\text{-}SiR_3 \; + \; H_2O$$

$$R_2SiCl_2 \; + \; 2\,H_2O \; \longrightarrow \; R_2Si(OH)_2 \; + \; 2\,HCl$$

$$2\,R_2Si(OH)_2 \; \longrightarrow \; 2\,[R_2SiO]_n \; + \; H_2O$$

But it must be said to the credit of Kipping, that he realized that the silicones that he isolated were complex structures which were probably built from the monomeric R_2SiO unit. The hydrolysis of organochlorosilanes by Kipping invariably led to the isolation of a lot of oils and gums. In some instances, he was able to isolate cyclic rings such as $[Ph_2SiO]_3$ or $[Ph_2SiO]_4$. Kipping was quite disappointed with the nature of organosilicon compounds and was quite pessimistic about the future of organosilicon chemistry [5, 7]. However, there is no doubt that his initial and pioneering work laid strong foundations for the future development of this branch of chemistry.

The other pioneer of organosilicon chemistry, Alfred Stock, excelled in the development of the synthesis of the volatile and inflammable silicon (and boron) hydrides. He developed new types of glassware and vacuum techniques to isolate and characterize these highly reactive compounds. Stock was also able to prepare dichlorosilane, H_2SiCl_2, which served as starting materials for dimethylsilane, $(CH_3)_2SiH_2$. Stock was also able to find out that dimethysilane hydrolyzes to give small quantities of an oily material. This oil was $[Me_2SiO]_n$. Thus, Stock was probably the first per-son to have prepared poly(dimethylsiloxane) [5,7] (see Eq. 6.4). He did not, however, proceed to characterize this compound any further or investigate its properties.

$$SiH_4 + 2HCl \xrightarrow{AlCl_3} SiH_2Cl_2 + 2H_2 \qquad (6.4)$$

$$SiH_2Cl_2 + (CH_3)_2Zn \longrightarrow (CH_3)_2SiH_2 + ZnCl_2$$

$$(CH_3)_2SiH_2 + 2H_2O \longrightarrow (CH_3)_2Si(OH)_2 + H_2$$

$$n(CH_3)_2Si(OH)_2 \longrightarrow [(CH_3)_2SiO]_n + nH_2O$$

Interest in organosilicon polymeric resins was at first aroused in industry because of the need to develop insulation materials that would be indefinitely stable at high temperatures [5]. It was envisaged that organosilicon polymers could provide a matrix for glass fibers and the entire matrix could be cured to afford the required insulation materials. Using Kipping's methodology, several organochlorosilanes such as $PhEtSiCl_2$ were synthesized by the organic chemist Hyde at the Corning Glass works company. These diorganochlorosilanes could be hydrolyzed to afford the polymeric derivatives $[PhEtSiO]_n$. It was also found that this polymer could be cured (crosslinked) by heating it at 200 °C. This crosslinked polysiloxane material could be used continuously at 160 °C without significant degradation [5].

The above important result in polysiloxanes was followed by the work of Rochow at General Electric Corp. towards development of polymers of the type $[Me_2SiO]_n$. Rochow was motivated in avoiding C-C bonds in his polymers (because of their thermal instability at high temperatures) and this led him to investigate methylorganosilanes. Initially, he also used the Kipping route and was able to synthesize methylchlorosilanes such as $MeSiCl_3$ and Me_2SiCl_2. Hydrolysis of this mixture gave a polymer which could be cured by thermal treatment. Some such crosslinked polymers (with Si/CH_3 ratio around 1.3-1.5) were very stable and underwent oxidation very slowly even at 300 °C. However, in spite of the attractive nature of the methylsilicone polymer there was a major difficulty in commercializing this material. The difficulty was in extending the Grignard route to the commercial scale. Among other factors the cost of magnesium metal was an important impediment. The breakthrough came in the way of a *direct process* of preparing the methylchlorosilanes without the need to go through the Grignard reagent. Rochow realized that the reaction of HCl with silicon affords chlorosilanes such as $HSiCl_3$ or H_2SiCl_2 as demonstrated by Alfred Stock. Rochow applied this procedure to react CH_3Cl with silicon. In one classic experiment he found that this reaction proceeded to afford methylchlorosilanes in the presence of copper as the catalyst. A mixture of chlorosilanes with close boiling points were obtained: $MeSiCl_3$, 57.3 °C; Me_2SiCl_2, 70.0 °C; Me_3SiCl, 57.3 °C, $SiCl_4$, 57.6 °C [5]. The process could later be optimized to obtain maximum yields of Me_2SiCl_2 (see Eq. 6.5).

$$2 \text{ MeCl } + \text{ Si } \xrightarrow[300^\circ C]{\text{Cu powder}} Me_2SiCl_2 \qquad (6.5)$$

The direct process is also quite versatile and can be fine-tuned to prepare other types of chlorosilanes also. Ethylchlorosilanes could also be similarly prepared analogous to the synthesis of methylchlorosilanes. Preparation of phenylchlorosilanes required a slight modification of the catalyst. Addition of a mixture of HCl and MeCl to silicon affords mixtures of methylchlorosilanes along with $MeSiHCl_2$. Simultaneously, along with Rochow, but independently, Richard Mueller in Germany had also come out with a direct process, initially for preparing $HSiCl_3$, and later for methylchlorosilanes [7].

The discovery of the direct process of the synthesis of organochlorosilanes took these compounds from the realm of laboratory curiosity to commercially important materials. The chemistry of the polymerization methods for the assembly of polysiloxanes is discussed in the subsequent sections. Polysiloxanes are synthesized by two principal methods (a) ring-opening polymerization of cyclosiloxanes and (b) condensation polymerization involving a hydrolysis/condensation reaction of diorganodichlorosilanes or condensation reaction between two difunctional diorganosilanes.

6.3 Ring-Opening Polymerization of Cyclosiloxanes

Ring-opening polymerization of cyclosiloxanes is a very effective method for obtaining linear polymers. Before we consider this aspect let us digress slightly and have a look at the synthesis and structures of a few representative examples of cyclosiloxanes. This will help us in understanding the relationship that these cyclic heterocyclic rings have with their linear counterparts. In this regard this relationship is similar to what we have encountered so far in the case of cyclophosphazenes, heterocyclophosphazenes and the linear polymers derived from them.

6.3.1 Cyclosiloxanes

Cyclosiloxanes are inorganic heterocylic rings that contain alternate silicon and oxygen atoms in their cyclic structures [10]. While silicon carries two substituents and is therefore tetracoordinate, oxygen has none and is dicoordinate. In this regard the cyclosiloxanes are similar to cyclophosphazenes. However, unlike cyclophosphazenes, cyclosiloxanes do not contain a valence unsaturated skeleton.

Several types of cyclosiloxanes are known in literature. The high temperature reaction of $SiCl_4$ with oxygen results in the formation of both open-chain compounds $Si_nO_{n-1}Cl_{n+2}$ and cyclic products $[Cl_2SiO]_n$ (n = 3,4 etc.) [11]. These compounds are extremely moisture-sensitive and have not found commercial applications as monomers for polymer synthesis. These six- and eight-membered chlorocyclosiloxanes $[Cl_2SiO]_3$ and $[Cl_2SiO]_4$ are isoelectronic analogues of the corresponding chlorocyclophosphazenes, $N_3P_3Cl_6$ and $N_4P_4Cl_8$ (Fig. 6.10).

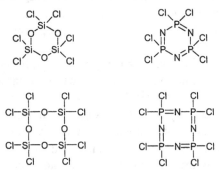

Fig. 6.10. The chlorocyclosiloxanes $[Cl_2SiO]_3$ and $[Cl_2SiO]_4$ along with their isoelectronic chlorocyclophosphazenes $[N_3P_3Cl_6]$ and $[N_4P_4Cl_8]$, respectively

Cyclosiloxanes with other substituents on silicon are also known. Various types of cyclosiloxanes containing different types of alkyl and aryl substituents on the silicon are known. The four-membered $[R'RSiO]_2$ cyclic rings are quite rare. Thus, disiloxanes of this type have been prepared by the oxygenation of the corresponding disilenes, $R'RSi=SiRR'$ (for example: R = R'=2,4,6-Me_3-C_6H_2-; R = 2,4,6-Me_3-C_6H_2, R'= *t*Bu) [12] (Fig. 6.11). The six- and eight-membered ring compounds, on the other hand, are generally more common. Three-dimensional silsesquioxanes $[RSiO_{1.5}]_8$ are obtained in the hydrolysis reaction of the trichlorosilane, $RSiCl_3$ [13].

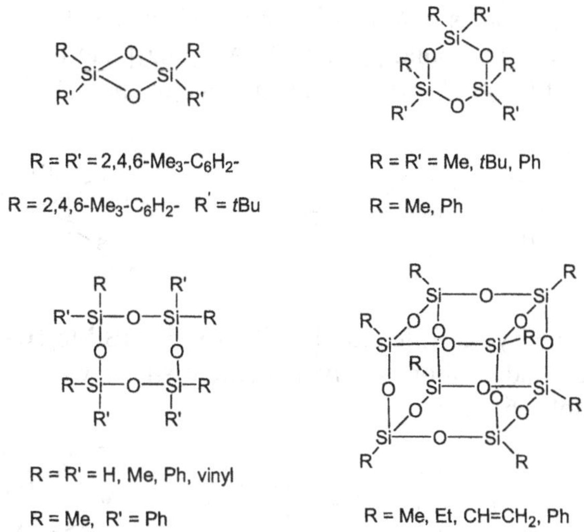

$R = R' = 2,4,6\text{-Me}_3\text{-C}_6\text{H}_2\text{-}$

$R = 2,4,6\text{-Me}_3\text{-C}_6\text{H}_2\text{-} \quad R' = t\text{Bu}$

$R = R' = \text{Me}, t\text{Bu, Ph}$

$R = \text{Me, Ph}$

$R = R' = \text{H, Me, Ph, vinyl}$

$R = \text{Me, } R' = \text{Ph}$

$R = \text{Me, Et, CH=CH}_2\text{, Ph}$

Fig. 6.11. Different kinds of organosiloxanes in ring or cage structures. The cage siloxanes are referred to as silsesquioxanes

The ring size in cyclosiloxanes varies a lot. The smallest sized ring is of course the four-membered ring mentioned above, while larger rings are also known. In a study of the hydrolysis of H_2SiCl_2 carried out with stoichiometric or a slight excess amount of water, Seyferth and coworkers have found the formation of $[H_2SiO]_n$ cyclosiloxanes. Cyclic oligomers with n = 4-23 were detected by gas chromatography [14]. However, the most common cyclic products in the hydrolysis of diorganodichlorosilanes, $RR'SiCl_2$, are the six- and the eight-membered cyclosiloxanes $[R'RSiO]_3$ and $[R'RSiO]_4$ [5, 7].

Although the hydrolysis of $RR'SiCl_2$ leads to the formation of several products including many linear products (both oligomers and polymers) it is also possible to control the reaction (for example, by proper dilution) to ensure that cyclic products predominate. For example, hydrolysis of Me_2SiCl_2 (in a scale of about 200 g) using a very large excess of water (600 mL) can lead to predominantly the eight-membered ring, $[Me_2SiO]_4$ [15] (see Eq. 6.6).

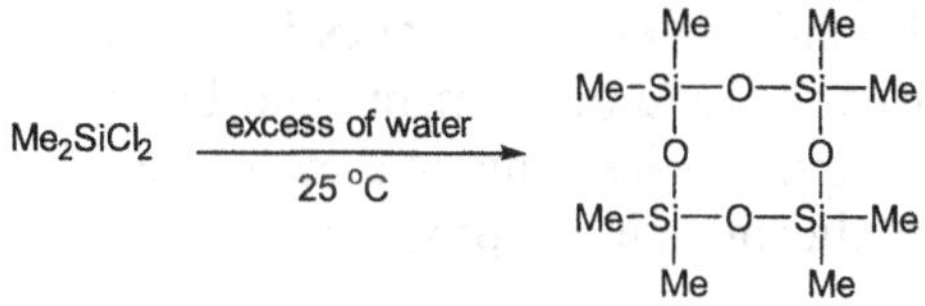

$$(6.6)$$

Some cyclosiloxanes can also be prepared by condensation reactions. For example, the compound $[\{SiPh_2\}\{SiMe_2\}_2O_3]$ can be prepared by the condensation of Me_2SiCl_2 with tetraphenyldisiloxane-1,3-diol [10, 13] (see Eq. 6.7).

$$(6.7)$$

Similarly, the cyclotrisiloxane $[\{Si(CH=CH_2)_2\{SiMe_2\}O_3]$ can be prepared by the condensation of divinyldichlorosilane with tetramethyldisiloxane-1,3-diol [16] (see Eq. 6.8).

$$(6.8)$$

The X-ray crystal structures of a number of cyclosiloxanes have been determined [13]. The structural data for a few selected cyclotrisiloxanes and cyclotetrasiloxanes are summarized in Table 6.1. The structural features of cyclotrisiloxanes can be summarized as follows.

1. Most of the six-membered rings are planar while the eight-membered rings are nonplanar. The Si-O bond distances depend on the substituents on silicon. The shortest distances are found for chlorocyclosiloxanes (see Table 6.1, entries 1 and 8).

2. The bond angles at silicon are very close to ideal tetrahedral angles. The bond angles at oxygen show considerable variation. Invariably the angles are all quite wide and range from 132 – 170°. The lowest bond angles are seen for the six-membered rings. This may be related to some amount of ring strain that these rings have in comparison to the eight-membered rings.

3. It is notable that in $[Cl_2SiO]_4$ two types of Si-O-Si bond angles are seen (148.5 and 170.7°). It is interesting to compare the Si-O-Si bond angles of cyclotrisiloxanes with acyclic derivatives. For example, in Cl_3Si-O-$SiCl_3$ the bond angle is 146°. This variation of bond angles in siloxanes is a reflection of the ease with which bond-angle adjustments can be made at oxygen to suit the structural requirements of a molecule. This property has important implications in the properties of polysiloxanes.

Table 6.1. Metric parameters of some cyclosiloxanes

SNo.	Cyclosiloxane	Bond length (Å)	Bond angle (°)	Bond angle (°)
		(Si-O)	(Si-O-Si)	(O-Si-O)
1	$[Cl_2SiO]_3$	1.617(2)	132.3(2)	109.6(1)
2	$[Me_2SiO]_3$	1.61(4)	136	104
3	$[tBu_2SiO]_3$	1.654(1)	134.9(1)	105.1(1)
4	cis-$[MePhSiO]_3$	1.65(1)	132(1)	107(1)
5	trans-$[MePhSiO]_3$	1.65(1)	132(1)	107(1)
6	$[Ph_2SiO]_3$	1.640(6)	132-133	107.0(3)
7	$(H_2SiO)_4$	1.628(4)	148.6(12)	112.0(9)
8	$[Cl_2SiO]_4$	1.592(2)	148.5(1) 170.7(1)	109.8(1)
9	$[Me_2SiO]_4$	1.65	142.5	109
10	$[MePhSiO]_4$	1.622(2)	150.2(2)	109.6(2)
11	$[Ph_2SiO]_4$	1.613(7)	167.4(5), 153.2(4)	109.1(4)

4. Finally it is interesting to note that the bond angles at oxygen in cyclotrisiloxanes are much wider in comparison to the angles found at nitrogen in the corresponding cyclophosphazenes. Thus, for example, the Si-O-Si bond angle in the six-membered ring $[Cl_2SiO]_3$ is 132.3°. In comparison, the P-N-P bond angle found in the isoelectronic cyclotriphosphazene, $N_3P_3Cl_6$, is

121.4°. Also in the latter the P-N-P and the N-P-N bond angles are nearly equal in contrast to the large difference between Si-O-Si and O-Si-O bond angles found in cyclotrisiloxanes.

6.3.2. Ring-opening Polymerization of Cyclosiloxanes by Ionic Initiators

Cyclosiloxanes, $[R_2SiO]_n$, (n = 3,4) can be polymerized by ROP methods to the corresponding linear polymer $[R_2SiO]_n$ (R = alkyl, aryl). Two different types of initiators are used for this reaction [17-19]. (1) Cationic initiators and (2) anionic initiators. Before we look at the individual methods of ROP let us look at the main features of the ROP of cyclosiloxanes.

1. There is virtually no ring-strain in the eight-membered cyclosiloxanes, while the ring strain in the six-membered derivatives has been estimated to be about 38 kJ mol^{-1}. Since Si-O bonds are being broken and re-formed during the process of ROP and since no ring strain is involved (at least in the case of eight-membered rings), this polymerization is *not enthalpy driven*.
2. However, there is an increase in the entropy because the skeletal mobility in the polysiloxane chains is far greater than that observed in the cyclic compounds. Therefore, the ROP of cylclosiloxanes is *entropy driven*.
3. The ring-opening polymerization of cyclosiloxanes is a complex equilibrium polymerization and detailed studies have shown that a series of competing reactions exist that allow formation of cyclic as well as polymeric products. The overall polymerization is a chain-growth polymerization.
4. The linear polymers can be reconverted to the cyclic compounds particularly if initiator concentration is too high and if the initiator is not removed or neutralized at the end of the reaction. However, it is the triumph of polymer chemists that in spite of these challenges polysiloxanes with molecular weights in the range of 2-5 million can be prepared quite routinely. The PDI's are not very narrow and no method seems to be yet available that allows synthesis of polysiloxanes with narrow molecular-weight distributions.

There is a special nomenclature that is used in polysiloxane literature which is not encountered elsewhere. Although we will not use this nomenclature in this book it is useful to know about it [5]. The monomeric structural unit of a poly(dimethylsiloxane) viz., Me_2SiO is termed **D** because this center is *difunctional* and can form two bonds with its neighbors. The Me_3SiO group is called **M** because it is *monofunctional* and can only bond to one neighbor. Similarly, the $MeSiO_3$ group could be termed **T** because it is *trifunctional*. Let us see how this nomenclature works. For example, $Me_3SiOSiMe_3$ will simply be called **MM**. $Me_3SiOMe_2SiOSiMe_3$ will be called **MDM**; $[Me_2SiO]_3$ will be called D_3; $[Me_2SiO]_4$ will be called D_4 and so on. If the substituents on silicon are not methyl groups the nomenclature can become cumbersome. For example, $[Ph_2SiO]_3$ will be referred to D_3^{Ph} with the superscript serving as the identifier for the substituent present on silicon.

With the above background let us first look at the cation initiated ROP of cyclosiloxanes.

6.3.2.1 Cationic Polymerization

Several cationic initiators such as Lewis acids or protic acids can be employed to effect the polymerization of cyclosiloxanes [17-19]. Among protic acids, H_2SO_4, $HClO_4$, CF_3SO_3H etc. have been widely used. Many Lewis acids have been used including $AlCl_3$ and $SnCl_4$. The mechanism of polymerization initiated by CF_3SO_3H is given below. Two broad types of mechanisms are envisaged 1) acidolysis/condensation and 2) generation of an active propagating center.

In the acidolysis/condensation mechanism the initiation of the polymerization is believed to generate the linear chain which contains a hydroxyl terminal group as well as a CF_3SO_2 end-group (see Eq. 6.9).

$$(6.9)$$

The propagation reaction proceeds in two ways. The first of these involves condensation of two molecules involving their hydroxyl end-groups. This leaves terminal triflate groups (see Eq. 6.10).

$$(6.10)$$

The second route of propagation involves condensation of a molecule containing one hydroxyl end-group with another molecule containing a CF_3SO_2 end-group (see Eq. 6.11). Further reactions lead to the formation of the high polymer.

$$(6.11)$$

$$\text{CF}_3\text{S(O)}_2\text{O}-\underset{\underset{\text{Me}}{|}}{\overset{\overset{\text{Me}}{|}}{\text{Si}}}\!\!-\!\!\left(\!\text{O}-\underset{\underset{\text{Me}}{|}}{\overset{\overset{\text{Me}}{|}}{\text{Si}}}\!\right)_{\!\!4}\!\!\text{O}-\underset{\underset{\text{Me}}{|}}{\overset{\overset{\text{Me}}{|}}{\text{Si}}}\!\!-\!\text{OH} \quad + \quad \text{CF}_3\text{S(O)}_2\text{O}-\underset{\underset{\text{Me}}{|}}{\overset{\overset{\text{Me}}{|}}{\text{Si}}}\!\!-\!\!\left(\!\text{O}-\underset{\underset{\text{Me}}{|}}{\overset{\overset{\text{Me}}{|}}{\text{Si}}}\!\right)_{\!\!4}\!\!\text{O}-\underset{\underset{\text{Me}}{|}}{\overset{\overset{\text{Me}}{|}}{\text{Si}}}\!\!-\!\text{OH}$$

$-CF_3SO_3H$

$$\text{CF}_3\text{S(O)}_2\text{O}-\underset{\underset{\text{Me}}{|}}{\overset{\overset{\text{Me}}{|}}{\text{Si}}}\!\!-\!\!\left(\!\text{O}-\underset{\underset{\text{Me}}{|}}{\overset{\overset{\text{Me}}{|}}{\text{Si}}}\!\right)_{\!\!4}\!\!\text{O}-\underset{\underset{\text{Me}}{|}}{\overset{\overset{\text{Me}}{|}}{\text{Si}}}\!\!-\!\text{O}-\underset{\underset{\text{Me}}{|}}{\overset{\overset{\text{Me}}{|}}{\text{Si}}}\!\!-\!\!\left(\!\text{O}-\underset{\underset{\text{Me}}{|}}{\overset{\overset{\text{Me}}{|}}{\text{Si}}}\!\right)_{\!\!4}\!\!\text{O}-\underset{\underset{\text{Me}}{|}}{\overset{\overset{\text{Me}}{|}}{\text{Si}}}\!\!-\!\text{OH}$$

This type of polymerization reaction can also be used to generate *short-chain* linear siloxanes. Thus, carrying out the acid-catalyzed ROP of $[Me_2SiO]_4$ in the presence of the disiloxane $Me_3Si-O-SiMe_3$ allows the formation of (predominantly) $Me_3SiO(SiMe_2O)_4Me_3Si$. The Me_3Si group functions as a chain blocker or a terminator group. This method of chain termination is also valid for the high polymers. In the case of the short chain generation, the reactions involved are summarized in Eqs. 6.12-6.14. Thus, the initiation is envisaged as the ring-opening of the eight-membered ring (see Eq. 6.12). This generates the difunctional linear siloxane (containing Si-OH and Si-OSO$_2$CF$_3$ end-groups).

$$(6.12)$$

$$\begin{array}{c} \text{Me} \quad\quad \text{Me} \\ \text{Me}-\underset{\underset{\text{O}}{|}}{\overset{\overset{}{|}}{\text{Si}}}\!\!-\!\text{O}-\underset{\underset{\text{O}}{|}}{\overset{\overset{}{|}}{\text{Si}}}\!\!-\!\text{Me} \\ \\ \text{Me}-\underset{\underset{\text{Me}}{|}}{\overset{\overset{}{|}}{\text{Si}}}\!\!-\!\text{O}-\underset{\underset{\text{Me}}{|}}{\overset{\overset{}{|}}{\text{Si}}}\!\!-\!\text{Me} \end{array} \quad\xrightarrow{\text{CF}_3\text{SO}_3\text{H}}\quad \text{HO}-\underset{\underset{\text{Me}}{|}}{\overset{\overset{\text{Me}}{|}}{\text{Si}}}\!\!-\!\text{O}-\underset{\underset{\text{Me}}{|}}{\overset{\overset{\text{Me}}{|}}{\text{Si}}}\!\!-\!\text{O}-\underset{\underset{\text{Me}}{|}}{\overset{\overset{\text{Me}}{|}}{\text{Si}}}\!\!-\!\text{O}-\underset{\underset{\text{Me}}{|}}{\overset{\overset{\text{Me}}{|}}{\text{Si}}}\!\!-\!\text{OS(O)}_2\text{CF}_3$$

Simultaneously, the disiloxane $Me_3SiOSiMe_3$ is also opened up by the protic acid (see Eq. 6.13).

$$Me_3SiOSiMe_3 \xrightarrow{\text{CF}_3\text{SO}_3\text{H}} Me_3SiOH + Me_3SiOS(O)_2CF_3 \quad\quad (6.13)$$

Subsequent combination of either Me_3SiOH or $Me_3SiOSO_2CF_3$ with the linear siloxane affords $Me_3SiO(SiMe_2O)_4Me_3Si$ (see Eq. 6.14). The reactions shown in Eqs. 6.12-6.14 represent the ideal situation; in practice both shorter and longer chain siloxanes are formed. This is because the reaction of triflic acid (or any other protic acid) with siloxanes essentially breaks up the Si-O-Si bonds to generate compounds containing Si-OH and/or Si-

OSO_2CF_3 end-groups. Thus, there is an opportunity for several types of products which eventually equilibrate to the most probable products. The important point to note is that the average molecular weight of such short-chain polysiloxanes $Me_3Si(OSiMe_2)_nOSiMe_3$ can be regulated by the initial ratio of $Me_3SiOSiMe_3$ and $[Me_2SiO]_4$. In the reactions shown in Eqs. 6.12-6.14 the ratio of $Me_3SiOSiMe_3$ and $[Me_2SiO]_4$ is 1:1. This can lead to the main product being $Me_3Si(OSiMe_2)_4OSiMe_3$.

$$2\ Me_3SiOS(O)_2CF_3 \quad + \quad HO\underset{Me}{\overset{Me}{-Si-}}O\underset{Me}{\overset{Me}{-Si-}}O\underset{Me}{\overset{Me}{-Si-}}O\underset{Me}{\overset{Me}{-Si-}}OS(O)_2CF_3 \tag{6.14}$$

$$\downarrow \text{-2 } CF_3SO_3H$$

$$Me_3SiO\underset{Me}{\overset{Me}{-Si-}}O\underset{Me}{\overset{Me}{-Si-}}O\underset{Me}{\overset{Me}{-Si-}}O\underset{Me}{\overset{Me}{-Si-}}OSiMe_3$$

Linear short-chain polysiloxanes $Me_3Si(OSiMe_2)_nOSiMe_3$ are used as *silicone oils*. Such oils are characterized by their high thermal stability and low *viscosity temperature coefficient* (VTC). A low VTC implies a small change in viscosity vis-à-vis temperature. Thus, many modern silicone oils retain their fluid-like characteristics even at temperatures as low as -80 °C.

The second mechanism of cationic polymerization involves the generation of an active propagation center. This could be an oxonium ion (oxygen centered cation) or a siliconium ion (silicon-centered cation). This type of ion has been detected in small molecule experiments. Thus, the reaction of Me_3SiH and $Me_3SiOSiMe_3$ with $[Ph_3C]^+[B(C_6F_5)_4]^-$ leads to the formation of $[(Me_3Si)_3O]^+$ which has been detected by NMR spectroscopic methods (see Eq. 6.15) [20].

$$\left[Ph_3C\right]^+\left[B(C_6F_5)_4\right]^- \quad + \quad Me_3SiH \quad + \quad Me_3Si-O-SiMe_3 \tag{6.15}$$

$$\downarrow \text{-}Ph_3CH$$

$$\left[\{SiMe_3\}_3O\right]^+\left[B(C_6F_5)_4\right]^-$$

Another example of an oxonium ion generation involves the reaction of Me_3SiH with $[Me_2SiO]_3$ with $[Ph_3C]^+[B(C_6F_5)_4]^-$ to generate the cyclotrisiloxane-centered oxonium ion (Fig. 6.12) [21].

In the oxonium ion-mediated polymerization it is envisaged that the attack of the protic acid on a cyclosiloxane such as $[Me_2SiO]_4$ leads to protonation of an oxygen and the formation of an oxonium ion. This is followed by ring-cleavage and generation of a propagating oxonium ion (Fig 6.13). The polymerization is terminated by any reagent that can quench the propagating species.

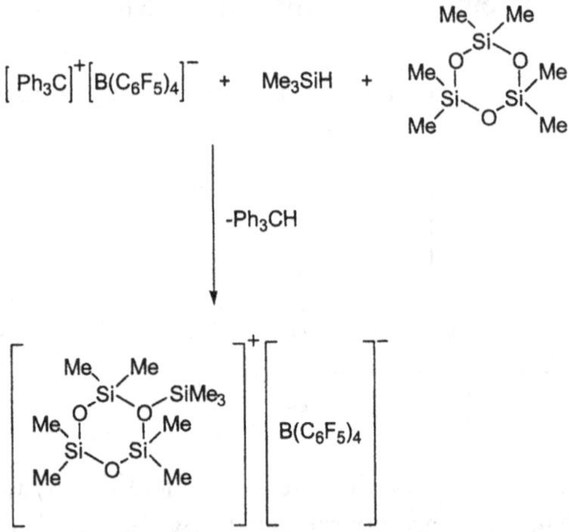

Fig. 6.12. Generation of a cyclotrisiloxane-based oxonium ion

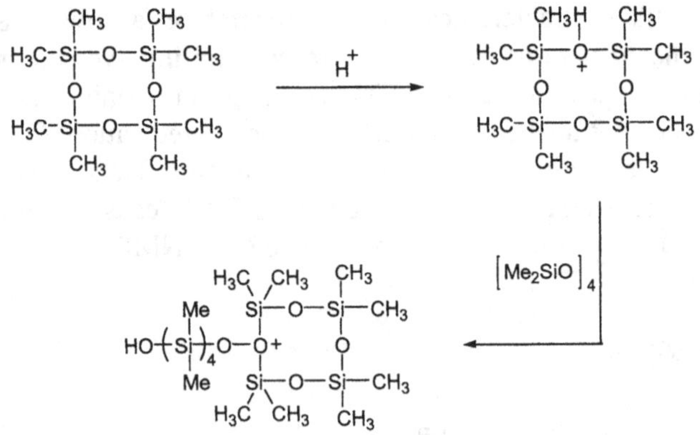

Fig. 6.13. Oxonium ion-mediated polymerization mechanism of $[Me_2SiO]_4$

An alternative proposal for the active propagating center involves the formation of the silicon-centered cation, the silylenium ion. In this proposal the initiation is similar to what has been proposed in the oxonium ion mechanism except that the positive charge is centered on the silicon to generate a silylenium ion (see Eq. 6.16).

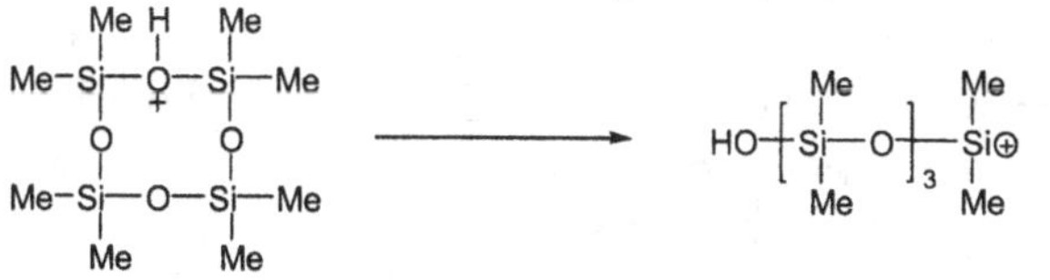

(6.16)

Propagation of the polymerization occurs by the nucleophilic attack of the oxygen center of another cyclosiloxane molecule on the silylenium ion. Evidence for stable silylenium (also called silylium) ions has been recently presented by Lambert and Reed [22]. They were able to synthesize and characterize the free silylenium ion $[Mes_3Si]^+$ (See Eq. 6.17). The proof for the existence of the silylenium ion was obtained from its ^{29}Si NMR (chemical shift of $\delta = +220$) as well as by single crystal X-ray analysis.

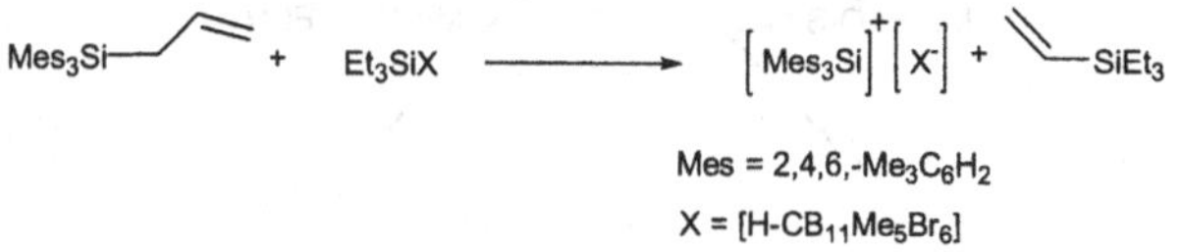

(6.17)

$$Mes = 2,4,6,-Me_3C_6H_2$$
$$X = [H-CB_{11}Me_5Br_6]$$

6.3.2.2 Anionic Polymerization

Anionic polymerization is more effective than the cationic polymerization in generating high-molecular-weight polysiloxanes. Generally, the molecular weights obtained in this method are in the order of $2\text{-}5 \times 10^6$. The initiators could be metal hydroxides such as KOH or even lithium salts of silanediols such as $Ph_2SiO_2Li_2$ [3, 21].

The anionic polymerization is initiated by the reaction of the base with a cyclosiloxane such as $[Me_2SiO]_4$ to generate the open-chain compound with a Si-O$^-$K$^+$ end-group (See Eq. 6.18).

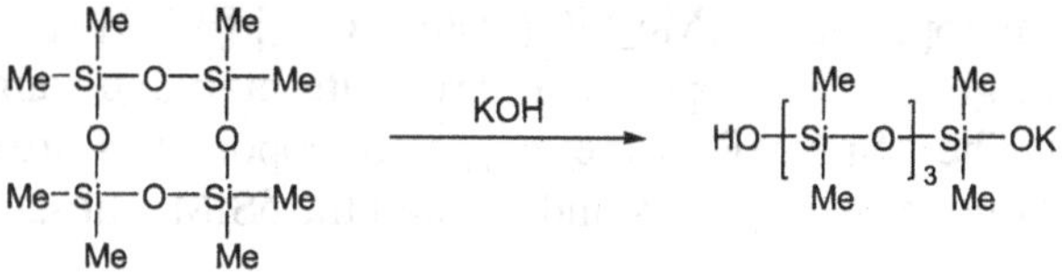

(6.18)

The propagation reaction is similar to that encountered in many conventional anionic polymerizations using organic monomers. Thus, the incoming monomer can *insert* between the ion-pair. This leads to chain propagation (Fig. 6.14).

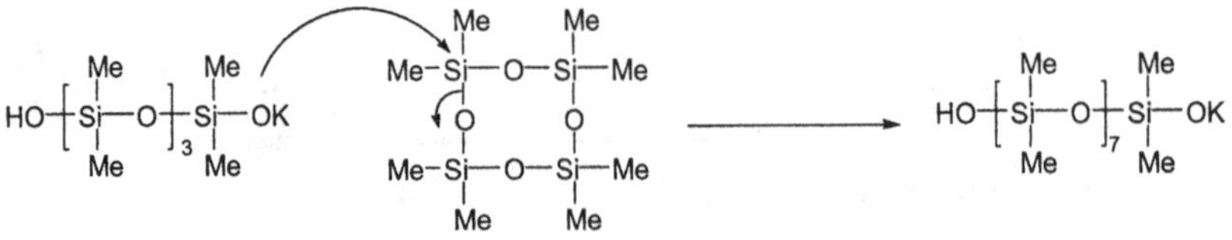

Fig. 6.14. Propagation reaction in the anionic polymerization of [Me$_2$SiO]$_4$

The termination of the reaction can be accomplished by the use of Me$_3$Si end-capping reactions (Fig. 6.15).

Fig. 6.15. Termination reactions of the anionic polymerization

6.3.3 Preparation of Copolymers by Ring-Opening Polymerization

Copolymers of the type [(SiR$_2$O)$_n$(SiR'$_2$O)$_m$] can be prepared by the polymerization of mixtures of cyclosiloxanes. Thus, for example, heating a mixture of [Me$_2$SiO]$_4$ and [Ph$_2$SiO]$_4$ in the presence of KOH as the catalyst affords the random copolymer [{Me$_2$SiO}$_n${Ph$_2$SiO}$_m$] [4, 23] (Fig. 6.16). The molecular weights of the copolymers are quite high and range from 100,000-200,000. The ^{29}Si NMR of these types of copolymers are characterized by signals due to δSiPh$_2$ at around -47 and the δSiMe$_2$ at -20.

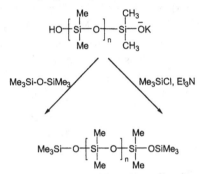

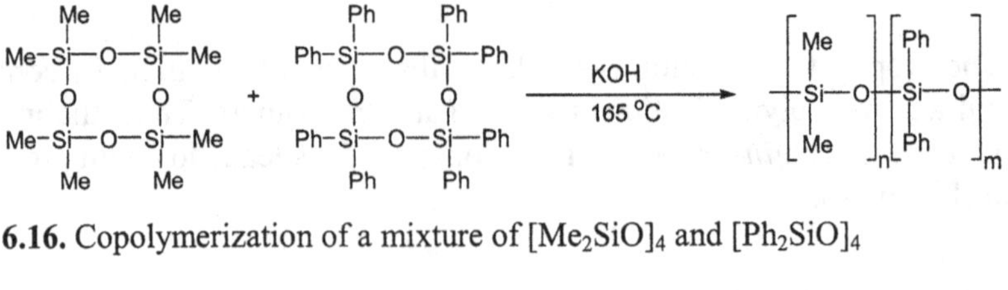

Fig. 6.16. Copolymerization of a mixture of [Me$_2$SiO]$_4$ and [Ph$_2$SiO]$_4$

Similarly, copolymerizing a mixture of $[Me_2SiO]_4$ and $[Si(Me)(CH=CH_2)O]_4$ affords poly(dimethylsiloxane-*co*-methylvinylsiloxane). This polymer has a random distribution of the $[Me_2SiO]$ and $[Me(CH_2=CH)SiO]$ units [24].

Copolymers containing a regular sequence of structural units can be prepared by polymerizing heterogeneously substituted siloxanes such as $[\{Si(CH=CH_2)(CH_3)(SiMe_2)_2\}O_3]$ and $[\{Si(CH=CH_2)_2-(SiMe_2)_2\}O_3]$ by anionic initiators [25, 16] (Fig 6.17). This method generates polysiloxanes with a regular sequence as revealed by their ^{29}Si NMR. Thus, for example, the polymer $[\{Si(CH=CH_2)_2(SiMe_2)_2\}O]_n$ shows signals at -49.17 ($\delta Si(CH=CH_2)_2$ of a pentad unit) and at -20.20 ($\delta Si(CH_3)_2$ of a pentad unit) [16]. Interestingly, polymerization of $[\{Si(CH=CH_2)_2 (SiMe_2)_2\}O_3]$ by CF_3SO_3H affords a random copolymer [16].

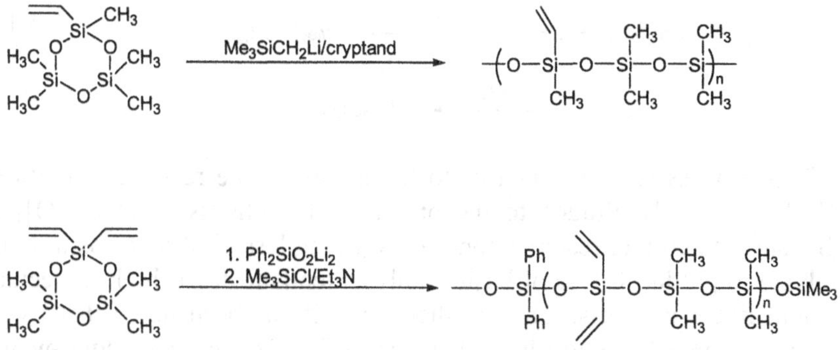

Fig. 6.17. Anionic polymerization of $[\{Si(CH=CH_2)(CH_3)(SiMe_2)_2\}O_3]$ $[\{Si(CH=CH_2)_2 (SiMe_2)_2\}O_3]$ to afford regular copolymers

6.3.4 Summary of the Ring-Opening Polymerization of Cyclosiloxanes

1. Ring-opening polymerization of cyclosiloxanes can be carried out by either cationic or anionic initiators.
2. Polymerization by cationic initiators occurs either by protic or Lewis acids. Different types of mechanisms of polymerization have been proposed such as acidolysis/condensation, oxonium ion-mediated polymerization or silylium ion-mediated polymerization.
3. Anionic polymerization can be carried out by metal hydroxides or even by lithium salts of silanols. This method of polymerization affords high-molecular-weight polymers.

4. The polymers have to be terminated by Me_3Si groups to afford stable products.
5. Copolymers can be prepared either by polymerization of a mixture of cyclosiloxanes or by the polymerization of heterogeneously substituted cyclosiloxanes.

6.4 Condensation Polymerization

The hydrolysis of $RR'SiCl_2$ leads to the formation of a diorganosilanediol which undergoes self-condensation to afford a variety of cyclic and linear products (see Eq. 6.19) [1, 3, 5]. This is the original route to silicones and still continues to be a process that is important.

$$Me_2SiCl_2 \ + \ 2\,H_2O \ \xrightarrow{\text{-2 HCl}} \ Me_2Si(OH)_2 \qquad (6.19)$$

$$nMe_2Si(OH)_2 \ \xrightarrow{\text{-}nH_2O} \ [Me_2SiO]_n$$

This process is very sensitive to the nature of the reaction conditions. We have already alluded to the preferential synthesis of $[Me_2SiO]_4$ by choosing the correct reaction conditions. Using less dilution leads to linear polymeric products. Use of basic catalysts tends to favor high-molecular-weight linear products. On the other hand, it has been noted that use of acidic catalysts tends to tilt the balance in favor of low-molecular-weight polymers or cyclic products.

Apart from the hydrolysis route to polysiloxanes there have been efforts to prepare appropriate difunctional silicon derivatives that can be used for step-growth polymerization. For example, diaminosilanes are good synthons for condensation with diorganosilanediols. This method is quite effective because of the lability of the Si-N bond. Thus, the reaction of $Me_2Si(NMe_2)_2$ with diphenylsilanediol affords a random copolymer [26] (see Eq. 6.20). The randomization occurs because of the catalytic action of the liberated amine to cleave Si-O bonds.

$$Me_2SiCl_2 \ + \ LiNMe_2 \ \xrightarrow{\text{-2 LiCl}} \ Me_2Si(NMe_2)_2 \qquad (6.20)$$

$$nMe_2Si(NMe_2)_2 \ + \ nPh_2Si(OH)_2 \ \xrightarrow{\text{-2}nMe_2NH} \ \left[\begin{array}{c} Me \quad\quad Ph \\ | \quad\quad\quad | \\ -Si-O-Si-O- \\ | \quad\quad\quad | \\ Me \quad\quad Ph \end{array} \right]_n$$

The reaction of $Ph_2Si(OH)_2$ with NMe_2-$[Si(Me_2)O]_3$-NMe_2 also affords a copolymer; however, this process also leads to random copolymers [26] (see Eq. 6.21).

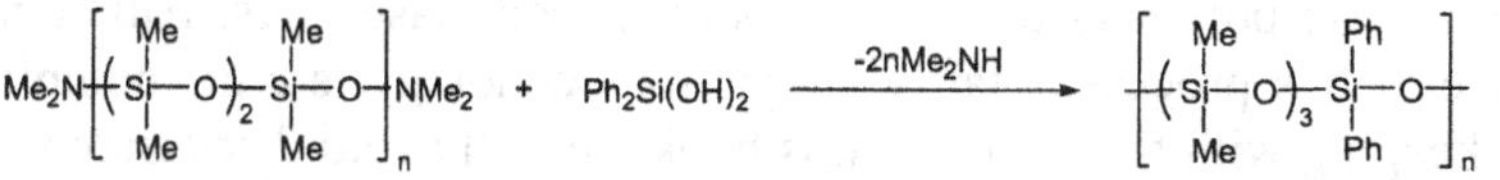

Alternating copolymers can, however, be prepared by choosing a bis-ureidosilane and allowing it to condense with diphenylsilanediol [26] (Fig. 6.18). The monomer bis-ureidosilane is prepared by a step-wise reaction involving first a reaction of Me_2SiCl_2 with pyrrolidine. The pyrrolidino derivative upon reaction with PhNCO affords the bis-ureidosilane. The choice of this precursor allows the formation of a nonreactive by-product. Thus, the condensation reaction of the bis-ureidosilane with diphenylsilanediol affords urea as the by-product which precipitates out of the reaction mixture and therefore does not participate in further reactions. Because of this a *nearly alternating* copolymer containing $[Me_2Si-O-SiPh_2O]_n$ is obtained (Fig. 6.18).

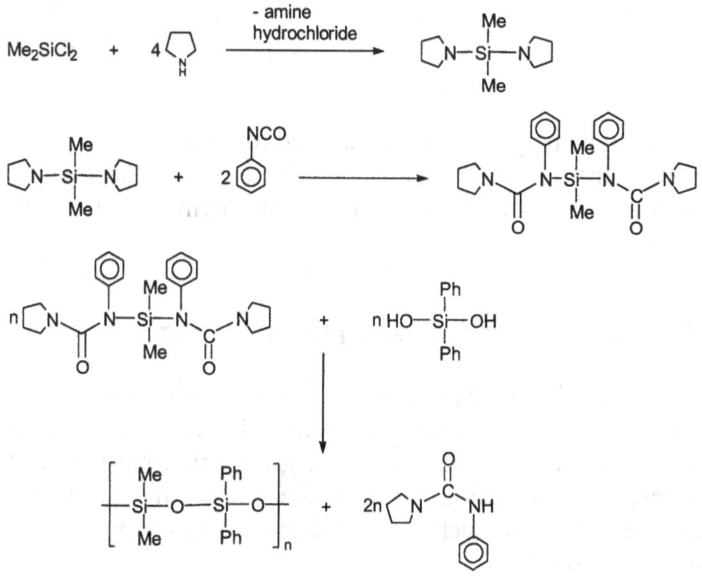

Fig. 6.18. Synthesis of poly(dimethylsiloxane-*co*-diphenylsiloxane)

6.5 Crosslinking of Polysiloxanes

Crosslinking or curing of polysiloxanes is the most important procedure that allows these polymers to be used as advanced elastomers. Although the science of this process is fairly simple its technology is quite complicated. In keeping with the nature of this book we will avoid discussing the details of the technology. The main aspects of the curing process of polysiloxanes are as follows.

High-molecular-weight polysiloxanes, along with reinforcing fillers such as fumed high-surface-area silica (as opposed to silica obtained from hydrolysis) are heated together with organic peroxides (along with some amount of cyclosiloxanes and some coloring pigments) to afford crosslinked products via $-CH_2CH_2-$ links [3-5] (see Eq. 6.22). This crosslinking is different from that applicable in organic rubber materials where sulfur cross-links are quite effective.

$$(6.22)$$

The siloxane polymers can be tailored to contain a few percent of vinyl groups and these can be crosslinked with polymers that contain Si-H units. We will look at these types of functional polymers in the following section.

6.5.1 Other Ways of Achieving Crosslinking

Apart from the peroxide curing method as discussed above, there are several other ways to crosslink siloxane polymers. These rely on coupling reactions between reactive functional groups that have been deliberately introduced either as end-groups or in the side-chain of the polymer. Representative examples of such functional-group- containing polysiloxanes are shown in Fig. 6.19.

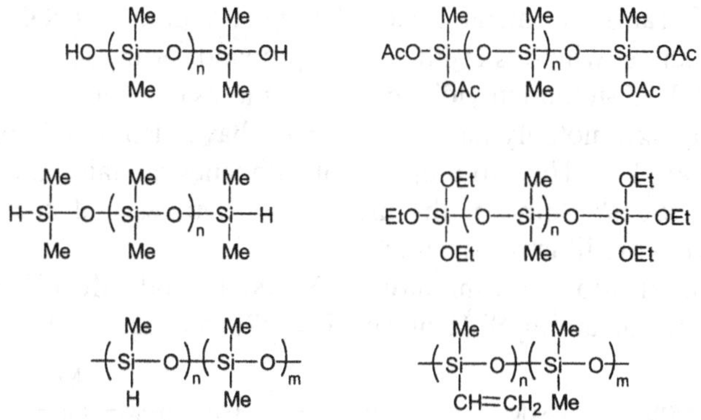

$$HO-\left(\underset{\underset{Me}{|}}{\overset{\overset{Me}{|}}{Si}}-O\right)_n-\underset{\underset{Me}{|}}{\overset{\overset{Me}{|}}{Si}}-OH \qquad AcO-\underset{\underset{OAc}{|}}{\overset{\overset{Me}{|}}{Si}}-\left(O-\underset{\underset{Me}{|}}{\overset{\overset{Me}{|}}{Si}}\right)_n-O-\underset{\underset{OAc}{|}}{\overset{\overset{Me}{|}}{Si}}-OAc$$

$$H-\underset{\underset{Me}{|}}{\overset{\overset{Me}{|}}{Si}}-O-\left(\underset{\underset{Me}{|}}{\overset{\overset{Me}{|}}{Si}}-O\right)_n-\underset{\underset{Me}{|}}{\overset{\overset{Me}{|}}{Si}}-H \qquad EtO-\underset{\underset{OEt}{|}}{\overset{\overset{OEt}{|}}{Si}}-\left(O-\underset{\underset{Me}{|}}{\overset{\overset{Me}{|}}{Si}}\right)_n-O-\underset{\underset{OEt}{|}}{\overset{\overset{OEt}{|}}{Si}}-OEt$$

$$\left(\underset{\underset{H}{|}}{\overset{\overset{Me}{|}}{Si}}-O\right)_n\left(\underset{\underset{Me}{|}}{\overset{\overset{Me}{|}}{Si}}-O\right)_m \qquad \left(\underset{\underset{CH=CH_2}{|}}{\overset{\overset{Me}{|}}{Si}}-O\right)_n\left(\underset{\underset{Me}{|}}{\overset{\overset{Me}{|}}{Si}}-O\right)_m$$

Fig. 6.19. Polysiloxanes with functional groups placed in the side chain or at the termini

Some of the important condensation reactions that have been very successful in bringing about crosslinking reactions are shown in Fig. 6.20. In each one of these reactions the main emphasis is on the fact that if any side-product emerges from the reaction it should not disturb the main-chain in terms of skeletal fragmentation or any other undesired reactions [1-2].

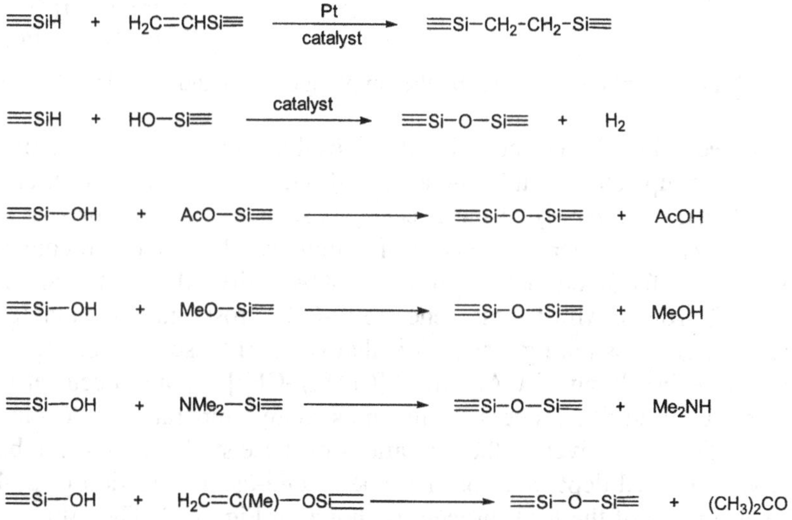

$$\equiv SiH \;+\; H_2C\!\!=\!\!CHSi\!\!\equiv \xrightarrow[\text{catalyst}]{Pt} \equiv Si-CH_2-CH_2-Si\equiv$$

$$\equiv SiH \;+\; HO-Si\equiv \xrightarrow{\text{catalyst}} \equiv Si-O-Si\equiv \;+\; H_2$$

$$\equiv Si-OH \;+\; AcO-Si\equiv \longrightarrow \equiv Si-O-Si\equiv \;+\; AcOH$$

$$\equiv Si-OH \;+\; MeO-Si\equiv \longrightarrow \equiv Si-O-Si\equiv \;+\; MeOH$$

$$\equiv Si-OH \;+\; NMe_2-Si\equiv \longrightarrow \equiv Si-O-Si\equiv \;+\; Me_2NH$$

$$\equiv Si-OH \;+\; H_2C\!\!=\!\!C(Me)-OSi\equiv \longrightarrow \equiv Si-O-Si\equiv \;+\; (CH_3)_2CO$$

Fig. 6.20. Several ways of preparing crosslinked polysiloxanes

The most important reaction is the coupling between a Si-H functional group and a Si-CH=CH$_2$ group to generate a Si-CH$_2$-CH$_2$-Si crosslink. This reaction is known as the *hydrosilylation* reaction and is catalyzed by variety of transition metal ions. The most common catalysts are the Speier's catalyst which is H_2PtCl_6 in isopropanol or the Karstedt's catalyst (a Pt(0)-1,1,3,3-tetramethyldisiloxane complex). More recently, other metal complexes notably those of Ru or Rh have also found application in this reaction [1-2]. How to prepare polysiloxanes containing Si-H and Si-CH=CH$_2$ groups? Some of the methods of synthesis of such functional polysiloxanes are illustrated below.

Copolymerization of a mixture of [Me$_2$SiO]$_4$ and [MeHSiO]$_4$ affords a polysiloxane containing Si-H side chains [27] (see Eq. 6.23).

$$x(Me_2SiO)_4 \;+\; y(MeSiHO)_4 \longrightarrow \left(\!O\!-\!\underset{Me}{\overset{Me}{Si}}\!\right)_x\!\!\left(\!O\!-\!\underset{H}{\overset{Me}{Si}}\!\right)_y \qquad (6.23)$$

A similar approach can also be used for preparing vinyl side-group-containing polysiloxanes. These can then be condensed together in the presence of a catalyst to afford crosslinking (Fig. 6.21).

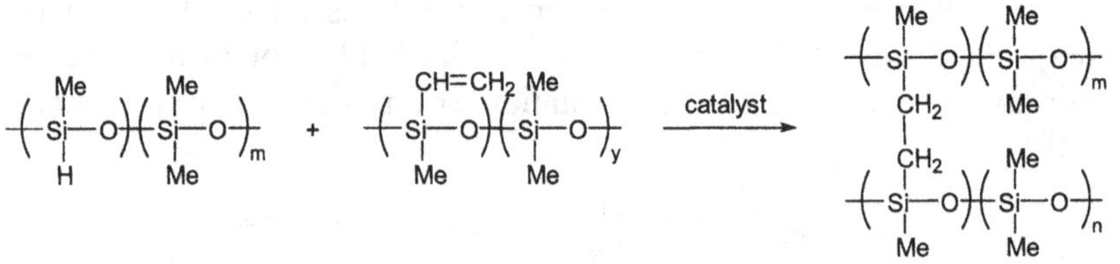

Fig. 6.21. Crosslinking siloxanes by the coupling of Si-H and Si-CH=CH$_2$ groups

The second method of coupling involves the reaction of a Si-H unit with a Si-OH group. This would generate a Si-O-Si crosslink. This reaction also is catalyzed by a variety of metal catalysts such as ferric, tin or zinc chlorides [1, 2]. The other methods of effecting crosslinks include condensation of a Si-OH functional group with Si-OAc, Si[C(Me)=CH$_2$], Si-NMe$_2$, or Si-OR groups. All of these generate Si-O-Si cross-links. A notable feature of these crosslinking reactions is that some of these reactions (particularly those involving Si-OAc and Si[C(Me)=CH$_2$] groups) occur at room temperature and this process is known as room-temperature-vulcanization (RTV). There are several other variations of these strategies and can be selectively utilized depending on the type of end-use application [1-2, 4-5]. The chemistry of the RTV process is shown in Fig. 6.22. The difunctional polymer, α-ω-dihydroxypolysiloxane can be reacted with MeSi(OAc)$_3$ to generate an acetoxy terminated polysiloxane. Exposure of this material to moisture hydrolyzes the acetoxy-groups with the formation of the Si-(OH)$_2$ end groups. These units self-condense rapidly to afford crosslinked systems. Such systems are known as single-component RTVs (Fig. 6.22).

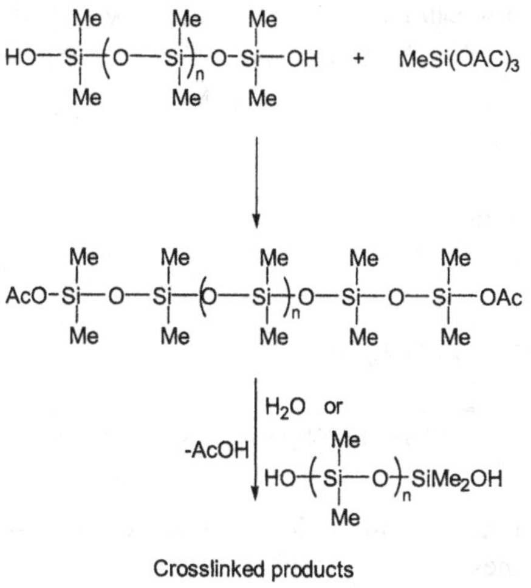

Fig. 6.22. Condensation of an α-ω-dihydroxypolysiloxane with MeSi(OAc)₃ to afford crosslinked products

A similar approach as above is also applicable for ethoxy-terminated siloxanes (Fig. 6.23).

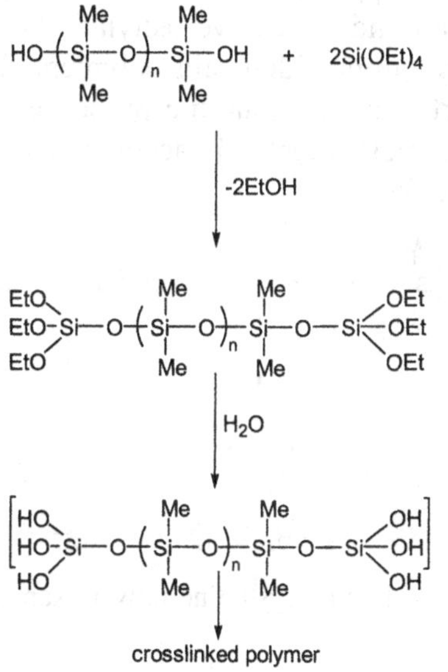

Fig. 6.23. Utilization of ethoxy end-groups for crosslinking reactions

Other type of end-groups such as amino-terminated and epoxy-terminated ones can also be prepared (Fig. 6.24).

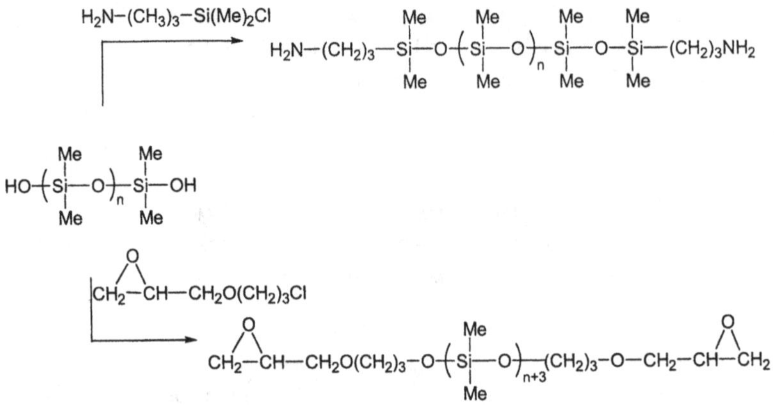

H$_2$N—(CH$_3$)$_3$—Si(Me)$_2$Cl

H$_2$N—(CH$_2$)$_3$—Si—O—(Si—O)$_n$—Si—O—Si—(CH$_2$)$_3$NH$_2$ (with Me groups on each Si)

HO—(Si—O)$_n$—Si—OH (with Me groups)

CH$_2$—CH—CH$_2$O(CH$_2$)$_3$Cl (epoxide)

CH$_2$—CH—CH$_2$O(CH$_2$)$_3$—O—(Si—O)$_{n+3}$—(CH$_2$)$_3$—O—CH$_2$—CH—CH$_2$ (with Me groups, epoxides)

Fig. 6.24. Modification of α-ω-dihydroxypolysiloxane to α-ω-didiamino- and α-ω-bis-epoxypolysiloxanes

The examples shown above illustrate the fact that polysiloxanes can be easily modified to incorporate various reactive functional groups which in subsequent reactions can be effectively utilized for generating cross-linked matrices.

Many other types of applications are also recently being exploited keeping these reactions in mind. Thus, hydrosilylation reactions have been used to prepare liquid-crystalline side-chain polymers. The idea is to utilize a mesogen that imparts liquid-crystalline properties as a side-chain of the polysiloxane. This is most effectively accomplished by the hydrosilylation reaction [28] (Fig. 6.25).

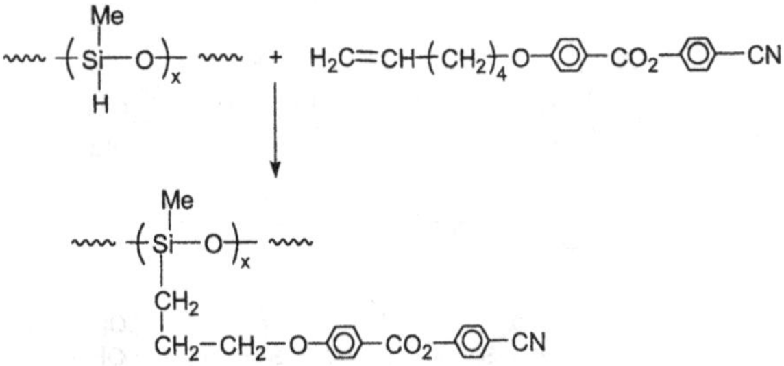

~~~(Si—O)$_x$~~~ + H$_2$C=CH—(CH$_2$)$_4$—O—⟨O⟩—CO$_2$—⟨O⟩—CN

~~~(Si—O)$_x$~~~ with side-chain CH$_2$—CH$_2$—CH$_2$—O—⟨O⟩—CO$_2$—⟨O⟩—CN

Fig. 6.25. Preparation of liquid-crystalline polysiloxanes by placing mesogens in the side-chains

6.6 Hybrid Polymers

Polymers containing polysiloxane segments and other types of backbones are of interest in generating new hybrid polymers. These polymers are expected to possess new properties which could be an increase in glass transition temperatures, increase in the thermal stability etc. There have been efforts to prepare such polymers by incorporation of rigid structural segments adjacent to the flexible siloxane units. A successful approach in this regard is the assembly of *exactly alternating* silarylene-siloxane copolymers [29-31]. These polymers are synthesized by using a step-growth condensation polymerization method. The silarylene unit is introduced by the monomer OH-Si(Me$_2$)-C$_6$H$_4$-p-Si(Me$_2$)-OH (Fig. 6.26) while the siloxane is introduced by the diueridosilane, Me$_2$Si(Ureido)$_2$. Condensation of these reagents in a solvent such as chlorobenzene leads to the elimination of urea as an insoluble by-product leaving the exactly alternating copolymer in solution (Fig. 6.27). High molecular weights (M_w's 100,000-300,000) and good polydispersity indices (1.70 - 1.90) are a feature of this reaction. Other types of silarylene monomers such as OH-Si(Me$_2$)-C$_6$H$_4$-p-O-C$_6$H$_4$-p-Si(Me$_2$)-OH have also been used [32].

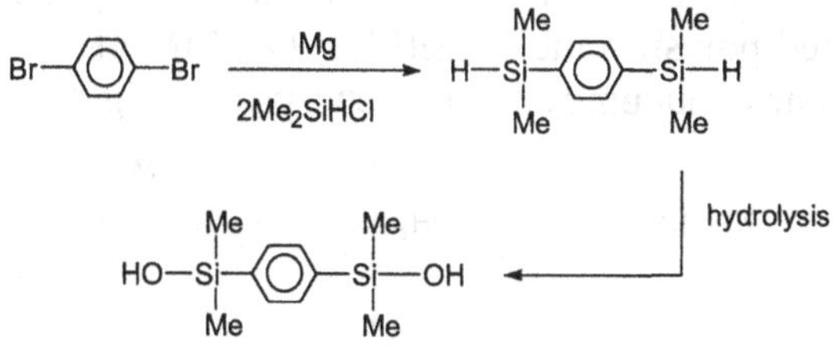

Fig. 6.26. Preparation of the monomer OHSiMe$_2$-C$_6$H$_4$-p-SiMe$_2$OH

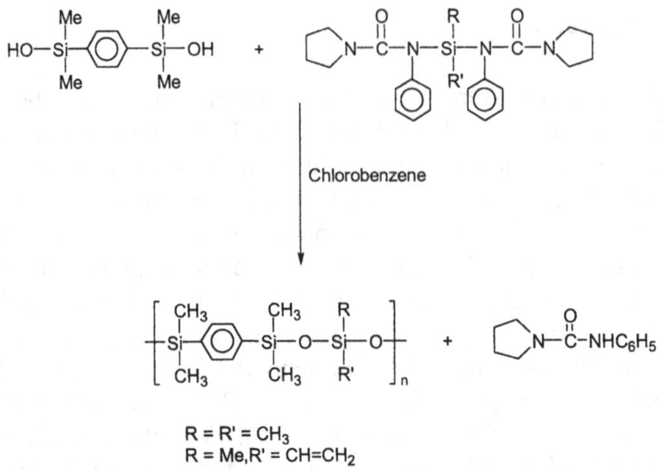

Fig. 6.27. Condensation of OHSiMe$_2$-C$_6$H$_4$-p-SiMe$_2$OH with RR'Si(ureido)$_2$ to afford exactly alternating hybrid copolymers

There have been other types of approaches also for the synthesis of hybrid polymers. Block copolymers containing polysiloxane segments and polyphosphazene segments have been synthesized by the reaction of hydride-terminated polysiloxane H-[Si(Me$_2$)-O]$_n$-SiMe$_2$H with the telechelic polyphosphazene containing an amino end-group (Fig. 6.28) [33].

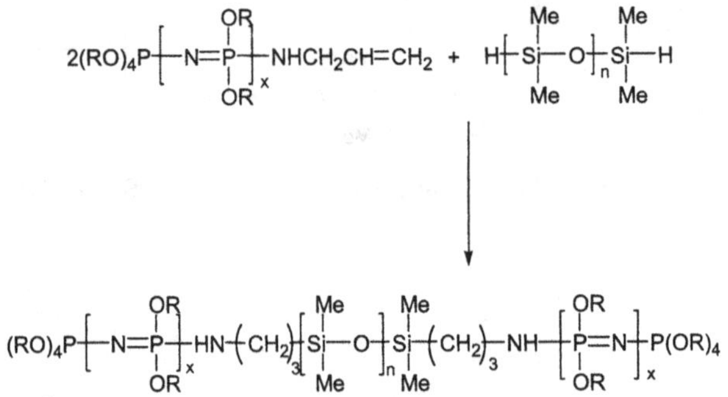

Fig. 6.28. Poly(phosphazene-siloxane) block copolymers

6.7 Properties of Polysiloxanes

6.7.1 Glass-Transition Temperatures and Conformational Flexibility

Polysiloxanes are among the most flexible polymers known and this is reflected by the very low glass-transition temperatures of this polymer family. Data for some representative examples of polysiloxanes along with some related polymers are given in Table 6.2.

Table 6.2. Glass-transition temperatures of polysiloxanes and some related polymers

S.No.	Polymer	T_g (°C)	S.No.	Polymer	T_g (°C)
1	$[Me_2SiO]_n$	-123	7	$[Me_2SiOSiMe_2\text{-}Ar\text{-}SiMe_2]_n$	-61 to -65
2	$[MePhSiO]_n$	-33	8	$[(Me_2SiO)(Ph_2SiO)]_n$	-30
3	$[MeHSiO]_n$	-138	9	$[NPCl_2]_n$	-66
4	$[Et_2SiO]_n$	-139	10	$[NP(CH_3)_2]_n$	-46
5	$[nBu_2SiO]_n$	-116	11	$[C(CH_3)_2CH_2]_n$	-70
6	$[Ph_2SiO]_n$	+49	12	Natural rubber	-72

The most well-studied polysiloxane viz., $[Me_2SiO]_n$ has a T_g of -123 °C [1]. This is one of the lowest values for any polymer. As mentioned earlier in Chapter 2 the glass-transition temperature of a polymer corresponds to a description of its amorphous state. The T_g of a polymer can be taken as a measure of the torsional freedom of polymer chain segments. Above its T_g a polymer has reorientational freedom of motion of its chain segments, while below its T_g this is frozen. Usually elastomeric materials have low glass-transition temperatures. For example, natural rubber has a glass-transition temperature of -72 °C. Similarly, polyisobutylene has a glass-transition temperature of -70 °C. Thus, poly(dimethylsiloxane) has considerably lower T_g than either of these well-known elastomers. It is also informative to compare the T_g of poly(dimethylsiloxane) with the isoelectronic polyphosphazene, $[NP(CH_3)_2]_n$. The latter has a T_g of -46 °C.

The effect of replacing the methyl substituents in PDMS by other rigid substituents has the expected effect. For example, poly(diphenylsiloxane) has a T_g of +49 °C. Replacement of only one methyl group per silicon in PDMS in the polymer $[MePhSiO]_n$ also increases the glass-transition temperature (entry 2, Table 6.2). This effect is also seen in random copolymers (entry 8, Table 6.2) as well as exactly alternating copolymers (entry 7, Table 6.2). Interestingly, replacing the methyl group in PDMS by the smaller substituent 'H' as in the polymer $[MeHSiO]_n$ lowers the T_g even further (-139°C, entry 3, Table 6.2) [34]. What are the main reasons for the unusual skeletal flexibility of PDMS? If we examine the structural characteristics of this polymer this would become clear.

The Si-O bond distance in PDMS is about 1.64 Å and the Si-C distance is about 1.87-1.90 Å. These bond distances are much longer than what are normally encountered in organic polymers (average C-C distance, 1.54 Å). This means that the substituents on silicon are farther away from the main chain in comparison to the situation present in organic polymers. Secondly, unlike in the latter in polysiloxanes only alternate atoms in the backbone (viz., silicon) have the substituents. This feature also allows considerable mobility and torsional freedom for the backbone. We might recall that polyphosphazenes also have a similar structural feature. Additionally, in polysiloxanes the Si-O-Si bond angle is quite adjustable and can be varied over a wide range. This feature is also found in silicate minerals. In linear polysiloxanes the observed Si-O-Si values are between 140-150°. This may be contrasted with the nearly rigid tetrahedral values found at silicon. Further, studies carried out on determining the shape and size of polysiloxane coils reveal that these macromolecules have a virtually unrestricted rotation about the main-chain bonds and can be approximated to ideal *freely rotating chains* [35]. Lastly, intermolecular interactions in PDMS are very low. This has been accounted in terms of the molecular structure of PDMS. The latter is believed to have an *all-trans* conformation (Fig. 6.29) although some recent studies also favor consideration of a *trans-syn* structure [36].

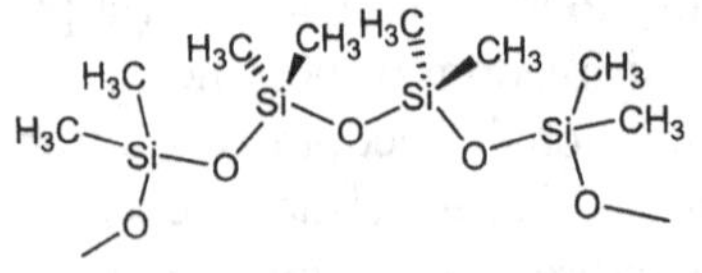

Fig. 6.29. The *all-trans* conformation of polydimethylsiloxane

Irrespective of the subtle difference in the *all-trans* or *trans-syn struc-tures* that PDMS possesses at a local level, its overall structure is believed to have a regular coiled helical conformation where the backbone Si-O bonds are effectively shielded from outside influence by the umbrella-type protection afforded by the lipophilic sheath of methyl groups [35]. Since the methyl groups are nonpolar, intermolecular interactions are minimum and this also accounts for the low melting temperature seen for poly(dimethylsiloxane) (-40 °C).

The unusual conformational behavior of polysiloxanes and its implica-tion on the polymer properties can be summarized as follows.

1) Many polysiloxanes have very low T_g values.
2) The Si-O and Si-C bond lengths are considerably longer than those (C-O or C-C) found in organic polymers.
3) Although the O-Si-O bond angles are rigid and are close to ideal tet-rahedral values, the Si-O-Si angles are very wide. Further, the capa-bility of the siloxane oxygen to adapt itself to various geometric situations is documented by the wide variation in Si-O-Si bond an-gles. Thus, Si-O-Si bond angles can be considered as soft and allow facile torsional freedom.
4) Only alternate atoms in the backbone of the siloxane polymer have side-groups. Thus, while silicon atoms have two side-groups, oxy-gen has none. This feature is not encountered in organic polymers and probably accounts for less steric congestion which also contrib-utes to the skeletal flexibility.
5) Conformational studies on poly(dimethylsiloxane) suggests that it adopts an *all-trans* (or *trans-syn*) local structure and that it has an overall regular helical coiled conformation. Within these coils the inner Si-O backbone is effectively shielded by the outer lipophilic methyl side-groups. Lack of intermolecular interaction between the methyl groups also accounts for low crystallinity in these polymers.

These characteristics of polysiloxanes are also reflected in their many other properties such as their high water repellency (hydrophobicity), low surface tension etc. Among the other desirable properties of polysiloxanes are their high gas permeability, thermal stability, oxidative stability and low volatility. In addition, their fire points are quite high and so they do not present a fire hazard [1, 5, 7]. Also, polysiloxanes have high biocom-patibility and very low toxicity. These properties of polysiloxanes have helped them to carve out a niche for themselves in the area of polymers and have helped their ever expanding utility some of which are outlined in the following section.

6.8 Applications of Polysiloxanes

It is quite difficult to enumerate all the possible applications of polysiloxanes. The reader is directed to other sources that are devoted to the technology and applications of polysiloxanes [1-2, 4-5]. The following serves to acquaint the reader with the large range of applications that these polymers possess.

Conventional utility of silicone oils has been commented upon already. Thus, these possess a low VTC and this enables their utility as hydraulic fluids for machinery that operate at very low temperatures. Silicone oils are used routinely as heat-transfer fluids and dielectric insulating materials.

The low surface tension of silicone oils have allowed their use as anti-foam agents in various applications such as textile dyeing, fermentation, fruit juice processing, antacid tablets and so on.

Silicone elastomers have exceptional stability to both high as well as low temperatures. Thus, silicone rubbers are quite flexible at temperatures as low as -60 to -70 °C. They do not seem to suffer much degradation up to 300 °C. Silicone resins find electrical and non electrical applications. Thus, these are used in the insulation of electrical equipment and for laminating printed circuit boards.

The biocompatibility and low toxicity of polysiloxanes have allowed their utility in artificial body organs, artificial skin, soft-contact lenses etc.

There are many emerging applications of silicone elastomers. Although these are also numerous we will present two examples that will serve to illustrate the versatility of polysiloxanes.

The utility of polysiloxanes that contain functional groups has already been mentioned. These functional groups can be used in many ways. Recently there have been efforts to use these as carriers for luminescent oxygen sensors. These are useful in various types of applications such as monitoring air pressures, visualizing flow particularly in aerospace applications or monitoring oxygen levels in vivo. The principle of the application is as follows. A luminescent inorganic complex is attached either covalently to a polymer support or is blended physically with a polymer. The important criterion is that the polymer should be permeable to oxygen. The triplet state of the oxygen quenches the excited luminophore (the inorganic complex or dye molecule) and reduction of intensity of the luminophore can be correlated quantitatively with the amount of oxygen present. If required, this data can be further correlated to the air pressure. Many types of luminophores are being tested for their applicability. These include phenanthroline ruthenium complexes; we have mentioned these in Chap. 5 in connection with the utility of poly(thionylphosphazene)s for a similar application. Recently, a cyclometalated iridium complex has been attached to a polysiloxane backbone through the hydrosilylation reaction. The design of the ligands around the metal complex has been carried out keeping in view that one of the ligands should be able to participate in a hydrosilylation reaction. This means that a suitable ligand should be present around the metal coordination sphere that would allow such a reaction. Use of vinylpyridine enables the Ir(III) complex to be attached to polysiloxane through the hydrosilylation reaction [37] (Fig. 6.30).

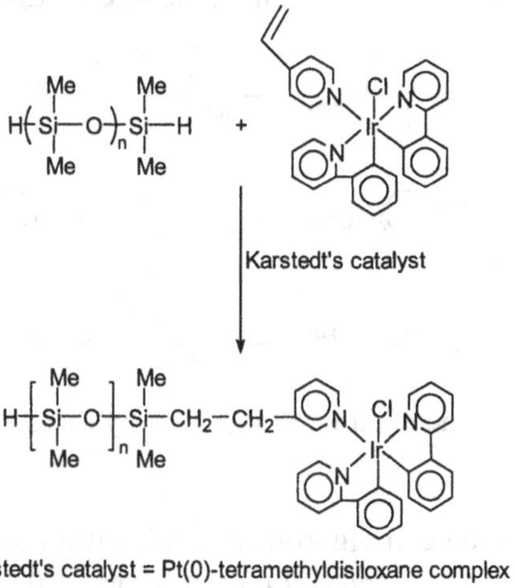
Karstedt's catalyst
Karstedt's catalyst = Pt(0)-tetramethyldisiloxane complex

Fig. 6.30. Anchoring a photoluminescent transition metal complex on the polysiloxane backbone

It was observed that the oxygen sensitivity of a thin film of the PDMS-Ir complex was very high. Blends of this polymeric complex with polystyrene have also been found to be quite effective [37].

The utility of poly(methylhydrosiloxane) $[MeHSiO]_n$ as a reducing agent in organic reactions is attracting attention [38-39]. Recently, combination of $[MeHSiO]_n$, and *in situ* generated metal hydrides (such as ZnH_2 generated *in situ* by the reaction of zinc 2-ethylhexanoate with $NaBH_4$) have been found to be quite effective as hydrosilylation/reduction reagents for a variety of organic compounds such as aldehydes, ketones, esters, triglycerides and epoxides (Fig. 6.31) [38].

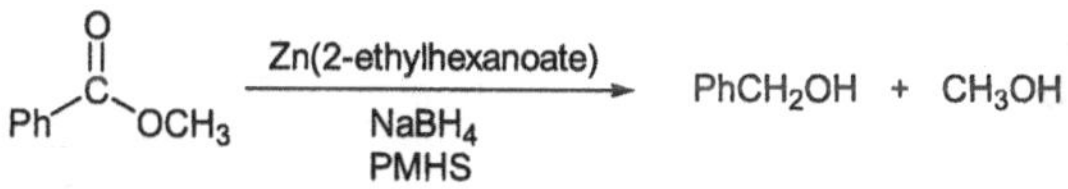
Zn(2-ethylhexanoate)
NaBH4
PMHS
PhCH2OH + CH3OH

Fig. 6.31. Reduction of esters by the use poly(methylhydrosiloxane) $[MeHSiO]_n$

In addition to mild reaction conditions, this procedure is also highly selective. For example, the reaction of methylbenzoate with this reagent combination affords benzyl alcohol exclusively (Fig. 6.31).

Utilization of a combination of Bu_3SnCl, KF and PMHS has been found to be effective to generate *in situ* the potent reducing agent Bu_3SnH. Thus, this combination of reagents can be used in a catalytic manner for the reduction of various organic substrates (Fig. 6.32). Although the role of KF is not understood very well, it is believed that it might activate the silicon center by forming an hypervalent silane intermediate [39].

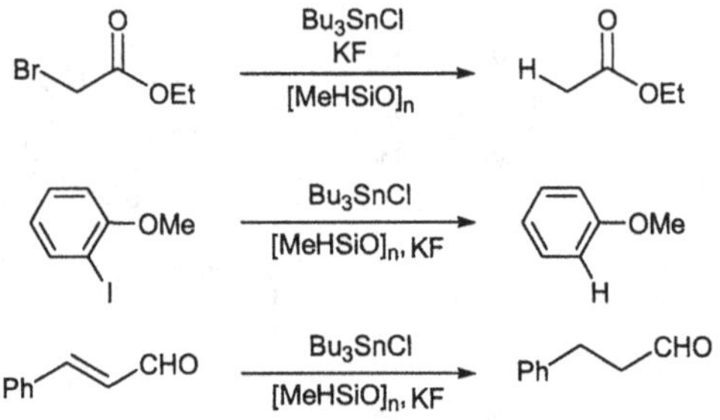

Fig. 6.32. *In situ* generation of triorganotin hydrides and their application as reducing agents

There have also been vigorous efforts to use poly(dimethylsiloxane) as a material for fabricating *microfluidic* devices [40]. This application is based on the following principle. RTV PMDS rubbers have excellent interfacial properties such as low surface energy. This allows these materials to be used as replicas for molds. Thus, molds containing various patterns can be prepared (masters) and PMDS is allowed to cure in the mold surface. Peeling off the PMDS from the mold is very facile. Conventionally, such a procedure has allowed the use PMDS rubbers for larger-scale replication of molds containing intricate designs. More recently there have been efforts to extend this principle on a smaller scale for generating rapid prototypes for microchannels. The limiting factor in this application is the resolution of the master itself. Resolutions of the order of 20 µm have been achieved. This methodology is believed to be useful for the generation of various types of microfluidic devices such as valves, pumps etc.

References

1. Clarson SJ, Semlyen JA (eds) (1993) Siloxane polymers. Prentice Hall, Englewood Cliffs

2. Jones RG, Ando W, Chojnowski J (eds) (2000) Silicon-containing polymers. Kluwer Acad., Dordrecht
3. Mark JE, West R, Allcock HR (1992) Inorganic polymers. Prentice-Hall, Englewood Cliffs
4. Noll W (1968) Chemistry and technology of silicones. Academic Press, New York
5. Rochow EG (1987) Silicon and silicones. Springer-Verlag, Heidelberg
6. Lebrun JJ, Porte H (1993). Polysiloxanes. In: Eastmond GC, Ledwith A, Russo S, Sigwalt P (eds) Comprehensive polymer science, Vol 4, Pergamon, Oxford, pp 593-609
7. Seyferth D (2001) Organometallics 20:4978
8. Greenwood NN, Earnshaw A (1984) Chemistry of the elements. Pergamon, Oxford
9. Wells AF (1975) Structural inorganic chemistry. 4th Ed. Oxford University Press, Oxford
10. Chvalovsky V (1987). In Haiduc I, Sowerby DB (eds) The chemistry of inorganic homo- and heterocycles. Academic Press, London, pp 287-348
11. Madagan RJAR, Nieuwenhuyzn M, Wilkins CJ, Williamson BE (1998) Dalton Trans 2697
12. Okazaki R, West R (1996) Adv Organomet Chem 39:231
13. Lukevics E, Pudova O, Sturkovich R (1989) Molecular Structure of organosilicon compounds. Ellis Harwood, Chichester
14. Seyferth D, Prud'hamme C, Wiseman GH (1983) Inorg Chem 22:2163
15. Braun D, Cherdron H, Ritter H (2001) Polymer synthesis: theory and practice. 3rd Ed. Springer-Verlag, Heidelberg, Berlin, New York
16. Weber WP, Cai G (2001) Macromolecules 34:4355
17. Kendrick TC, Parbhoo BM, White JW (1993). In: Clarson SJ, Semlyen JA(eds) Siloxane polymers. Prentice Hall, Englewood Clifs, pp 459-523
18. Chojnowski J (1993). In: Clarson SJ, Semlyen JA (eds) Siloxane polymers. Prentice Hall, Englewood Clifs, pp 1-71
19. Semlyen JA (1993). In: Clarson SJ, Semlyen JA (eds) Siloxane polymers. Prentice Hall, Englewood Clifs, pp 135-192
20. Wang Q, Zhang H, Surya Prakash GK, Hogen-Esch TE, Olah G (1996) Macromolecules 29:6691
21. Toskas G, Moraeu M, Masure M, Sigwalt P (2001) Macromolecules 34:4730
22. Kim KC, Reed CA, Elliott DW, Mueller LJ, Tham F, Lin L, Lambert JB (2002 Science 277:825
23. Kennan JJ (1993). In: Clarson SJ, Semlyen JA (eds) Siloxane polymers. Prentice Hall, Englewood Clifs, pp 72-134
24. Ziemelis MJ, Saam JC (1989) Macromolecules 22:2111
25. Rózge-Wijas KR, Chojnowski J, Zundel T, Boileau S (1996) Macromolecules 29:2711
26. Babu GN, Christopher SS, Newmark RA (1987) Macromolecules 20:2654
27. White JW, Treadgold RC (1993). In: Clarson SJ, Semlyen JA (eds) Siloxane polymers. Prentice Hall, Englewood Cliffs, pp 193-215

28. White MS (1993). In: Clarson SJ, Semlyen JA (eds) Siloxane polymers. Prentice Hall, Englewood Clifs, pp 245-308
29. Dvornic PR, Lenz RW (1982) J Polym Sci Polym Chem 20:951
30. Lai YC, Dvornic PR, Lenz RW (1982) J Polym Sci Polym Chem 20:2277
31. Dvornic PR, Lenz RW (1994) Macromolecules 27:5833
32. Dvornic PR, Lenz RW (1982) J Polym Sci Polym Chem 20:593
33. Allcock HR, Prange R (2001) Macromolecules 34:6858
34. Clarson SJ (1993). In: Clarson SJ, Semlyen JA (eds) Siloxane polymers. Prentice Hall, Englewood Clifs, pp 216-244
35. Dvornic PR, Lenz RW (1992) Macromolecules 25:3769
36. Darsey JA (1990) Macromolecules 23:5274
37. De Rosa MC, Mosher PJ, Yap GPA, Focsaneanu KS, Crutchley RJ, Evans CEB (2003) Inorg Chem 42:4864
38. Mimoun H (1999) J Org Chem 64:2582
39. Terstiege I, Maleczka Jr RE (1999) J Org Chem 64:342
40. Cooper McDonald J, Whitesides GM (2002) Acc Chem Res 35:491

7 Polysilanes and Other Silicon-Containing Polymers

7.1 Introduction

Polysilanes are polymers containing catenated silicon chains. These polymers have been attracting considerable attention in recent years because of their interesting electronic and photophysical properties [1-5]. Homopolymers of polysilanes can be of the type $[(n\text{-}C_6H_{13})_2Si]_n$, where similar side chains are present on silicon, or of the type $[PhMeSi]_n$, where dissimilar side chains are present. Copolymers of polysilanes can be of the type $[(PhMeSi)_x(Me_2Si)_y]$ where there is a random arrangement of the two different types of organosilicon units or of the type $[Me_2Si\text{-}Si(n\text{Hex})_2]_n$ where the two types of organosilicon units alternate with each other in the polymer chain (Fig. 7.1). Polysilanes are a large family of inorganic polymers. There is considerable diversity in their structure which is brought about by a variation of the side chain.

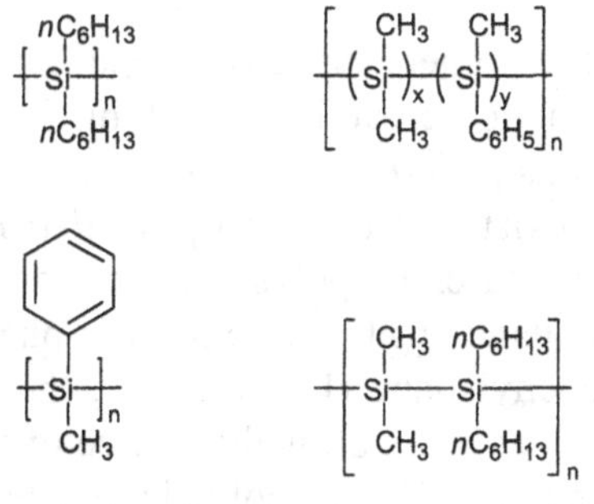

Fig. 7.1. Representative examples of polysilanes

Polysilanes are quite different from the other inorganic polymers that we have discussed in this book so far. Unlike polyphosphazenes, $[R_2P=N]_n$, or polysiloxanes, $[R_2Si\text{-}O]_n$ which contain two different elements that alternate with each other in their polymeric chains, polysilanes contain *only one element* viz., *silicon* in the main chain of the polymer. In some respects this is somewhat similar to what is found in many organic polymers such as polyethylene. However, even in many organic polymers which contain catenated carbon atoms, the adjacent atoms in the main chain usually have *different* side chains. Thus, for example, while polymers such as polyethylene, polytetrafluoroethylene, and polyacetylene are built of repeat units where *every* adjacent carbon atom has the *same* substituent or side chain (this is at least the situation in *perfectly linear* chains) most of the other organic polymers contain *dissimilar* side chains on adjacent carbon atoms as, for example, in poly(isobutylene), polystyrene and poly(vinylchloride) (Fig. 7.2).

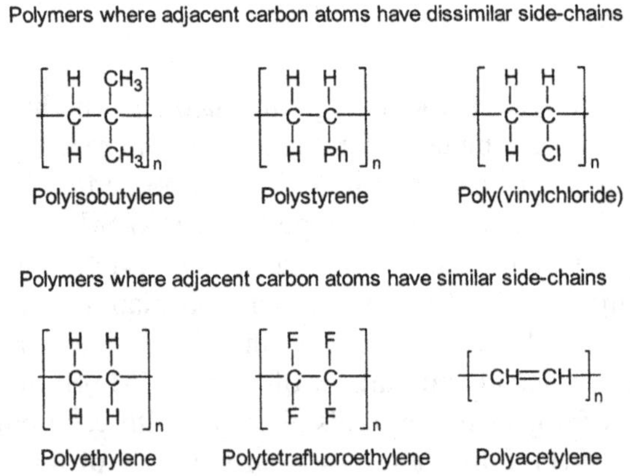

Polymers where adjacent carbon atoms have dissimilar side-chains

Polyisobutylene   Polystyrene   Poly(vinylchloride)

Polymers where adjacent carbon atoms have similar side-chains

Polyethylene   Polytetrafluoroethylene   Polyacetylene

Fig. 7.2. Examples of organic polymers containing similar and dissimilar side chains on adjacent backbone atoms

A small note on the nomenclature of polysilanes: There seems to be a general consensus among researchers working in this area to call these catenated silicon chains *polysilanes*. However, alternative descriptions of these polymers also exist and the term *polysilylenes* is also used. Since most preparative procedures of polysilanes utilize organosilanes such as dicyclohexyldichlorosilane $[(nC_6H_{13})_2SiCl_2)]$, phenylmethyldichlorodisilane $(PhMeSiCl_2)$, phenylsilane $(PhSiH_3)$ etc., the term *polysilanes* is an appropriate name and has as its basis the source of the polymer [6]. This is similar to referring to $[CH_2CH_2]_n$ as polyethylene because its source is ethylene (and not polymethylene although it *does* have a methylene repeat unit). The *repeat unit-based* nomenclature of $[R_2Si]_n$ viz., polysilylene would base the name on the repeat unit R_2Si which is a *silylene* (analogous to a carbene, R_2C). We will be using the term *polysilane* in this book to describe this family of polymers.

In addition to polysilanes we will also briefly look at some other polymer systems that contain silicon. These are known as polysilynes or network polysilanes, polycarbosilanes, and polysiloles.

7.2 Historical

The foundation of polysilane chemistry was laid by F. S. Kipping and his coworkers. Investigations by these researchers on the reactions of alkali metals with diorganodichlorosilanes in general and diphenyldichlorosilane in particular led them to isolate a number of cyclic rings. For example, the cyclotetrasilane $[Ph_2Si]_4$ was a prominent product that was isolated by Kipping and his coworkers (see Eq. 7.1) [1, 4].

$$Ph_2SiCl_2 \xrightarrow{Li} \text{(cyclotetrasilane structure)} \tag{7.1}$$

Research work carried out much later also indicated that cyclic products dominate the reactions between alkali metals (particularly lithium) and dialkyl/diaryldichlorosilanes. Even the reactions of Na/K alloy with dimethyldichlorosilane leads to the formation of the six-membered ring $[Me_2Si]_6$. If this reaction is carried out under controlled conditions using adequate dilution, a continuous range of cyclic products $[Me_2Si]_n$ (n = 5-35) can be obtained (see Eq. 7.2) [7].

$$Me_2SiCl_2 \xrightarrow[THF]{Na/K} \text{(ring structure)} + Cyclo\text{-}(Me_2Si)_n \qquad n = 5, 7\text{-}35 \tag{7.2}$$

The first example of a polymeric product involving catenated silicon atoms was described by Burkhard in 1949 [8]. He prepared poly(dimethylsilane) $[Me_2Si]_n$ by the following procedure. He took about 450 g of sodium metal along with 700 g of Me_2SiCl_2 and one liter of benzene, sealed them in an autoclave, and heated them. High temperatures (200 °C) and high pressures (1517 kPa) developed in this reaction. Most of the product from this reaction was an intractable solid which was identified as $[Me_2Si]_n$. It was not soluble in any common organic solvent and thus precluded a study of its solution properties.

Although not much work was carried out on polysilanes since the original research of Burkhard, interest in this area was revived by the disclosure of Yazima and coworkers that $[Me_2Si]_n$ could be transformed into silicon carbide fibers [9-11]. Thermal rearrangement of the insoluble $[Me_2Si]_n$ affords a soluble polymer, viz., polycarbosilane, $[MeHSiCH_2]_n$. The latter is crosslinked (by heating in air) and heated at 1300°C to afford silicon carbide fibers containing strength reinforcing by β-SiC crystallites (Fig. 7.3).

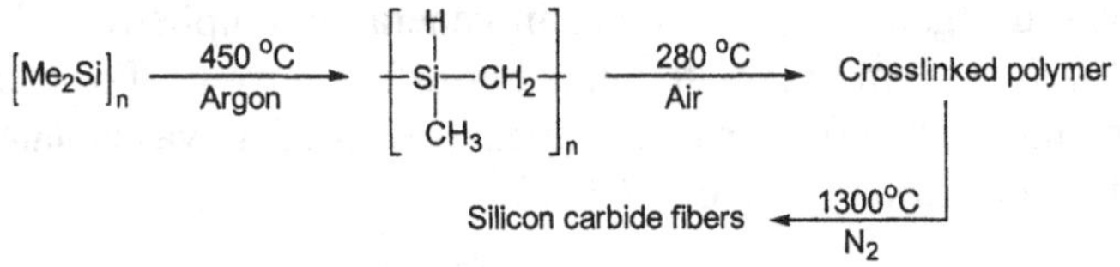

Fig. 7.3. Conversion of $[Me_2Si]_n$ into β-SiC

Yajima's trendsetting work has not only renewed interest in polysilanes but also in other polymeric materials that can function as precursors for ceramics such as silicon nitride, Si_3N_4 and boron nitride, BN [6].

Before we consider the synthesis of polysilanes in the next section, let us consider if addition polymerization involving disilene monomers can be used for the preparation of polysilanes. Robert West from the University of Wisconsin prepared the first stable disilene in 1981. His synthetic route consisted of photolyzing an acyclic trisilane. Loss of $Me_3Si-SiMe_3$ leads to a silylene intermediate that dimerizes to the stable disilene (Fig. 7.4) [12-13].

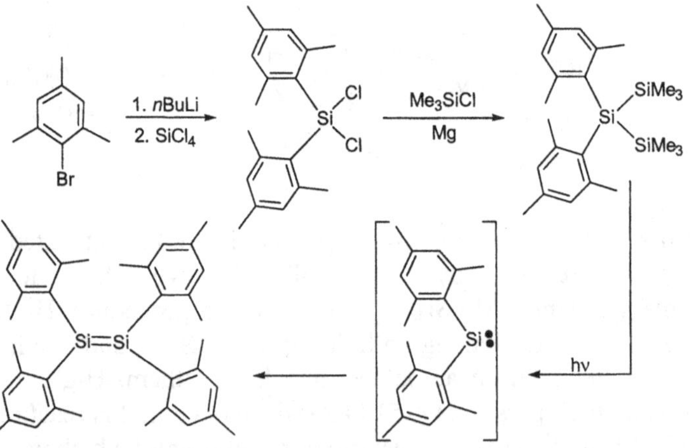

Fig. 7.4. Synthesis of a stable disilene by the photolysis of $[Mes_2Si(SiMe_3)_2]$

Other alternative ways of preparing disilenes are now known. Thus, Masamune and his coworkers found that the disilene $[(2,6\text{-}Et_2C_6H_3)_2Si]_2$ could be isolated in the photolytic reaction involving the cyclotrisilane $[(2,6\text{-}Et_2C_6H_3)_2Si]_3$ (Fig. 7.5) [12].

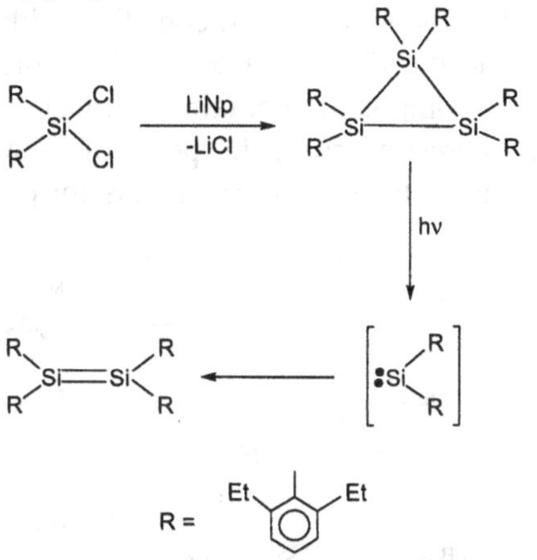

Fig. 7.5. Synthesis of a stable disilene from a cyclotrisilane

Using these and other strategies many other disilenes have been prepared. These include disilenes that contain two different substituents on silicon. The resulting E and Z isomers have also been isolated in some instances (Fig. 7.6) [12-13].

R= CH(SiMe₃)₂

Fig. 7.6. Some examples of E and Z disilenes

Although many disilenes have now been prepared and characterized they are not very good monomers for the preparation of polysilanes. The very method of preparing these kinetically stabilized disilenes defeats their use *as olefin-like monomers* for addition polymerization. However, as will be shown in the next section *trapped disilenes* or *masked disilenes* that contain appropriate substituents on silicon can be polymerized to polysilanes [5].

More recently, stable silylenes which are structurally the repeating unit of a polysilane chain have been prepared and characterized [14-15]. Their synthesis involves the reduction of silicon dichlorides. Two methods of synthesis are shown in the Fig. 7.7. Both of these procedures involve the preparation of cyclic compounds containing Si(II). Although these compounds are highly interesting from the point of view of their chemistry, they have not been shown to serve as precursors for polymers.

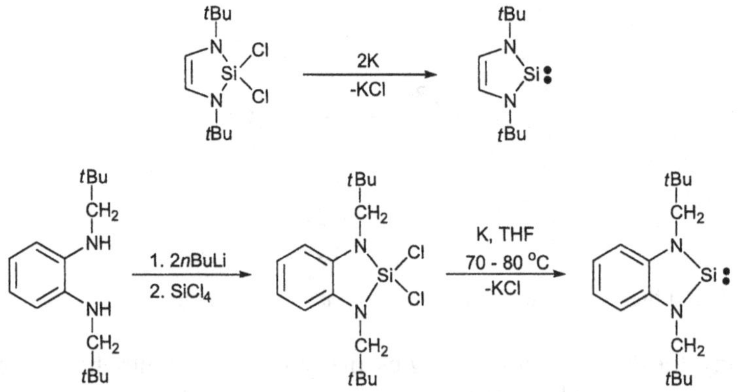

Fig. 7.7. Synthesis of stable silylenes

7.3 Synthesis of Polysilanes

The completely inorganic polysilane (containing only silicon and chlorine), perchloropolysilane $[SiCl_2]_n$, is prepared by a sublimation of the cyclic silane Si_4Cl_8 [16]. The polymerization was proposed to proceed through a diradical species generated through homolytic cleavage of the Si-Si bond and the consequent ring opening of the cyclic silane. Combination of the diradicals leads to the formation of the polymeric dichlorosilane $[SiCl_2]_n$ (Fig. 7.8).

The polymer $[SiCl_2]_n$ is not soluble in any organic solvent and is extremely moisture sensitive. However, it has been structurally characterized by X-ray crystallography and solid-state ^{29}Si NMR. Thus, the X-ray crystal structure of this polymer shows an *all-trans* conformation. The ^{29}Si-NMR spectrum of this compound in the solid state shows a resonance at δ 3.9. The polymer $[SiCl_2]_n$ has two reactive chlorine atoms on silicon and therefore the possibility of replacing these by other substituents by a nucleophilic substitution reaction is present. However, there have not been many reports on this. This may be due to the experimental difficulty in implementing the substitution reaction procedure.

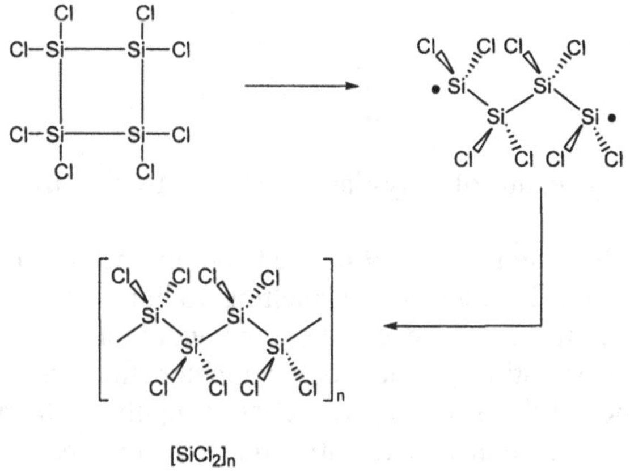

[SiCl₂]ₙ

Fig. 7.8. Synthesis of $[SiCl_2]_n$

Other polysilanes that contain organic side-groups can be prepared essentially by the following procedures.

1. Wurtz-type coupling reaction involving an alkali metal (usually sodium)-assisted dehalogenation of diorganodichlorosilanes, $RR'SiCl_2$. This is essentially the same procedure that Burkhard had originally employed in the preparation of $[Me_2Si]_n$ [1-5, 17].
2. Anionic polymerization of masked disilenes and strained cyclosilanes [18]
3. Catalytic dehydrogenation of organosilanes such as $PhSiH_3$ [19-21].

In addition to these general synthetic procedures, other methods of polysilane synthesis include electrochemical reduction of diorganodichlorosilanes [22], condensation reactions between appropriate acyclic difunctional reagents such as $Li(Ph_2Si)_5Li$ and $Cl(Me_2Si)_5Cl$ [4]. These latter methods are not discussed here.

7.3.1 Synthesis of Polysilanes by Wurtz-type Coupling Reactions

In spite of its many drawbacks (as will be evident during the course of this discussion), the alkali metal-assisted Wurtz-type coupling reaction remains by far the most widely used procedure for the preparation of polysilanes

[1-5, 17]. This is mainly because of its general applicability to al-kyl/aryldichlorosilanes. The reaction involved can be represented as fol-lows (see Eq. 7.3).

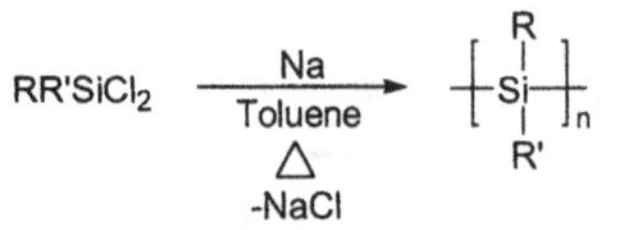

$$RR'SiCl_2 \quad \xrightarrow[\substack{\text{Toluene} \\ \triangle \\ \text{-NaCl}}]{\text{Na}} \quad \left[\begin{matrix} R \\ | \\ Si \\ | \\ R' \end{matrix} \right]_n \tag{7.3}$$

A typical procedure of polysilane synthesis by this methodology is as follows:

Usually a dispersion of finely divided sodium metal is required for the reaction. This can be achieved by melting sodium metal in high-boiling solvents such as toluene, xylene and decane under reflux conditions and by the use of efficient stirring. The diorganodichlorosilane is added to the so-dium dispersion at this stage. This reaction is highly exothermic and needs to be performed carefully. The entire reaction mixture is heated further (typically for 2-4 h), cooled and worked-up. One of the common ways of work-up involves addition of methanol or i-propanol to the reaction mix-ture to destroy the excess of sodium metal. This is followed by precipita-tion of the polymer by the addition of an excess of methanol or i-propanol. A large number of polysilanes such as $[R_2Si]_n$ and $[RR'Si]_n$ have been pre-pared by adopting this general procedure [1, 4, 17]. Representative exam-ples of some of the polysilanes prepared by the Wurtz-type of coupling re-action are shown in Figs. 7.9 and 7.10.

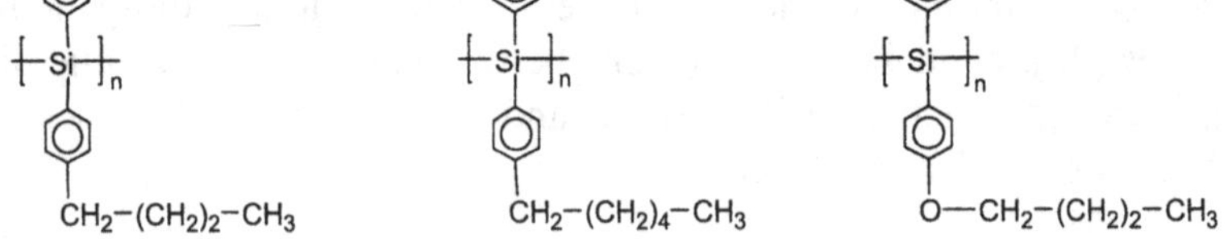

Fig. 7.9. Examples of polysilanes that contain similar substituents on silicon

Figs. 7.9 and 7.10 are self-explanatory. However, a few points are worth mentioning. Although $[Me_2Si]_n$ is a highly crystalline and intractable solid, other poly(dialkylsilane)s such as $[(n\text{-}C_5H_{11})_2Si]_n$ [23], $[(n\text{-}C_6H_{13})_2Si]_n$ [24] and polymers such as $[(n\text{-}C_{10}H_{23})_2Si]_n$ [25] which have long alkyl side-chains are soluble in many common organic solvents. Polymers containing aryl substituents can also be prepared [26]. A number of polymers with mixed substituents have also been synthesized (Fig. 7.10) [27]. Even polymers containing trifluoromethyl substituents such as [(CF$_3$CH$_2$CH$_2$)$_2$(Me)Si]$_n$ [28] or those that contain pyrrolyl substituents [Ph(C$_4$H$_4$N)Si]$_n$, [Me(C$_4$H$_4$N)Si]$_n$ and [nBu(C$_4$H$_4$N)Si]$_n$ have been prepared [29].

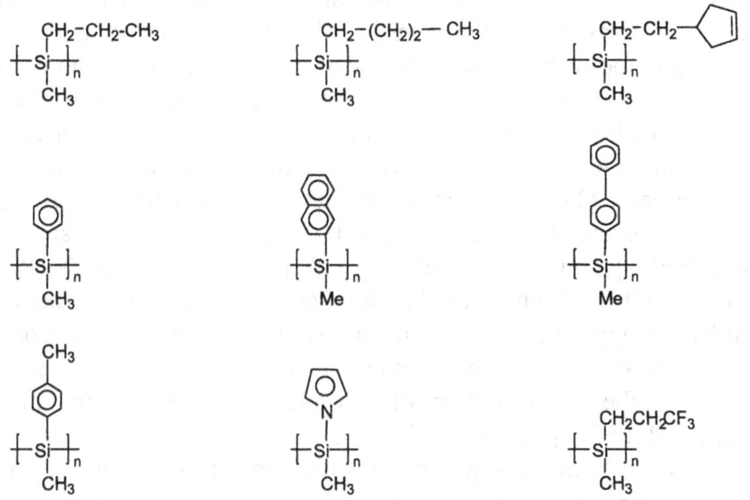

Fig. 7.10. Polysilanes of the type [RR'Si]$_n$ that contain different substituents on silicon

In addition to homopolymers, a number of copolymers can also be prepared by the Wurtz-type coupling methodology [30-34]. This involves the reaction of two diorganodichlorosilanes with sodium. Representative examples of some copolymers are shown in Fig. 7.11.

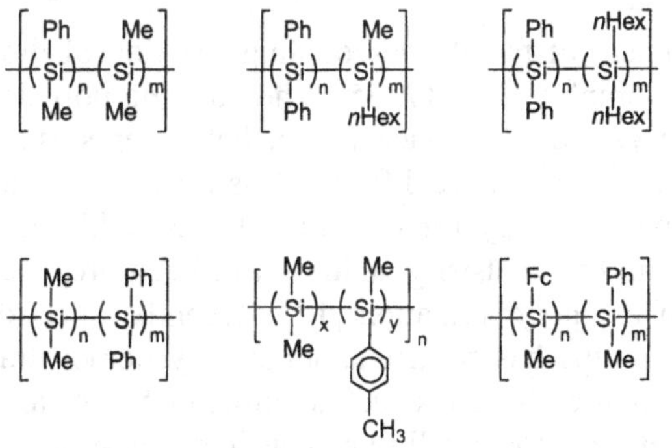

Fig. 7.11. Polysilane random copolymers

Many of the copolymers shown in Figure 7.11 probably have a random arrangement of the monomer segments. Some copolymers such as poly(methylphenylsilane-*co*-dimethylsilane) [{PhMeSi}$_x${Me$_2$Si}$_y$] and poly(methylphenylsilane-*co*-*n*hexylmethylsilane), [{PhMeSi}$_x${(*n*Hex)$_2$Si}$_y$], probably have a *block type* arrangement of the monomer segments. Thus, for example rate studies on the co-polymerization of methylphenyldichlorosilane and dimethyldichlorosilane reveals that the initial rate of disappearance of the former was approximately four times that of the latter [1, 4]. Another interesting observation of the copolymerization reactions is that monomers that cannot be otherwise homopolymerized can be incorporated into copolymers. For example, monomers such as ferrocenylmethyldichlorosilane [34] and sterically hindered compounds such as *t*-butylmethyldichlorosilane cannot be homopolymerized [1]. However, these can be copolymerized along with other monomers to afford co-polymers such as [{(Me)(*t*Bu)Si}$_x${(*n*C$_6$H$_{13}$)(Me)Si}$_y$]$_n$ and [{(Me)(Fc)Si}$_x${(Me)(Ph)Si}]$_n$. Additionally it has been found that the properties of the copolymers can be tuned by the appropriate choice of the nature and ratio of the monomers. Thus, the crystallinity of the homopolymer [Me$_2$Si]$_n$ can be considerably reduced by the incorporation of {(Me)(Ph)Si}units in the copolymer [(Me$_2$Si)$_x$(MePhSi)$_y$] ($x = y = 1$) [1, 30].

Most of the homopolymers and copolymers of polysilanes are usually obtained as air stable solids and are soluble in a range of organic solvents and are generally thermally stable [1]. However, they are sensitive to radiation and undergo scission upon exposure to various kinds of radiation including UV-visible light. This radiation-sensitivity is accentuated in solution. The molecular weights of the high-molecular weight fraction of the polymers can be quite high, ranging from 100,000 to orders of millions [1].

7.3.1.1. General Features of the Wurtz-Coupling Process

The mechanism of the Wurtz-coupling reaction of diorganodichlorosilanes is quite complex [1-4, 17, 35]. The growing polymerization requires access to reactive sodium surface and the heterogeneous nature of the reaction can only make this more difficult. It is for this reason that an efficient stirring is quite necessary for isolation of reasonable polymer yields. Involvement of silyl radicals/silyl anions have been proposed in the reaction mechanism of the polymerization [17]. Recently, spectroscopic evidence for Si-H end-groups has been unambiguously demonstrated [36]. It was further shown that end-groups such as Si-Cl or Si-OR are either absent or their concentration is too small to be detected. These authors propose a sequential anion-radical mode of propagation. The formation of the Si-H end groups supports the presence of silyl radicals that can abstract a proton from the solvent [36]. The possible involvement of silyl anions is supported by the experimental evidence that additives such as 15-crown-5 ether (specific complexing agent for Na^+) has a beneficial effect on the polymerization [4]. A tentative proposal for the process of polymerization is shown in Fig. 7.12. This indicates the formation of a silyl radical which can form a silyl anion and propagation can occur by further reactions with a chlorosilane [17].

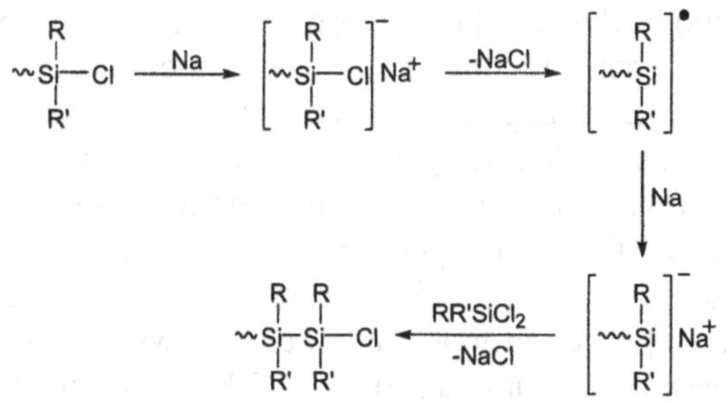

Fig. 7.12. Mechanism of polymerization of polysilanes by the Wurtz-type coupling reaction

The Wurtz-type coupling reaction of diorganodichlorosilanes is characterized by the appearance of an intense blue-purple color in the reaction mixture. Although the initial suggestions for the origin of the blue color related to the possibility of F-centered defects in the sodium chloride that was formed, careful experimentation by EPR and UV-Visible spectroscopy indicated that colloidal sodium formed in the reaction is responsible for the color [37].

The molecular weights of the *as-obtained* polysilanes (without any fractionation) are generally bimodal, many times trimodal [35]. The lowest molecular-weight compounds are usually cyclic compounds with molecular weights lower than 1500. The next higher fraction is composed of oligomers and medium-molecular-weight polymers whose M_w's lie between 4,000-30,000. The highest molecular weight species which are the long chain linear polysilanes have M_w's in the range of 10^5-10^6 with relatively high PDI's. In order to separate the cyclic compounds from the polymeric species repeated re-precipitation procedures are required. Fractionation techniques can separate the medium-and high-molecular-weight polymers. Because of this cumbersome protocol the yields of polysilanes are often quite low and generally range from 5-25%. Only in very few cases the yields are actually greater than 50% [1-4, 17].

7.3.1.2 Improvements to the Wurtz-Coupling Process

Several types of modifications have been tried in the basic Wurtz-coupling process to improve the yields of the high-molecular-weight polymers and also to reduce the polydispersity index. These include the use of additives such as ethyl acetate [38], diglyme [39], 15-crown-5-ether [40], using ultrasound [41-42], use of alternative reducing reagents such as K/18-crown-6-ether [43], graphite-potassium, C_8K [44-46] and so on. While the action of the additives seems to improve the polymer yields, this is at the expense of the higher molecular weights. Ultrasonic reactions have been carried out with a view to disperse the metal at lower temperatures. This seems to be effective but its general applicability remains to be demonstrated.

Use of alternative reducing agents such as C_8K allows the reaction to be carried out at lower temperatures. Thus, phenylmethyldichlorosilane can be polymerized by C_8K at temperatures below 0 °C to afford polymers in reasonable yields (up to 29%) with molecular weights ranging from 5,000-100,000 [44, 45]. Side-groups that are otherwise sensitive to Wurtz-coupling process by sodium can tolerate C_8K. Thus, etheroxy side-chain-containing diorganodichlorosilanes also can be readily polymerized by the use of graphite-potassium to afford water-soluble poly(4,7,10,13-tetraoxatetradecylmethylsilane (see Eq. 7.4) [46].

$$\underset{\underset{Me}{|}}{\overset{\overset{Ph}{|}}{Cl-Si-Cl}} \quad \xrightarrow[\substack{THF \\ T<0\ ^\circ C}]{C_8K} \quad \underset{\underset{Me}{|}}{\overset{\overset{Ph}{|}}{\left[Si\right]_n}} \tag{7.4}$$

$$\underset{\underset{R}{|}}{\overset{\overset{Me}{|}}{Cl-Si-Cl}} \quad \xrightarrow[\text{THF, 0 }^\circ C]{C_8K} \quad \underset{\underset{R}{|}}{\overset{\overset{Me}{|}}{\left[Si\right]_n}}$$

$$R = CH_2CH_2CH_2O(CH_2CH_2O)_3Me$$

Although several improvements have been effected in the basic Wurtz-type reaction it must be mentioned that there is still not a single universal synthetic protocol that allows a high yield, monomodal, high-molecular-weight polysilane synthesis. For every new system that needs investigation, the optimum reaction conditions have to be experimentally determined.

7.3.2 Polymerization by Anionic Initiators

Cyclosilanes can be used as monomers in a ROP to afford linear chains. Although, as mentioned earlier the three-membered cyclic compound

[(2,6-Et$_2$C$_6$H$_3$)$_2$Si]$_3$ affords a stable disilene upon photolysis [12], other strained rings that contain less sterically encumbered substituents have been used for polysilane synthesis. Matyjasjewski and coworkers have targeted strained cyclosilanes as potential candidates for ROP. However, it was observed that the strained four-membered ring octaphenyltetrasilane [Ph$_2$Si]$_4$ cannot be polymerized to the linear polymer. Attempts to polymerize this compound leads to a ring expansion and the formation of the five-membered ring [Ph$_2$Si]$_5$. This led to a modification of the four-membered ring [Ph$_2$Si]$_4$ in the following manner. The reaction of octaphenyltetrasilane with CF$_3$SO$_3$H leads to partial dearylation. This important reaction in organosilicon chemistry can be controlled stoichiometrically to effect the replacement of four *nongeminal* phenyl substituents from the starting cyclic ring [Ph$_2$Si]$_4$ and to afford the functional and reactive silyl triflate [Ph(OSO$_2$CF$_3$)Si]$_4$. Further reaction of this compound with methylmagnesiumbromide leads to 1,2,3,4-tetramethyl-1,2,3,4-tetraphenylcyclotetrasilane [PhMeSi]$_4$ which was shown to exist as four stereoisomers. Fractional crystallization of this compound allowed the isolation of the *all-trans* derivative which has been used for further ring-opening polymerization (Fig. 7.13) [47].

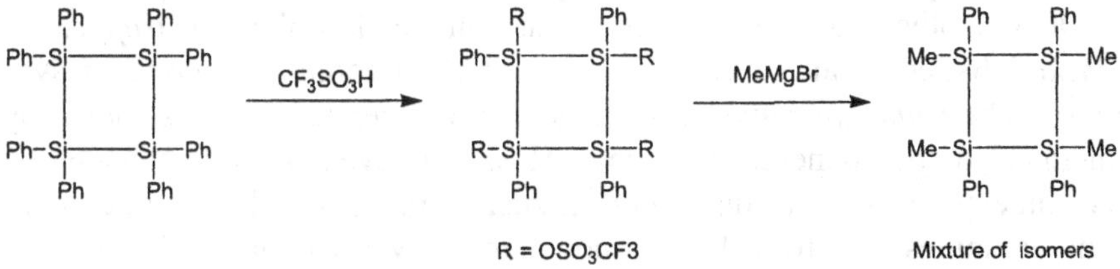

Fig. 7.13. Preparation of the mixed substituent cyclotetrasilane [PhMeSi]$_4$ from the homogeneously substituted derivative [Ph$_2$Si]$_4$

Thus, the *all-trans* isomer of 1,2,3,4-tetramethyl-1,2,3,4-tetraphenylcyclotetrasilane can be converted into the linear polymer in nearly quantitative yields by the use of anionic organolithium reagents such as n-butyllithium (see Eq. 7.5). Alternative types of initiators include 1,4-dipotassiooctaphenyltetrasilane, or [(PhMe$_2$Si)$_2$Cu(CN)Li$_2$]. It was observed from molecular weight analysis by gel-permeation chromatography (GPC) that high polymers were formed with average molecular weights of 100,000 and a M_w/M_n of about 2.0. Since two different substituents were present on silicon the possibility of the formation of stereoregular polymers by this methodology was investigated. It was found that the polymer isolated is not fully stereoregular. However, some amount of regularity in terms of syndiotactic and isotactic segments was detected by the use of NMR techniques [47].

$$\begin{array}{c}\text{Me}\quad\text{Me}\\ \text{Ph–Si—Si—Ph}\\ \text{Me–Si—Si—Ph}\\ \text{Ph}\quad\text{Me}\end{array}\quad\xrightarrow{\ [(\text{PhMe}_2\text{Si})_2\text{Cu(CN)Li}_2]\ }\quad\begin{bmatrix}\text{Ph}\\ \text{Si}\\ \text{Me}\end{bmatrix}_n$$

$$(7.5)$$

Poly(phenylmethylsilane)

Matyjasjewski's procedure can be applied to other types of cyclosilane rings. Thus, the ROP of the five-membered cyclosilane [(PhMeSi)(Me$_2$Si)$_4$] can be accomplished by silylanion initiators such as Me$_2$PhSiK (see Eq. 7.6) [48].

$$(7.6)$$

$$\begin{array}{c}\text{Me}\quad\text{Ph}\\ \text{Me}\!\!\diagdown\!\!\text{Si}\!\!\diagup\!\!\text{Si}\!\!\diagdown\!\!\text{Me}\\ \text{Me}\!\!\diagup\!\!\text{Si}\!\!\diagup\!\diagdown\!\!\text{Si}\!\!\diagdown\!\!\text{Me}\\ \text{Me}\ \text{Me}\ \text{Me}\ \text{Me}\end{array}\quad\xrightarrow[\substack{\text{HMPA}\\ -78\ ^\circ\text{C}}]{\text{Me}_2\text{PhSiK}}\quad\begin{bmatrix}\text{Me}\ \text{Me}\ \text{Me}\ \text{Me}\ \text{Ph}\\ \text{Si–Si–Si–Si–Si}\\ \text{Me}\ \text{Me}\ \text{Me}\ \text{Me}\ \text{Me}\end{bmatrix}_n$$

We have seen in the beginning of this chapter that methods of polymerizing stable disilenes have not yet been found primarily because the very process of stabilizing these compounds by the use of sterically hindered substituents prevents their controlled polymerization to high polymers.

On the other hand, compounds that can be described as *trapped* or *masked* disilenes can be viewed as potential monomers for polymer synthesis. The *trapped* or *masked* disilene can be liberated from its adduct by chemical or photochemical process. However, even if moderately bulky substituents are present on silicon in such compounds, the liberated disilene combines to afford the thermodynamically favorable cyclized products. Thus, the masked disilene containing isopropyl substituents on silicon affords the cyclotetrasilane [iPr$_2$Si]$_4$ (see Eq. 7.7) [49].

$$(7.7)$$

Masked disilene $\xrightarrow[\text{-anthracene}]{h\nu}$ [R$_2$Si=SiR$_2$] $\longrightarrow$

$$\begin{array}{c}\text{R}\quad\text{R}\\ \text{R–Si—Si—R}\\ \text{R–Si—Si—R}\\ \text{R}\quad\text{R}\end{array}$$

R = iPr

Cyclotetrasilane

Sakurai and coworkers have found that disilenes such as 1-phenyl-7,8-disilabicylo[2.2.2]octa-2,5-dienes can be polymerized by the use of anionic initiators such as *n*-butyllithium [18, 50, 51]. High-molecular-weight polymers with a narrow PDI (about 1.5-1.9) were obtained from this method. As mentioned above, the choice of substituents on silicon is quite critical. If all the substituents are sterically encumbered the reaction does not lead to polymeric products. The monomers required for this process were prepared by the reaction of 1,2-dichlorosilanes with lithium naphthalenide. Usually a mixture of isomeric products is obtained in this reaction although in some cases the reaction appears to lead to the formation of a single isomer (see Eq. 7.8) [18].

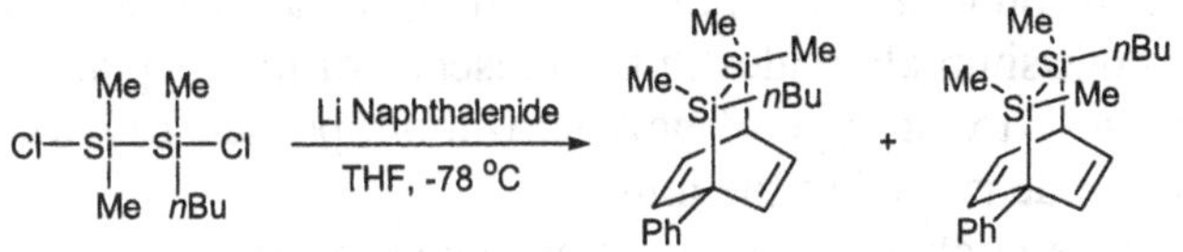

$$(7.8)$$

A number of such masked disilenes (as shown in Eq. 7.8) can be reacted with anionic initiators such as *n*-butyllithium to afford high-molecular-weight polymers (Fig. 7.14).

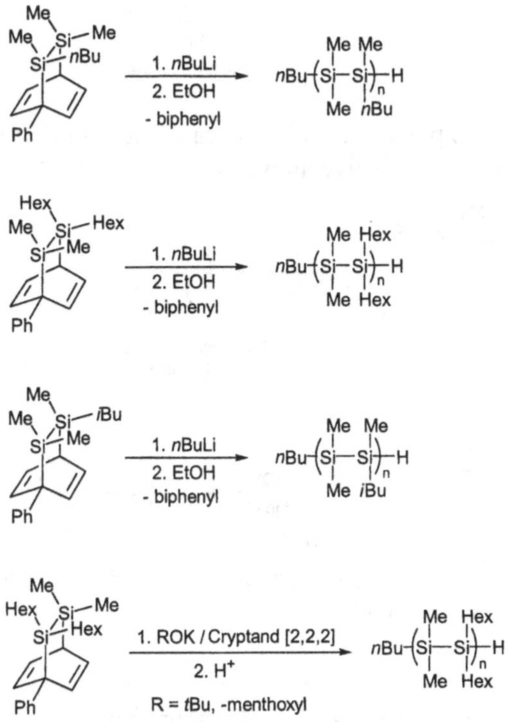

Fig. 7.14. Anionic polymerization of masked disilenes

The mechanism of polymerization of masked disilenes by anionic initiators has been suggested as involving a propagating silylanion with concomitant expulsion of the *trap* viz., biphenyl. The synthesis of polysilanes using the masked disilene method is useful in the preparation of polymers with a high amount of stereoregularity. Thus, polymers containing *alternate* segments of $(nC_6H_{13})_2Si$ and Me_2Si could be prepared [51]. In contrast, a polymer of a similar composition prepared by the Wurtz-type coupling reaction involving a 1:1 mixture of $(nC_6H_{13})_2SiCl_2$ and Me_2SiCl_2 has a *random* sequence of $(nC_6H_{13})_2Si$ and Me_2Si units. Sakurai and coworkers have also found that the polymerization of masked disilenes can be initiated by potassium alkoxides in the presence of the cryptand [2.2.2] [52]. This leads to a living type of polymerization, particularly when potassium(-)menthoxide is used as the initiator.

Masked disilenes containing amino substituents on silicon can also be prepared (see Eq. 7.9) [18].

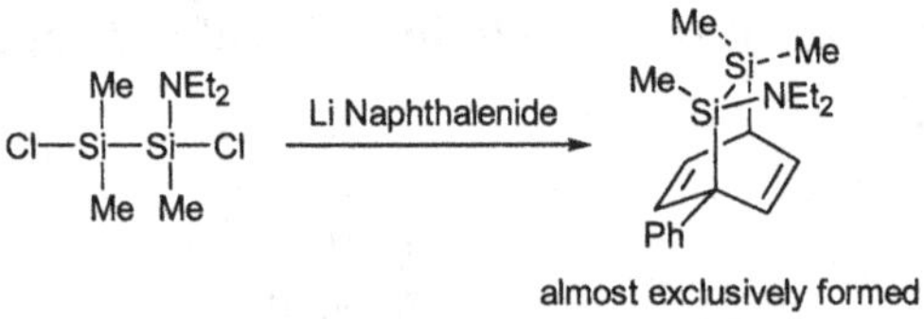

(7.9)

almost exclusively formed

These can then be polymerized by the catalytic action of *n*BuLi to afford polysilanes that contain reactive amino side-chains (see Eq. 7.10) [18].

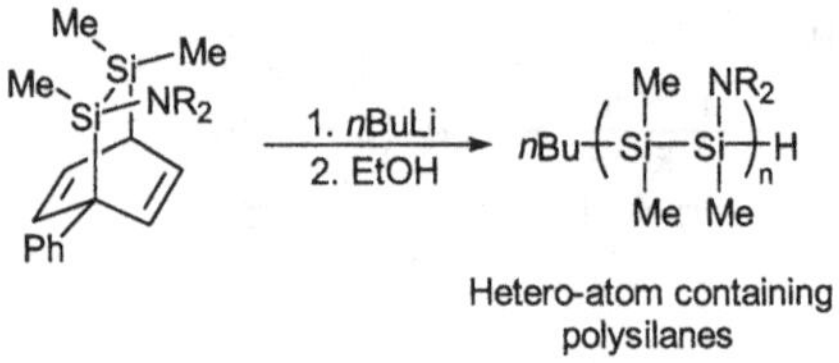

(7.10)

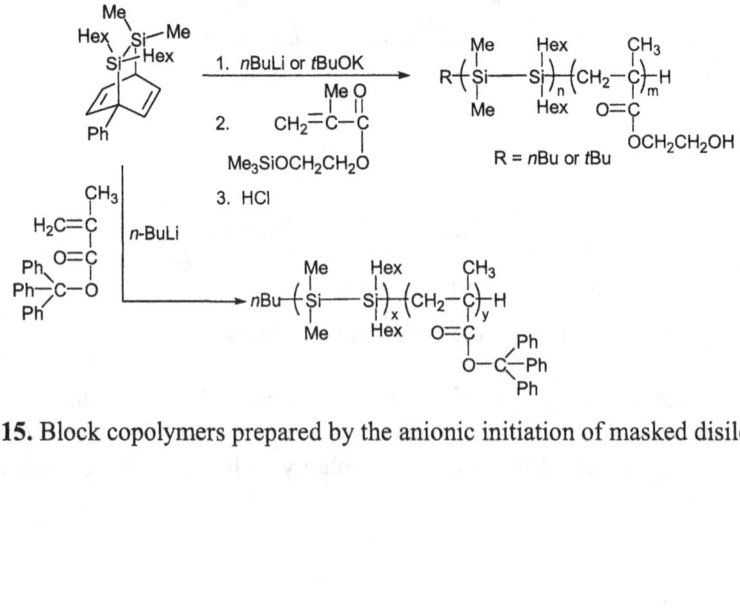

Hetero-atom containing polysilanes

Fig. 7.15. Block copolymers prepared by the anionic initiation of masked disilenes

The anionic polymerization strategy of masked disilenes allows the synthesis of block copolymers. Various kinds of organic blocks can be attached to polysilane segments (Fig. 7.15) [53-55]. Thus, a silyl-terminated organic monomer can be used for a block co-polymer formation. At the end of the polymerization the trimethylsilyl group can be removed to afford a functional polymer (Fig. 7.15) [53]. Another interesting example is in the utilization of this technique to build polymers that can be induced to have reversible helical conformations. Thus, for example, poly(1,1-dimethyl-2,2-dihexylsilene)-*p*-poly(triphenylmethyl methacrylate) was prepared by first carrying out the polymerization of the masked disilene (Fig.7.15) [54]. After this, (-)sparteine, an amine that induces optical induction was added. This was followed by the addition of the monomer triphenyl methacrylate. A 1:1 block copolymer was obtained with a M_n of about 10,000 and a M_w/M_n of about 1.5. Interestingly it was observed that in this polymer the helical sense of one of the blocks viz., the triphenyl methacrylate can be *transferred* to the polysilane block. This process is reversible and can be modulated with temperature. Strategies such as this allow new material synthesis with potential applications as chirooptical materials for switch/memory type of devices [54].

7.3.3 Polysilanes Obtained by Catalytic Dehydrogenation

Harrod and coworkers have found that metallocene catalysts such as Cp_2ZrMe_2 and Cp_2TiMe_2 have been found to catalyze the conversion of primary arylsilanes such as $PhSiH_3$ at room temperatures to afford short-chain polysilanes H-(PhHSi)$_n$-H which contain Si-H terminal groups (Fig. 7.16) [56-57]. These Group 4 metallocenes have bent structures in contrast to ferrocene where the iron is sandwiched between the two cyclopentadienyl rings.

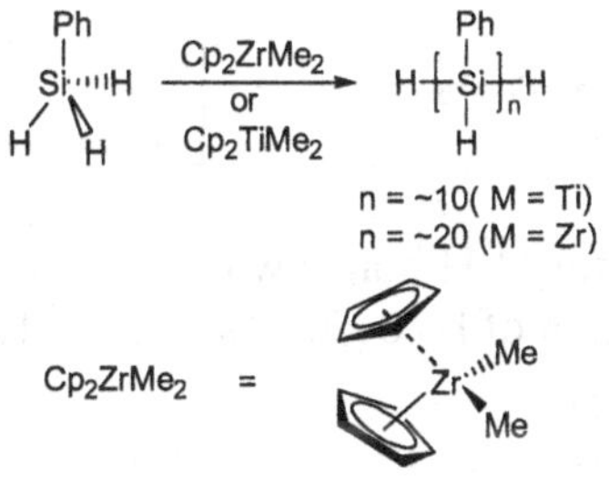

Fig. 7.16. Synthesis of short-chain poly(phenylsilane)s by the catalytic dehydrogenation of $PhSiH_3$

Improvements in catalyst design have led to an increase in chain length and polymers with a degree of polymerization of up to 80 have been realized [19-21, 58-60]. Although, these polymers still do not possess very high molecular weights, they are an interesting class of polysilanes which contain the reactive Si-H groups in the side-chain. This allows the opportunity for elaborating these polymers into other types by means of many reactions including hydrosilylation reactions.

The new generation dehydrogenative polymerization catalysts employed include [CpCp*Zr(H)$_2$](Cp* = pentamethylcyclopentadienyl), CpCp*M(Si(SiMe$_3$)$_3$R) (M = Zr, Hf; R = CH$_3$, Cl) [19-21]. Combinations of metallocene dichlorides and n-butyllithium such as Cp$_2$MCl$_2$/2nBuLi (M = Ti, Zr Hf) have also been employed (Fig.7.17) [21].

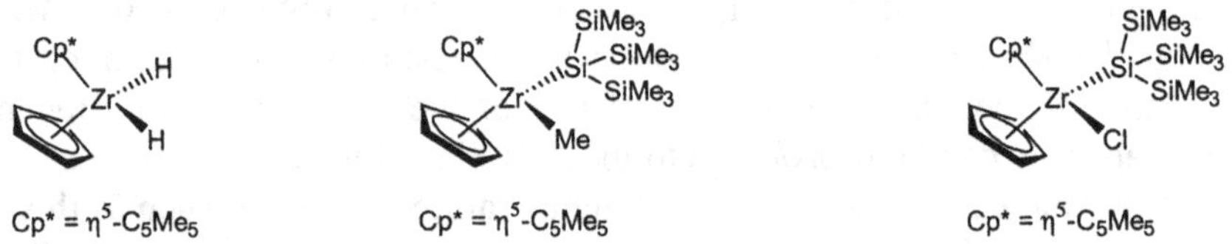

Cp* = η^5-C$_5$Me$_5$ Cp* = η^5-C$_5$Me$_5$ Cp* = η^5-C$_5$Me$_5$

Fig. 7.17. Some of the mixed-substituent zirconocene catalysts used for polymerization of RSiH$_3$ to H[HSiR]$_n$H

The mechanism of the catalytic dehydrogenation appears to involve the cleavage of the Si-H bonds which is activated by the catalyst [19-20]. It has been proposed by Tilley that the active catalytic species is a metal hydride of the type L$_n$M-H (where L is the other ligand or ligands such as cyclopentadienyl or pentamethylcyclopentadienyl that are present on the metal). The metal hydride interacts with a compound containing a Si-H bond to generate a four-centered transition state (see Eq.7.11)

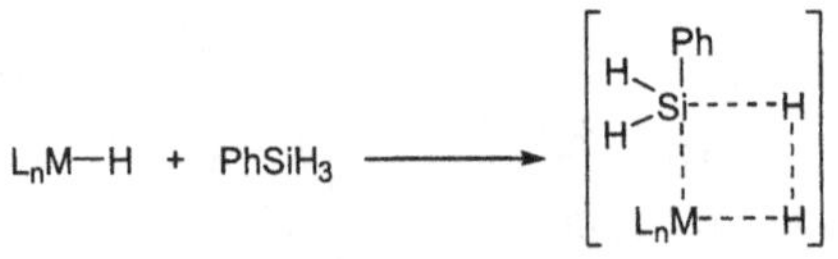

$$ \text{L}_n\text{M—H} + \text{PhSiH}_3 \longrightarrow \left[\begin{array}{c} \text{H} \searrow \overset{\text{Ph}}{\underset{|}{\text{Si}}} \text{- - - -H} \\ \text{H} \nearrow \quad \vdots \quad \vdots \\ \text{L}_n\text{M- - - -H} \end{array} \right] \tag{7.11} $$

Cleavage of Si-H and M-H bonds with the concomitant formation of Si-M bonds and elimination of H$_2$ occurs (see Eq. 7.12).

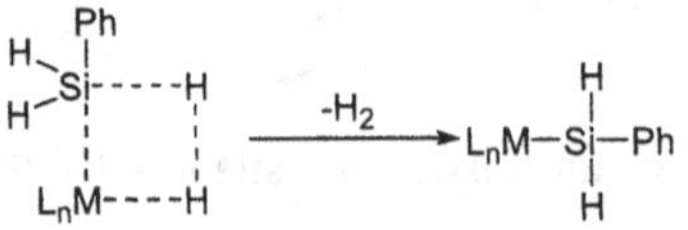

$$ \begin{array}{c} \text{H} \searrow \overset{\text{Ph}}{\underset{|}{\text{Si}}} \text{- - - -H} \\ \text{H} \nearrow \quad \vdots \quad \vdots \\ \text{L}_n\text{M- - - -H} \end{array} \xrightarrow{\text{-H}_2} \text{L}_n\text{M—}\overset{\text{H}}{\underset{\text{H}}{\text{Si}}}\text{—Ph} \tag{7.12} $$

The species $L_nM\text{-Si(H)}_2Ph$ interacts with another molecule of silane to form another four-membered transition state (see Eq. 7.13)

(7.13)

$$L_nM-\underset{H}{\overset{H}{Si}}-Ph \quad + \quad Ph-SiH_3 \quad \longrightarrow$$

Cleavage of another Si-H bond along with formation of Si-Si and M-H bonds leads to the formation of a disilane (see Eq. 7.14).

(7.14)

$$\longrightarrow L_nM-H \quad + \quad H-\underset{H}{\overset{Ph}{Si}}-\underset{H}{\overset{Ph}{Si}}-H$$

Notice that the catalytic species $L_nM\text{-H}$ is regenerated in this reaction sequence. This process can go on to produce a polymeric species. Although the precise reasons for not achieving high molecular weights through the dehydrogenation reaction are not entirely clear, there seem to be several factors that have a cumulative effect. 1) The catalyst $L_nM\text{-H}$ is also active towards insertion of Si-Si bonds and hence can cleave them to afford oligomeric or even cyclic products. 2) As the polymerization reaction proceeds, the entire reaction mixture becomes viscous and access of the catalyst active site to the growing polymer chain may become increasingly difficult. However, diluting the reaction mixture does not seem to dramatically improve the molecular weights.

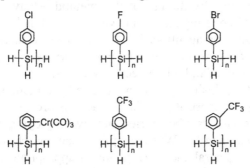

Fig. 7.18. Representative examples of polysilanes prepared by the dehydrogenation of the corresponding $RSiH_3$

In spite of the limitations of the dehydrogenation procedure, this method leads to the synthesis of functional polysilanes. Also interestingly, many reactive groups are quite tolerant to catalysis by the dehydrogenation reaction as evident by the type of polymers prepared by this method (Fig. 7.18). Although the catalytic dehydrogenation occurs with diorganosilanes $RR'SiH_2$ at elevated temperatures, this procedure seems to be best suited for monoorganosilanes, $RSiH_3$. Also, with the disilanes, $RR'SiH_2$, oligomeric and not polymeric products are formed in catalytic dehydrogenation. Furthermore, among the $RSiH_3$ the reactivity of arylsilanes appears to be greater than the corresponding alkyl compounds. The only exception to this is CH_3SiH_3 which has a comparable reactivity to that of $PhSiH_3$ [21].

7.3.4 Summary of the Synthetic Procedures for Polysilanes

1. Polysilanes can be prepared by (a) a Wurtz-coupling process involving a diorganodichlorosilane and an alkali metal (b) the use of anionic initiators and the use of strained cyclosilanes or masked disilenes (c) catalytic dehydrogenation of primary silanes, $RSiH_3$ and (d) miscellaneous methods including electrochemical reduction methods, condensation methods involving, for example, reactions between a monomer containing two reactive Si-Cl end groups and another that has silyllithium end groups.

2. The Wurtz-coupling is the most widely used synthetic method inspite of its obvious limitations such as harsh reaction conditions, intolerance to many reactive functional groups on the silicon monomers and so on. This process leads to polymers that usually have a bimodal distribution of molecular weights. The yields of polysilanes are low to moderate. Many poly(alky/arylsilane)s have been prepared by the use of this method. Many modifications of the Wurtz-coupling reaction have been tried including use of alternative reducing agents such as C_8K. However, there is still no general procedure that allows a high yield synthesis of polysilanes that have monomodal, high molecular weights.

3. Strained cyclosilanes can be utilized as monomers in a ring-opening polymerization. These reactions are catalyzed by anionic initiators and can be carried out by employing milder reaction conditions than the Wurtz-coupling reaction. However, this process is not general and construction of appropriate strained cyclosilane monomers is a prerequisite. Similarly, masked disilenes can be polymerized by anionic initiators. Several interesting polymers have been prepared by this method including stereoregular polymers. Design of the masked disilenes has to be carefully carried out and should avoid the use of sterically hindered groups.

4. Catalytic dehydrogenation is quite effective with monomers such as RSiH$_3$ (particularly when R = an aryl group). Various types of catalysts can be employed to effect the deydrogenation. Most well-used catalysts are metallocene catalysts containing Group 4 metals such as titanium, zirconium or hafnium. Although high molecular weights are not achieved in this process, the polysilanes obtained have reactive Si-H functional groups that can be elaborated further.

7.4 Modification of Polysilanes

Polysilanes can be elaborated, after their synthesis, by appropriate chemical methods [1-4, 61]. Representative methods of such possibilities are discussed in this section.

Polysilanes containing Si-H groups can be reacted further, in a variety of ways. Waymouth and coworkers have shown that free-radical-assisted addition of C=C or C=O bonds can be effected (Fig. 7.19) [62].

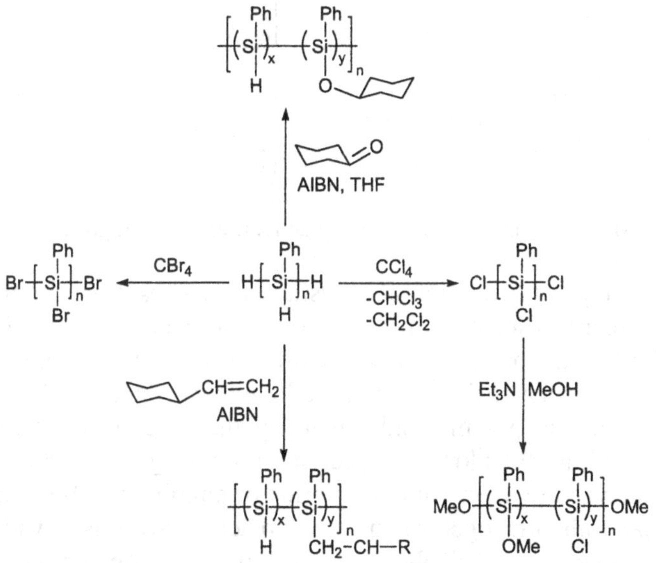

Fig. 7.19. Modification of polysilanes

Free-radical-assisted hydrosilylation can be carried out on poly(phenylsilane), H(PhHSi)$_n$H. Various types of reactive substrates such as 1-hexene, cyclohexanone etc., have been utilized and about 70-90% of substitution was observed without much chain degradation (Fig. 7.19).

Similarly, the conversion of the Si-H unit in poly(phenylsilane) into the more reactive Si-Cl and Si-Br functional group can be easily effected by the reaction with carbon tetrachloride and carbon tetrabromide, respectively (Fig. 7.19) [63]. The chlorination or bromination does not affect the Si-Ph substituent but only involves the Si-H bonds. In these reactions about 80% of the Si-H bonds can be converted into the corresponding Si-X bonds. These halogenated polysilanes are reactive polymers that can react with other types of nucleophiles such as alcohols to afford polysilanes containing alkoxy groups [63].

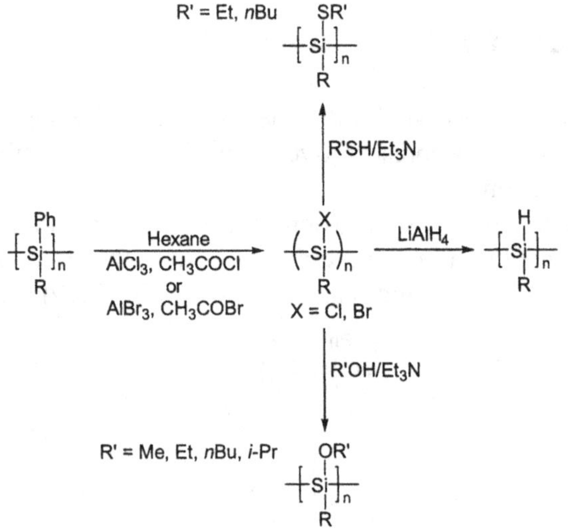

Fig. 7.20. Modification of polysilanes by activation of Si-Ph groups

Another type of activation of polysilanes involves dearylation reactions. We have seen earlier that Si-Ph bonds are susceptible to attack by triflic acid. West and coworkers have shown that it is possible to convert poly(alkyl/phenylsilane)s to poly(alkyl/halogenosilane)s. The halogen is introduced in the polymer side-chain by the reaction of the polysilane [PhRSi]$_n$ with acetylchloride or acetylbromide (Fig. 7.20) [64]. This reaction is catalyzed by aluminum trichloride or aluminum tribromide. The resulting poly(alkyl/halogenosilane)s are reactive towards a variety of reagents such as alcohols and thiols. Reduction of the poly(alkyl/halogenosilane)s also can be carried out by the use of LiAlH$_4$ to generate the corresponding polymers containing Si-H groups (Fig. 7.20) [64].

The phenyl substituent in poly(phenylmethylsilane)s can also be activated by its chlorination in a Sn(IV)-catalyzed reaction (see Eq. 7.15) [65].

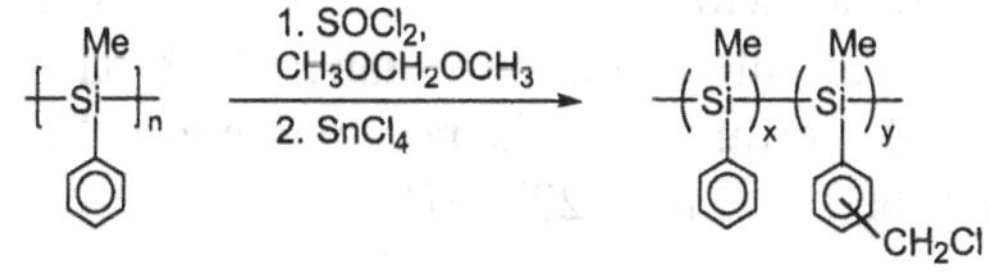

$$(7.15)$$

The chlorination of [MePhSi]$_n$ leads to a reactive group that is remote with respect to the main silicon chain. It is possible to modify the resulting functional polysilane in many ways. For example, recently Sakata and his research group have utilized this approach to anchor tetraphenylporphyrin units as pendant groups on the polysilane backbone (Fig. 7.21). It was observed that in this polymer an intramolecular charge transfer occurs from the silicon main-chain to the porphyrin side-chain [66].

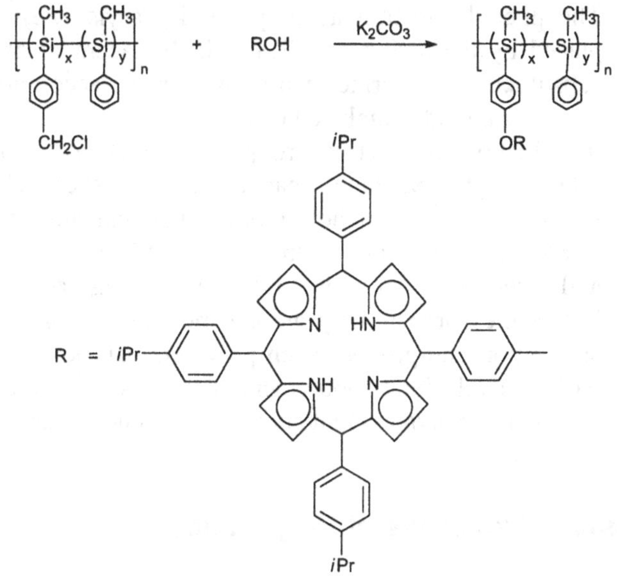

Fig. 7.21. Attachment of a tetraphenylporphyrin unit to a polysilane side-chain

Polysilanes with reactive silyl anions such as those prepared by the anionic polymerization method (which are not quenched) can be utilized to prepare blockcopolymers containing polysilane-organic polymer blocks. Mention of this was made in the previous section [53, 54]. A variation of this approach allows preparation of polysilanes with functional terminal groups that can be utilized in many novel ways. For example, a tripodal sulfur-containing coordinating ligand was anchored to the silyl end group by adopting the procedure shown in Fig. 7.22 [67]. Thus, the reaction of a polysilane containing a silyl anion terminal group (prepared by the anionic

ROP of the masked disilene) with an alkyl bromide that contains the tripodal sulfur ligand leads to an interesting polysilane which has three coordinating sulfur units at its terminus. This polymer could be utilized for
chemisorption on gold surface (Fig. 7.22) [67].

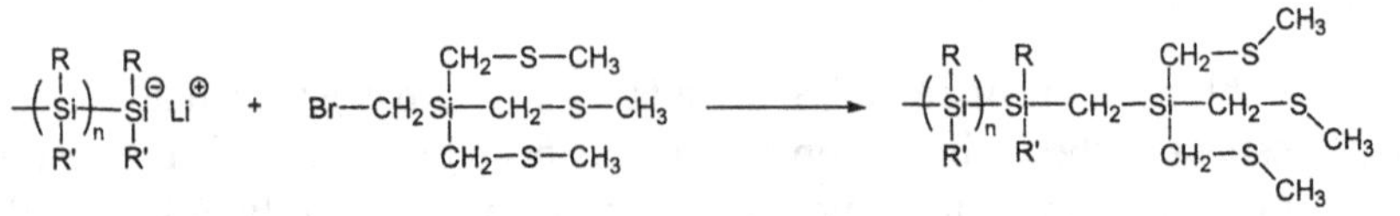

Fig. 7.22. Anchoring a tripodal sulfur ligand at the terminus of a polysilane

Thus, polysilanes can be elaborated (after their primary synthesis is accomplished) by many ways. These can be summarized as follows:

1. The phenyl substituents in poly(alkyl/arylsilane)s can be replaced by the action of CF_3SO_3H. Replacement of the phenyl
   substituent by chlorine or bromine can be accomplished by an
   aluminum halide-catalyzed reaction.
2. Introduction of reactive groups on the phenyl substituents of
   poly(phenyl/alkylsilane)s can be accomplished. These can be
   further utilized for anchoring other chemical entities of choice.
3. The Si-H group in polysilanes can be utilized as reactive centers
   and several types of hydrosilylation reactions can be carried out.
4. Polysilanes prepared by anionic polymerization have reactive
   silyl anion end-groups. Such polymers can be utilized for preparing novel block copolymers. They also can be used for preparing new polymers that contain novel end-groups.

7.5 Physical Properties of Polysilanes

Many of the physical properties of polysilanes depend on the actual substituents present on silicon. However, polysilanes have some distinct features in comparison to other polymers which is a direct result of the unique
characteristics that a catenated chain of silicon atoms provide. These can
be summarized as follows:

1. Most polysilanes are soluble polymers and dissolve in common
   organic solvents. The solubility decreases with increasing crystallinity of the polymer. The classic example of this feature is
   Burkhard's polymer $[Me_2Si]_n$. Other examples include $[Et_2Si]_n$,
   $[Ph_2Si]_n$ etc., [1-4]. However, disrupting the crystallinity by means
   of copolymerization such as that found in $[(Me_2Si)_n(PhMeSi)_m]$ or

in homopolymers that contain longer alkyl chains makes the polymers soluble [1-4, 30].

2. Polysilanes have unique electronic properties and this results from the electron delocalization in the σ-bonded chain. Although the Si-Si bond energy is comparable to that of C-C bond energy, silicon has a lower ionization potential and a greater electron affinity [4, 68-69]. The combination of these factors leads to electron delocalization in the σ-bonded chain of polysilanes. Thus, polysilanes can be compared with organic polymers that have conjugated skeleton such as polyacetylene. The important difference is that in the latter the conjugation occurs as a result of delocalized π-electrons. Because of the σ-delocalization of the electrons, polysilanes show strong absorption in the region between 300-400 nm [1-4, 69]. Further, since the σ-σ* transitions are *allowed*, the extinction coefficients of the optical absorption spectra are very high.

3. Polysilanes are radiation-sensitive polymers and degrade upon exposure to UV light. This process is accentuated in solution [1-4].

4. The glass-transition temperatures of polysilanes range from very low values of -75 °C to high values up to +120 °C[1]. Thus, polymers such as [(Me)(n-Hex)Si]$_n$ has the lowest T_g of about -75 °C. As the hexyl group is replaced by shorter chain alkyl substituents as in [(Me)(CH$_2$CH$_2$CH$_3$)Si]$_n$ (-28 °C) the T_g increases. Interestingly, the related polymer [(Me)(CH$_2$CH$_2$CF$_3$)Si]$_n$ (-3 °C) has a higher T_g [28]. In general, polysilanes that contain long alkyl chains as substituents are elastomeric. Crystallinity is introduced when the alkyl groups are smaller or when substituents such as phenyl groups are present on the silicon. Polysilanes are thermally quite stable. Most polymers are stable up to 200-300 °C without decomposition. It has been already mentioned that poly(dimethylsilane) undergoes a rearrangement without decomposition at 450 °C to afford the polycarbosilane [(H)(Me)SiCH$_2$]$_n$ which is converted to the ceramic silicon carbide only above 1000 °C [9-11].

5. Many polysilanes possess helical main-chain structures although the details of these helical structures vary from case to case [5].

7.5.1 NMR of Polysilanes

The ^{29}Si-NMR spectra of polysilanes show that most of the chemical shifts are in the negative region [4, 72]. Representative examples of polymers and their ^{29}Si-NMR chemical shifts are given in Table 7.1.

Table 7.1. Solution ^{29}Si-NMR chemical shifts of polysilanes

S.No	Polymer	δ
1	$[(n\text{-}C_5H_{11})_2Si]_n$	-29.2
2	$[(n\text{-}C_6H_{13})_2Si]_n$	-23.0
3	$[(Me)(Ph)Si]_n$	-39.4, -40.1, -41.4
4	$[(nPr)PhSi]_n$	-30.2, -32.4, -39.2
5	$[(nC_6H_{13})PhSi]_n$	-32.0, -35.0, -38.9
6	$[\{OCH_3(CH_2)_6\}_2Si]_n$	-24.0
7	$[Si(CH_3)_2\text{-}Si(n\text{-}C_6H_{13})_2]_n$	-26.0 (Hex$_2$Si)
		-36.0 (Me$_2$Si)
8	$[(Me)(CH_2CH_2CH_3)Si]_n$	-32.5
9	$[(Me)(CF_3CH_2CH_2)Si]_n$	-32.1
10	$H[(H)(Ph)Si]_nH$	-59 to -63

If the same substituents are present on silicon (in polymers of the type $[R_2Si]_n$) a single resonance is seen in the ^{29}Si-NMR spectrum (for example, see entries 1 and 2, Table 1). On the other hand, if the substituents are different on the silicon, there is a possibility of fully ordered polysilanes. Thus, if all the substituents are on the same side it would be an isotactic polymer (Fig. 7.23). However, such an ordered isotactic polysilane has not yet been prepared.

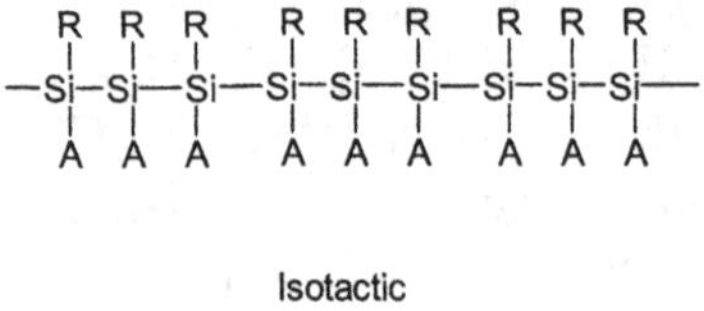

Isotactic

Fig. 7.23. Arrangement of side-groups in an isotactic polysilane

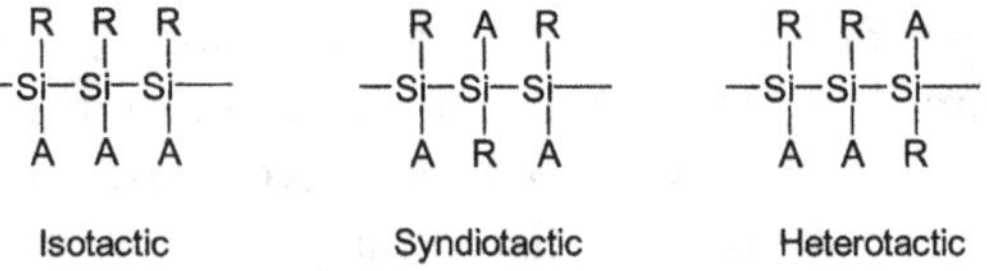

Isotactic   Syndiotactic   Heterotactic

Fig. 7.24. Arrangement of substituents in a triad of successive silicon atoms in an atactic polysilane

Even in atactic polymers where the organization of the side groups is perfectly random, if one considers short segments, say a triad of three successive silicon atoms there are three possibilities of a short-range order that could be present. These are termed *isotactic, syndiotactic* and *heterotactic* (Fig. 7.24). It should be emphasized that the terms isotactic, syndiotactic and heterotactic as applied in this context represent the statistical possibility of the organization of side-groups in a successive three-silicon segment within an atactic polymer chain. These stereochemical arrangements will resonate separately in the ^{29}Si NMR and three signals should be seen for each of these three triad sequences. Thus, for example, in atactic [PhMeSi]$_n$ three clear signals are seen for the isotactic, syndiotactic and heterotactic segments of the polymer [4]. It is possible to observe more signals in the ^{29}Si NMR for polymers of the type [RR'Si]$_n$ if stereochemical sequences are expressed and detected beyond the triad to extend into tetrad and pentad sequences.

Many copolymers of polysilanes show signals in the ^{29}Si NMR that indicates a random arrangement. However, if ordered copolymers can be prepared this can be detected by ^{29}Si NMR. For example, West and co-workers prepared an ordered copolymer by the Wurtz-coupling reaction involving the monomer BrSiMe$_2$Si(nHex)$_2$SiMe$_2$Br (Fig. 7.25) [73]. Theoretically this reaction should lead to the formation of [SiMe$_2$Si(nHex$_2$)SiMe$_2$]$_n$ where there is a clear ordering of the SiMe$_2$ groups vis-a vis the Si(nHex$_2$) groups. While in the high-molecular-weight portion of the polymer some randomization was detected the lower-molecular-weight portion was determined to be an ordered copolymer by the observation of sharp signals at -36 ppm (SiMe$_2$) and -26 ppm (Sin-Hex$_2$).

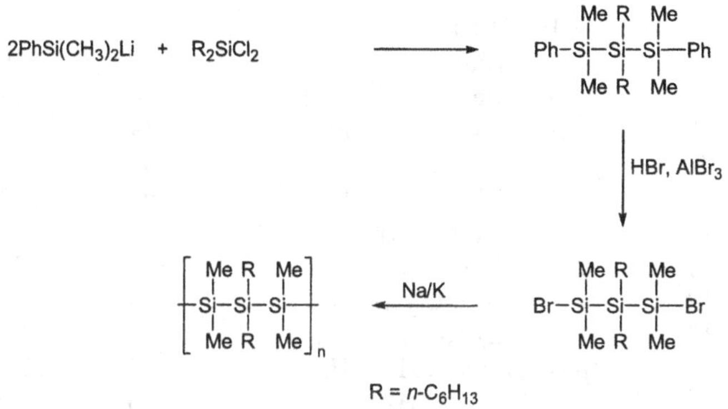

$$2PhSi(CH_3)_2Li + R_2SiCl_2 \longrightarrow$$

Me R Me
| | |
Ph—Si—Si—Si—Ph
| | |
Me R Me

HBr, AlBr$_3$

Me R Me
| | |
Br—Si—Si—Si—Br
| | |
Me R Me

Na/K

$$\left[\begin{array}{c} Me\ R\ Me \\ | \ | \ | \\ Si-Si-Si \\ | \ | \ | \\ Me\ R\ Me \end{array} \right]_n$$

R = n-C$_6$H$_{13}$

Fig. 7.25. Preparation of an ordered copolymer

Another way of preparing the ordered copolymer consists of using the masked-disilene method. If the monomer (masked disilene) is designed in such a manner that one of the silicon atoms has two methyl groups and the other silicon has two n-hexyl substituents, it would be expected that the polymer obtained from it should have a stereo regular arrangement. This has been accomplished and in the ^{29}Si NMR of this polymer two distinct sharp signals were obtained indicating an ordered sequence of $SiMe_2$ and $Si(n\text{-Hex})_2$ units (see Eq. 7.16) [51].

(7.16)

7.5.2 Electronic Spectra of Polysilanes

Most polysilanes absorb in the region between 300-400 nm [1-4, 26, 74-77]. The electronic absorption data of representative examples are summarized in Table 7.2.

Table 7.2. Electronic spectroscopic data for representative examples of polysilanes

S.No	Polymer	λ_{max}	ϵ Si-Si
1	$[(nC_6H_{13})_2Si]_n$	317	9500
2	$[(nC_{10}H_{21})_2Si]_n$	324	8940
3	$[(nPr)(Me)Si]_n$	306	5600
4	$[(nC_{10}H_{13})(S\text{-}CH_2CH(CH_3)(C_2H_5))Si]_n$	323	55,000
5	$[(Ph)(Me)Si]_n$	341	9300
6	$[(p\text{-}OMe\text{-}C_6H_4)(Me)Si]_n$	346	8180
7	$[(p\text{-}tBu\text{-}C_6H_4)(Me)Si]_n$	339	7400
8	$[\{p\text{-}OMe\text{-}C_6H_4\text{-}(CH_2)_3\}(Me)Si]_n$	308	6735
9	$H[(Ph)(H)Si]_nH$	294	2026
10	$H[(Ph)(nC_6H_{13})Si]_n H$	325	4321
11	$H[(Ph)(OC_6H_{13})Si]_nH$	353	2349
12	$[(p\text{-}nBu\text{-}C_6H_4)_2Si]_n$	395	26,600
13	$[(p\text{-}C_2H_5\text{-}C_6H_4)_2Si]_n$	390	10,000
14	$[\{p\text{-}(CH_3)_2CH\text{-}CH_2\text{-}C_6H_4\}_2Si]_n$	390	16,200
15	$[(p\text{-}tBu\text{-}C_6H_4)_2Si]_n$	376	3400
16	$[(p\text{-}nC_6H_{13}\text{-}C_6H_4)_2Si]_n$	397	23,300
17	$[(p\text{-}OBu\text{-}C_6H_4)_2Si]_n$	326	9000
18	$[(m\text{-}OBu\text{-}C_6H_4)_2Si]_n$	403	18,300

Studies on model compounds [70-71] reveal that the energy of absorption shifts to longer wavelengths upon increase of the number of silicon atoms: Si_2Me_6 (190 nm); Si_3Me_8 (217 nm); Si_4Me_{10} (233 nm); Si_5Me_{12} (250 nm); Si_6Me_{14} (260 nm); Si_7Me_{16} (269 nm); Si_8Me_{18} (276 nm); Si_9Me_{20} (278 nm); $Si_{10}Me_{12}$ (284 nm); $Si_{16}Me_{34}$ (312 nm). A similar feature has also been noticed for the corresponding oligomers containing the n-hexyl substituents: Si_3Hex_8 (217 nm); Si_4Hex_{10} (233 nm); Si_5Hex_{12} (250 nm); $Si_{10}Hex_{22}$ (290 nm). Such a progressive shift of electronic absorption to the longer wavelength is attributed to a σ-delocalization. Thus, a delocalized band-model is more appropriate to describe the polysilane electronic structure. This arises primarily because of the low ionization potential of silicon coupled with its greater electron affinity. This electronic picture is unlike the carbon backbone polymers where electron delocalization occurs only in polymers where there are alternate double bonds in the backbone. The classic example of such a polymer is polyacetylene.

The main features of the electronic spectra of polysilanes can be summarized as follows:

1. The electronic transitions in polysilanes are of the σ-σ* type. The high values of the extinction coefficients of these absorptions is consistent with the σ-σ* nature of these transitions. One of the highest ε values are observed for $[(nBu-p-C_6H_4)_2Si]_n$ (ε = 26,600) [2,26] and $[(nC_{10}H_{13})(S-CH_2CH(CH_3)(C_2H_5)Si]_n$ (ε = 55,000) [74]

2. The $λ_{max}$ and the ε values depend on the chain length but seem to reach an optimum value for a chain length equaling $[RR'Si]_n$ (n=30). The $λ_{max}$ value also depends on the type of substituents present on the silicon. Thus, the highest energy absorption is observed for $[(CF_3CH_2CH_2)(CH_3)Si]_n$ at 285 nm [28].

3. Poly(dialkylsilane)s have a $λ_{max}$ between 305-315 nm. Introduction of aryl groups causes a bathochromic shift. Thus, for poly(alkyl/arylsilane)s the $λ_{max}$ increases to 325-350 nm. The highest bathochromic shifts are observed for poly(diarylsilane)s [26,75]. This progressive shift of absorption to lower energy upon introduction of aryl groups has been attributed to σ-π mixing.

4. Poly(dialkylsilane)s show fluorescence with a small Stokes shift. In contrast, polymers such as $[PhMeSi]_n$ show two emission bands: (a) a weak emission band with a small Stokes shift and (b) an intense emission with a substantial Stokes shift (for $[PhMeSi]_n$ this emission occurs at 460 nm). The latter has been attributed to the presence of chain branching defects [75].

Most polysilanes also show thermochromic behavior in solution as well as in the solid state [1-4, 75, 78-82]. The most well-studied example in this regard is the poly(di-*n*-hexylsilane), $[(n-Hex)_2Si]_n$ which shows an abrupt thermochromic transition (generally observed over a range of about 10 °C [4, 75]. Thus, at ambient temperature this polymer shows a $λ_{max}$ at 315 nm. On cooling this transition moves to lower energy. Thus, by about -56 °C a new band at around 355 nm is fully developed at the expense of the high energy band. These changes are reversible with changes in temperature.

More recent studies on poly(di-*n*-hexylsilane) seem to indicate that this polymer can exist in three different conformations in solution [82]. Detailed studies on various types of polysilanes reveal that the above type of thermochromic behavior is typical of polysilanes where similar (or identical) alkyl groups are present on the silicon. In polymers with widely different types of alkyl groups such as [(Me)(*n*-Hex)Si]$_n$, although no abrupt transition occurs, a gradual shift to a lower energy transition is observed. An intermediate behavior is also seen for polymers such as [(*n*Pr)(*n*Hex)Si]$_n$ [75]. Bathochromic shifts (upon lowering temperature) are not observed in polysilanes that contain sterically hindered substituents or if both the substituents are aryl groups. The causes of thermochromic behavior are quite complex and the most simplistic explanation seems to suggest the dominance of an ordered conformation (such as a helical rod-like conformation) at lower temperatures leading to better delocalization and consequently to the σ-σ* transition moving to lower energy. For a detailed discussion on the various models of this interesting phenomenon in polysilanes the reader is directed to authoritative sources [1-4, 75, 79, 83].

Thermochromism is exhibited by many polysilanes in the solid state as well but is different from that shown in solution. Polymers with short alkyl side-chains such as [Et$_2$Si]$_n$, do not show thermochromism. Even polymers such as [*n*Bu$_2$Si]$_n$, [*n*Pent$_2$Si]$_n$ do not show any dramatic thermochromism. However, polymers with longer alkyl side-chains starting from [(*n*Hex)$_2$Si]$_n$ show thermochromism in the solid state. Thus, [(*n*Hex)$_2$Si]$_n$ undergoes a phase transition at 43 °C. This is coincident with the shifting of the electronic transition from 375 nm (below 43 °C) to 317 nm (above 43 °C) [75]. Recent studies on the thermochromic behavior in the solid state reveal that in polymers such as [(*n*C$_{10}$H$_{23}$)$_2$Si]$_n$ multiple low-energy bands are seen [25].

7.6 Optically Active Polysilanes

Although most polysilanes adopt main-chain helical structures, most of them do not show a preferred screw sense. Thus, if a helical polymer shows a preferred screw sense it can be either left-handed or right-handed type. These possibilities can be detected by CD (circular dichroism) spectroscopy. Thus, in regular optical absorption spectroscopy the extinction coefficient or molar absorptivity (ε) is related to the absorbance (A), concentration (c) and path length of the detecting cell (l) by the famous Beer-Lambert's expression (see Eq. 7.17).

$$A = \varepsilon cl \qquad (7.17)$$

However, when plane-polarized light passes through a solution containing an optically active substance, the left and right circularly polarized components are absorbed to different extents. This is reflected in CD spectroscopy by the positive or negative Cotton effects ($\Delta\varepsilon$) [84]. The appropriate expression in CD spectroscopy is shown in Eq. 7.18.

$$\Delta\varepsilon = \varepsilon_L - \varepsilon_R = (A_L - A_R)/cl \qquad (7.18)$$

The ε_L, ε_R, A_L and A_R represent the respective extinction coefficients and absorbances of the absorption of the left and right circularly polarized light, respectively. Polysilanes that have equal segments of the opposite screw sense that absorb in the same wavelength will not show Cotton effect. On the other hand, if the polymer has a preferred screw sense it will show a Cotton effect.

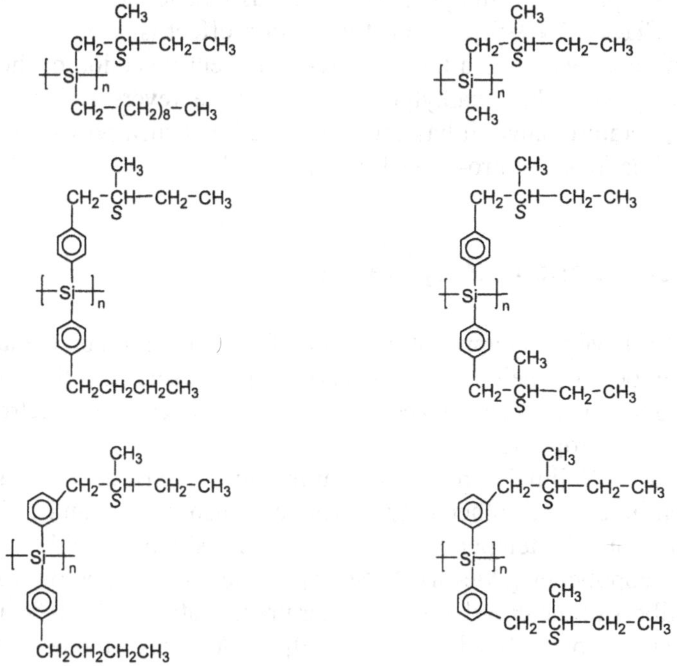

Fig. 7.26. Representative examples of optically active polysilanes

Recently, many helical polysilanes with a preferred screw sense have been synthesized mainly by Fujiki and coworkers [74, 85-92]. These contain either one or two chiral side chains. Representative examples of such polymers are shown in Fig. 7.26.

Optically active polysilanes are synthesized by the conventional Wurtz synthesis starting from the appropriate dichloro derivatives. The polymer, poly(n-decyl-(S)-2-methylbutylsilane) shows a rod-like helical chromophore and shows a positive Cotton absorption at 323 nm [74]. Fujiki has observed that the rigid-rod-like helical structure was maintained in the series of polymers $[(R)(S\text{-}2\text{-methylbutyl})Si]_n$ as the R is varied from n-propyl to n-dodecyl [74]. However, when R = Me a CD band with both positive and negative Cotton effects are seen [85]. This is interpreted as being consistent with the polymer possessing a diastereoisomeric block structure with the opposite helical segments coexisting in the same main chain. Fujiki and coworkers have also observed a dependence on the type of preferred screw sense adopted by the polymers. Thus, in the polymer poly(n-butyl-(S)-2-methylbutylsilane) which contains only one chiral unit per silicon a positive Cotton effect is seen. On the other hand, if two chiral substituents are present in the polysilane such as that found in $[\{C_6H_4\text{-}p\text{-}(S)\text{-}CH_2\text{-}CH(CH_3)(C_2H_5)\}_2Si]_n$, a negative Cotton effect is seen [87]. Another interesting observation is a temperature-dependent inversion of the helical screw conversion. For dialkylpolysilanes such a reversal occurs over a small temperature range. It has been suggested that such polymers are useful in applications as chiro-optical switches [88].

7.7 Applications of Polysilanes

Polysilanes have many applications [1-5]. These are based on the ability of some members of their family to serve as precursors to silicon carbide. Other applications of polysilanes stem from their interesting electronic and photophysical properties.

Apart from Yajima's procedure of utilizing $[Me_2Si]_n$ as a precursor for silicon carbide, many other polysilanes have also been found to be good precursors for high-temperature processing into silicon carbide. These include the copolymer $[(Me_2Si)_x(PhMeSi)_y]$. The latter is photo-crosslinked and the fibers of the crosslinked polymer upon heating at 1100 °C in argon atmosphere to afford SiC [30]. A high-molecular-weight poly(methylsilane) of composition $[(CH_3SiH)_x(CH_3Si)_y(CH_3SiH_2)_z]$ was prepared by the sonochemical/Wurtz-coupling condensation of CH_3SiHCl_2. This polymer upon pyrolysis in argon affords silicon carbide. Silicon carbide in the form of fibers, films as well as solid monoliths could be obtained by this method [93].

Other applications of polysilanes include their utility as photoinitiators for vinyl polymerization. This application is based on the phenomenon of generating silyl radicals upon photolysis. Although polysilanes are not efficient photoinitiators they have unique properties such as being not susceptible to oxygen. Many organic monomers such as acrylates, and styrene have been polymerized by this method [4].

Polysilanes have been examined as materials for photoresist applications involving microlithography which has applications in microchip fabrication [1-4, 94]. This application is based on the sensitivity of polysilanes towards Si-Si bond scission upon photolysis.

Polysilanes are now being looked at from the point of view of electroluminescent materials as these polymers have high-energy HOMO's [95]. Because of this it is easy to oxidize polysilanes. Consequently, polysilanes have been predominantly shown to be *hole transport* materials. These features have enabled these materials to be considered as possible electroluminescent materials particularly for emission in the near ultraviolet (NUV) or ultraviolet (UV). These types of light sources are difficult to obtain from conventional organic polymers such as poly(phenylenevinylene) which have been vigorously examined as light-emitting materials. A functional electroluminescent device would consist of a thin film of polysilane sandwiched between a hole and an electron-injecting electrode. Electroluminescence is exhibited by the recombination of holes and electrons which are injected by electrodes. Typically the electrodes that are used are indium-tin oxide (ITO) for hole injection and metals such as Ca, Al, or alloys of magnesium and silver for electron injection. Polymers such as $[PhMeSi]_n$ have been found to show electroluminescence which can be tuned over the range of colors from blue to red by the use of dye materials. However, the electroluminescence of $[PhMeSi]_n$ is exhibited only at low temperatures and is seen from 230-110 K. In contrast, the diaryl polymer $[(n\text{-}C_4H_9\text{-}p\text{-}C_6H_4)_2Si]_n$ shows electroluminescence peaking at 407 nm at room temperature. The successful implementation of these materials in EL devices would have to overcome the weakness of the polysilane backbone under UV light irradiation [95].

Polysilanes are also attracting attention in application as photoreceptors in copiers and printers because of their high carrier mobility [95].

7.8 Polysilynes

Reduction of organotrichlorosilanes $RSiCl_3$ can be utilized to prepare polymers with network structures known as polysilynes [4]. The presence of three reactive chlorines means that every silicon can be connected to *three other* silicon centers. If the substituent R is bulky and is sterically hindered, discrete molecules in various structures such as cubes, trigonal prisms or tetrahedra can be prepared [96]. The type of cluster isolated depends on the nature of the substituent present on silicon and also on the nature of reductant apart from the reaction conditions. Thus, a cubic silane [RSi]$_8$ was isolated in the reduction of RSiCl$_3$ with Mg/MgBr$_2$ (R = 2,6-Et$_2$-C$_6$H$_3$) (Fig. 7.27).

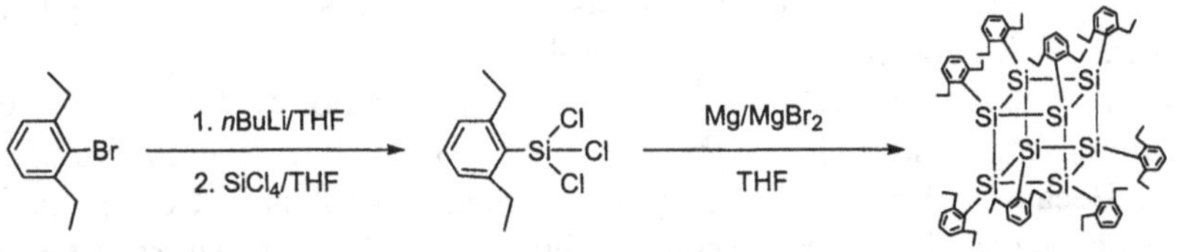

Fig. 7.27. Synthesis of a cubic silane [RSi]$_8$

On the other hand, trigonal prismatic [RSi]$_6$ is obtained in the reduction of RSiCl$_3$ or [RSiCl$_2$]$_2$ (R=2,6-i-Pr$_2$-C$_6$H$_3$) (Fig. 7.28).

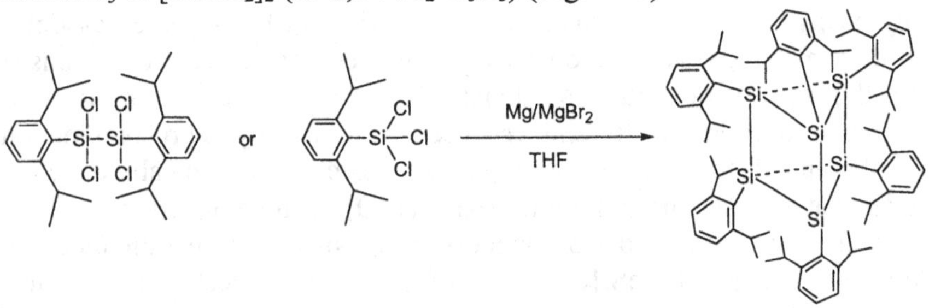

Fig. 7.28. Synthesis of the hexameric [RSi]$_6$ in the trigonal prism structure

The tetrahedral cluster [RSi]$_4$ is prepared by the reduction of [RSiBr$_2$]$_2$ (R = tBu$_3$Si) (Fig. 7.29). This cluster synthesis is made possible by the use of the sterically hindered tris-t-butylsilyl substituent on silicon. It is noticed that in these reduction reactions the substituent on silicon has a crucial and structure-directing role. Although it is not possible *apriori* to predict the type of product that may be formed in the reduction of a given RSiCl$_3$, it has been experimentally found, as shown above, that the nuclearity of the cluster decreases with an increase of the steric bulk of the substituent on silicon.

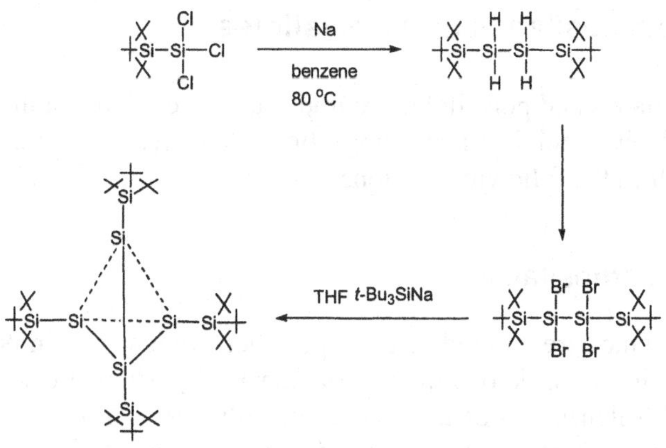

Fig. 7.29. Synthesis of tetrahedral [RSi]₄ by the *t*Bu₃SiNa-mediated reduction of [*t*Bu₃SiSiBr₂]₂

In order to prepare polymeric derivatives from RSiCl₃, the steric bulk of the substituent on silicon has to be reduced. Thus, reduction of RSiCl₃ (R= an alkyl group such as *n*-propyl, *n*-butyl, *n*-hexyl, cyclohexyl etc., or where the R = sterically unhindered aryl group such as phenyl, *p*-tolyl, *p*-NMe₂-C₆H₄- etc.) afforded polysilynes. A typical synthesis exemplified by the preparation of [*n*HexSi]ₙ consists of reacting *n*-HexSiCl₃ with sodium/potassium alloy under high-intensity ultrasound conditions at ambient temperature [97-98]. This polymer showed a ^{29}Si-NMR chemical shift at -57 ppm which is about 30 ppm upfield shifted in comparison to poly(di-*n*-hexylsilane), [*n*-Hex₂Si]ₙ (-24.8 ppm). This chemical shift is indicative of a tetrahedral silicon environment where each silicon is attached to three other silicon centers. Based on this, these polymers have been suggested to have an open sheet-like structure containing fused cyclosilane rings. The poly(*n*-hexylsilyne) [*n*HexSi]ₙ has a M_w of 2.4 × 10⁴ (monomodal, PDI = 2.1). The electronic spectra of these polysilynes reveal a broad absorption tailing into the visible region. Recently, some of these types of polymers have been shown to have enhanced electrical conductivity upon doping with I₂. They also have been shown to form strong charge-transfer complexes with electron acceptors such as 9,10-dicyanoanthracene, *p*-chloranil and tetracyanoquinodimethane [99].

7.9 Polycarbosilanes and Polysiloles

Polycarbosilanes and polysiloles are important silicon-containing polymers that have been receiving increasing attention in recent years. These are briefly dealt in the following sections.

7.9.1 Polycarbosilanes

Polycarbosilanes are a broad class of polymers which contain both silicon and carbon in the backbone of the polymer [100]. There are several types of polycarbosilanes. All of these types of polymers will not be dealt with here. The reader is referred to a more detailed review article on this subject [100]. The most important member of this family is of course the one prepared by Yazima by the thermolysis of poly(dimethylsilane), $[Me_2Si]_n$ at 450 °C to afford $[(H)(CH_3)SiCH_2]_n$ (see Eq. 7.19). The importance of this polymer arises from the fact that it can be spun into fibers, crosslinked by heating in air at around 300 °C and transformed by pyrolysis at high temperatures to silicon carbide fibers impregnated with silicon carbide crystallites [9-11]. This pioneering work has in fact inspired much of the research on polysilanes in general as well as polycarbosilanes in particular.

$$[Me_2Si]_n \xrightarrow[\text{Argon}]{450\ °C} \left[\begin{matrix} H \\ | \\ Si\!-\!CH_2 \\ | \\ CH_3 \end{matrix} \right]_n \qquad (7.19)$$

Poly(carbosilane)

An elegant synthetic route has been recently developed for polymers containing alternate R_2Si and CH_2 units by Interrante and coworkers. These polymers are called poly(silylenemethylene)s or polysilaethylenes [101-102]. The preparative route to these polymers consists of a platinum (H_2PtCl_6)-catalyzed ring-opening polymerization of four-membered 1,3-disilacyclobutanes. These monomeric inorganic rings can be prepared in many ways. Thus, the reduction of $Cl(OEt)_2SiCH_2Cl$ with magnesium affords an alkoxydisilacyclobutane. This can be converted into the corresponding tetrachloro derivative by treatment with acetylchloride. The chlorine atoms on the silicon can then be replaced by alkyl groups by the corresponding Grignard reagent (Fig. 7.30). These disilacyclobutanes have been found to be ideal substrates for ring-opening polymerization [100-101].

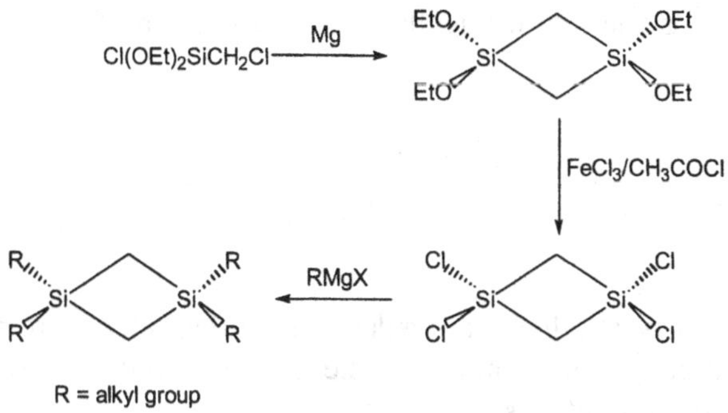

Fig. 7.30. Preparation of disilacyclobutanes

Tetrachlorodisilacyclobutane can be ring-opened to afford the polymeric derivative $[Cl_2SiCH_2]_n$. This polymer can be reduced with $LiAlH_4$ to afford $[H_2SiCH_2]_n$ which has a M_w of about 80,000 (see Eq. 7.20). Notice that this polymer is a monosilicon analogue of polyethylene where every alternate CH_2 unit is replaced by the SiH_2 unit (Fig. 7.31) [102].

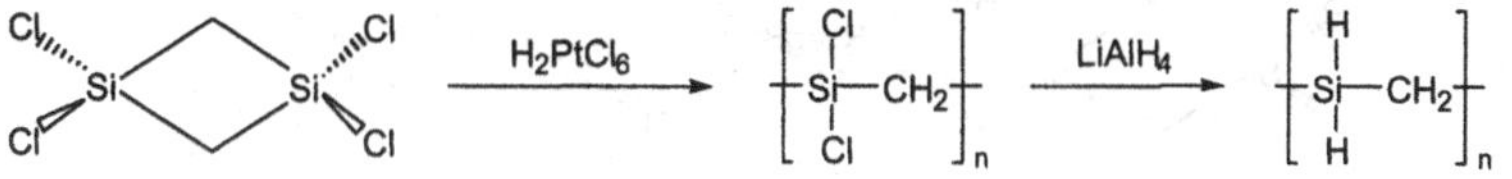

$$(7.20)$$

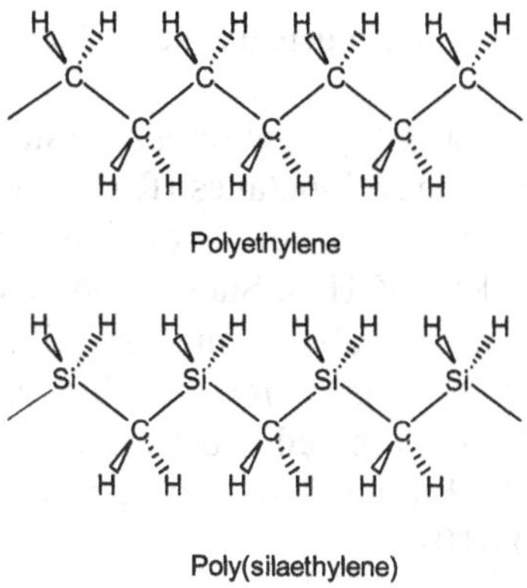

Polyethylene

Poly(silaethylene)

Fig. 7.31. Relationship between polyethylene and polysilaethylene

Poly(silaethylene) also has been shown to possess an *all-trans* structure such as polyethylene [100-101].

The fluoro derivative $[F_2SiCH_2]_n$, has been prepared by a slightly different method [103]. Unlike the dichloro compound which is prepared by a direct ROP reaction, the difluoro derivative is obtained in a ROP of the diethoxy derivative. The diethoxy polysilaethylene upon treatment with

BF$_3$.Et$_2$O affords the hydrolytically sensitive difluoro derivative (see Eq. 7.21).

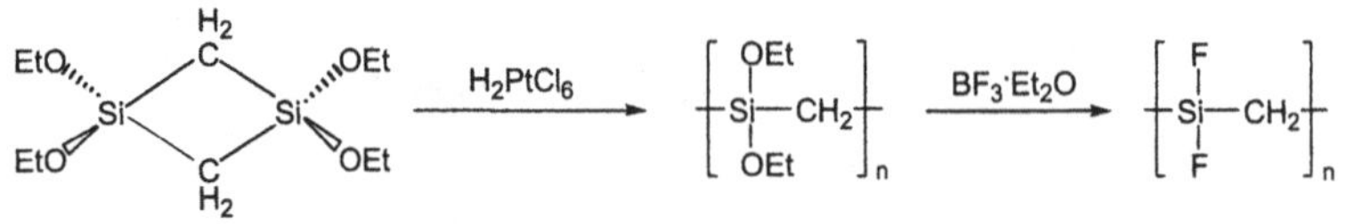

(7.21)

Other dialkyl or alkyl/aryl polysilaethylene derivatives are prepared by the ring-opening polymerization of the *cis-trans* mixtures of the corresponding disilacyclobutanes.

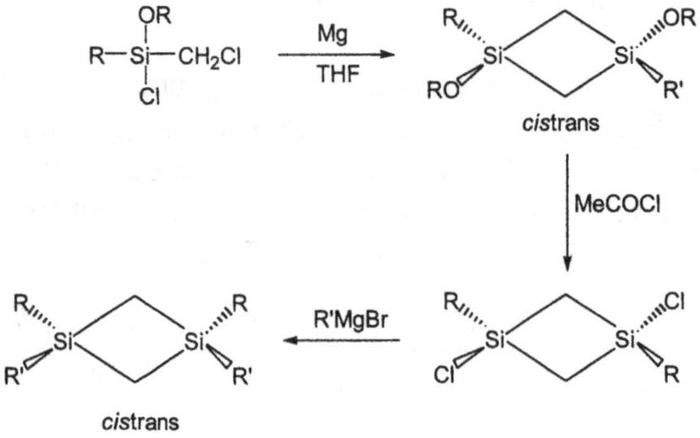

Fig. 7.32. Synthesis of mixed substituent disilacyclobutanes

Reduction of RSi(OR′)ClCH$_2$Cl with magnesium affords a *cis-trans* mixture of the dialkoxydisilacyclobutanes [R(OR′)SiCH$_2$]$_2$. These on reaction with acetylchloride afford the corresponding mixtures of chloro/alkyl, chloro/aryl derivatives, [RClSiCH$_2$]$_2$. Such compounds are ideal precursors for the preparation of mixed substituent derivatives. Thus, [RClSiCH$_2$]$_2$ can be further alkylated to give mixed-substituent disilacyclobutanes [RR′SiCH$_2$]$_2$ (Fig. 7.32). Such mixed-substituent disilacyclobutanes can be used as substrates for ROP in a chloroplatinic acid-catalyzed reaction (see Eqs. 7.22 and 7.23) [100-101].

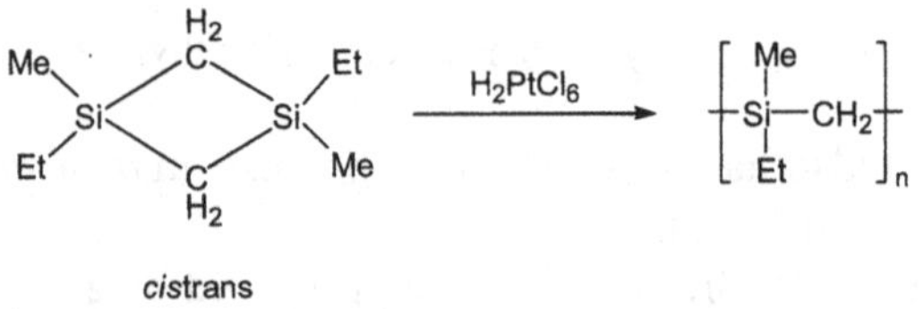

(7.22)

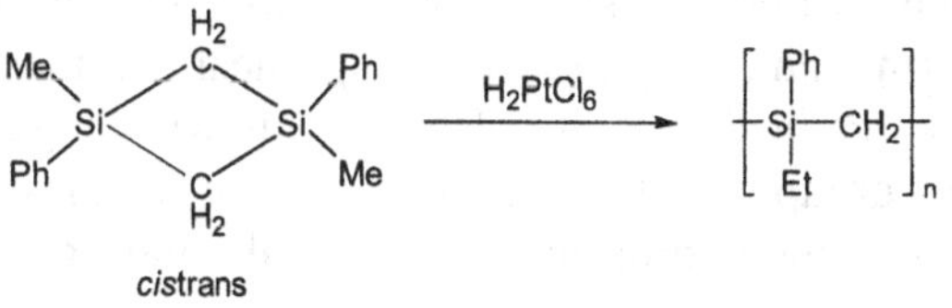

$$(7.23)$$

cistrans

Thermal properties of poly(silaethylene)s follow predictable lines. Thus, one of the lowest T_g containing polymer is $[H(n\text{-Hex})SiCH_2]_n$ (-91.4 °C). Introduction of the aryl groups increases the T_g, $[H(Ph)SiCH_2]_n$ (-37.6 °C). Presence of methyl and phenyl groups on the silicon increases the T_g further , $[MePhSiCH_2]_n$ (+28.6 °C) [100-101].

The most important application of this family of polymers is as precursors for silicon carbide. Poly(silaethylene), $[H_2SiCH_2]_n$, has been converted into silicon carbide at 1000 °C [101]. This conversion has been studied in some detail and has been found to proceed in a step-wise manner. Thus, between 300-450 °C poly(silaethylene) starts to lose H_2 to afford a crosslinked polymer. This is further converted into a fully crosslinked network by about 800 °C. The conversion of this network material into silicon carbide occurs above 1000 °C [101].

Polymers containing a $R_2Si\text{-}CH_2CH_2$ repeat unit can be prepared by the hydrosilylation reaction of the difunctional monomer $[CH=CH_2SiH_2Cl]$. This leads to the formation of poly [(dichlorosilylene)ethylene]. Reduction of this polymer with lithiumaluminumhydride affords poly(silyleneethylene),$[H_2SiCH_2CH_2]_n$ (Fig. 7.33) [104].

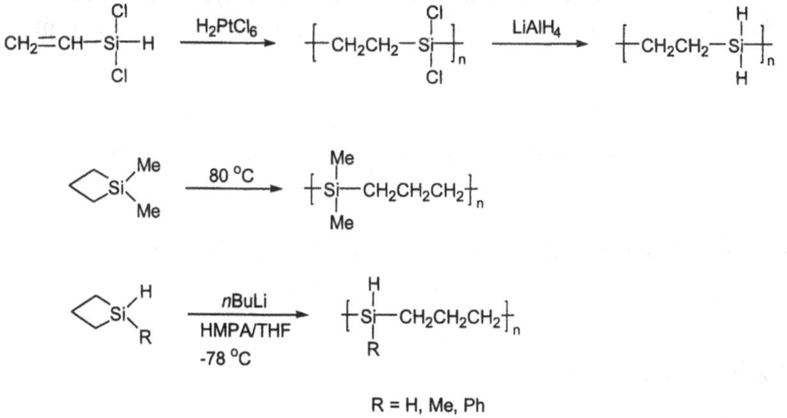

R = H, Me, Ph

Fig. 7.33. Preparation of polymers with a Si-C-C repeat unit

Polymers containing higher alkylene units alternating with silylene units can be prepared by a thermal ROP of the corresponding monosilacyclobu-

tanes. Anionic polymerization of silacyclopropanes [105] or silacyclobutanes leads to the formation of the corresponding linear polymers [106].

Other types of polycarbosilanes are also now known. Polycarbosilanes
containing aromatic groups in the backbone can be synthesized by the
Wurtz-type coupling of the appropriate dichloro/dibromo precursors (Fig.
7.34) [107-109].

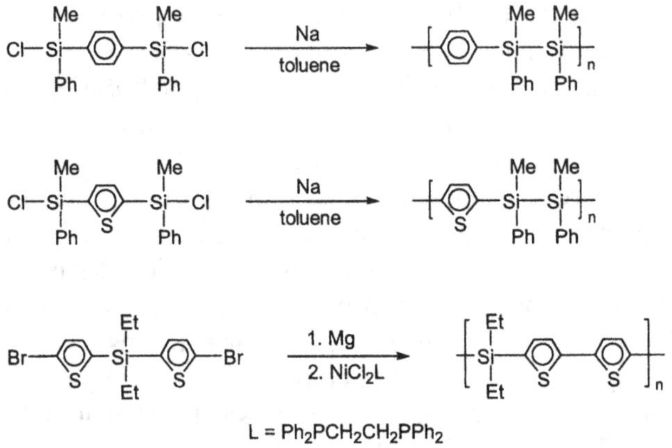

Fig. 7.34. Synthesis of polycarbosilanes containing aromatic spacer groups

Rigid-rod polymers interrupted by silicon atoms are another group of
polymers that belong to the family of polycarbosilanes (Fig. 7.35) [110].
These type of polymers are of interest because the introduction of the silicon units in the rigid-rod structure allows the polymer to become more
soluble apart from modifying the electronic properties of the parent polymers.

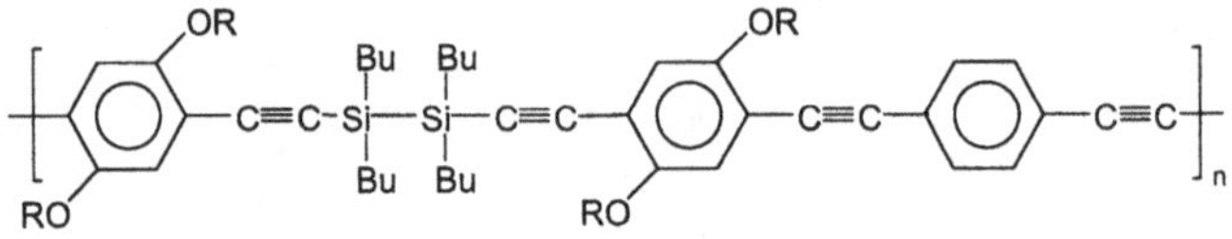

Fig. 7.35. Example of a rigid-rod polymer containing disilane spacer groups

7.9.2 Polysiloles

Polysiloles are a relatively new class of silicon-containing polymers which
contain as their building blocks the *silacyclobutadiene* [111]. Interest in
these compounds is because of the relatively low energy gap between their

HOMO and LUMO and the expected unusual electronic behavior due to σ-π conjugation.

One of the basic starting materials for the preparation of polysiloles is 1,1-dichlorotetraphenylsilole. This compound is readily prepared by the reaction of the dilithium salt of tetraphenylbutadiene with silicon tetrachloride (see Eq. 7.24) [112].

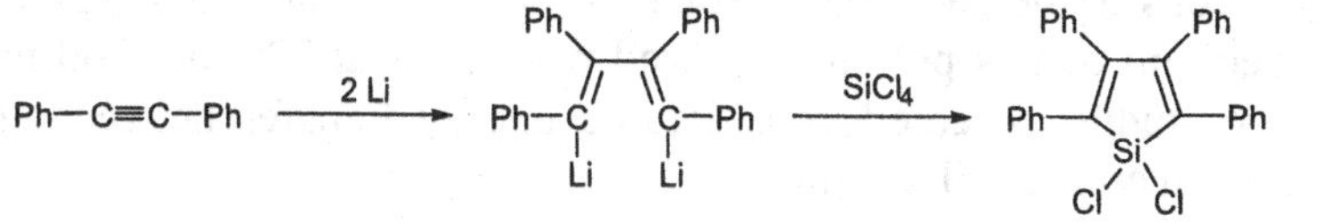

$$(7.24)$$

Reduction of 1,1-dichlorotetraphenylsilole with lithium, sodium or potassium in tetrahydrofuran affords the polysilole of a moderate molecular weight, M_n = 5500 with a M_w/M_n of 1.1 (see Eq. 7.25) [113]. The ^{29}Si NMR shows a major resonance at δ = -40.8 and has a red-shifted emission at 520 nm. This polymer is also electroluminescent. It has been suggested that the emission properties of the polysilole results from excimer formation in solution and/or the formation of a twisted intramolecular charge-transfer state. Such a conformation brings the polymer into a *trans* state which brings the silole rings into a face-to-face relationship with each other [113].

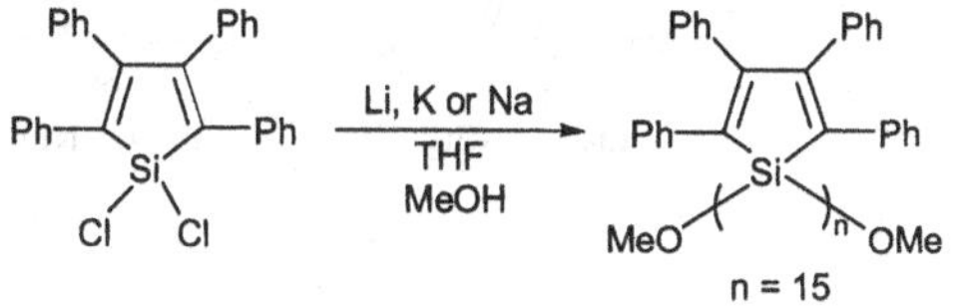

$$(7.25)$$

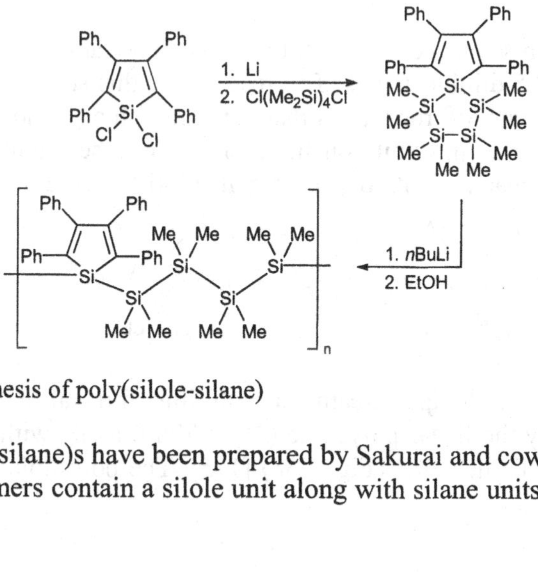

Fig. 7.36. Synthesis of poly(silole-silane)

Poly(silole-silane)s have been prepared by Sakurai and coworkers [114]. These copolymers contain a silole unit along with silane units that are connected to each other (Fig. 7.36). The reaction of 1,1-dichlorotetraphenylsilole with lithium metal along with 1,4-dichlorooctamethyltetrasilane leads to the formation of a spirocyclic compound. This serves as a monomer for an anionic ROP to afford poly(silole-silane). This polymer has a M_n of 17,000 (M_w/M_n=1.3). The ^{29}Si NMR shows three types of signals. The silole silicon resonates at δ = -27.0, while the other silicons in the backbone resonate at δ = -32.5 and -33.0. The absorption spectrum of this polymer showed two bands at 320 and 360 nm. Fluorescence studies revealed considerable energy transfer between the polysilane chain and the silole ring.

Alternative ways of preparing polysiloles involves the use of diaminosiloles. These are prepared by ring-closure reaction of bis(diethylamino)bis(phenylethynylsilane) (Fig. 7.37).

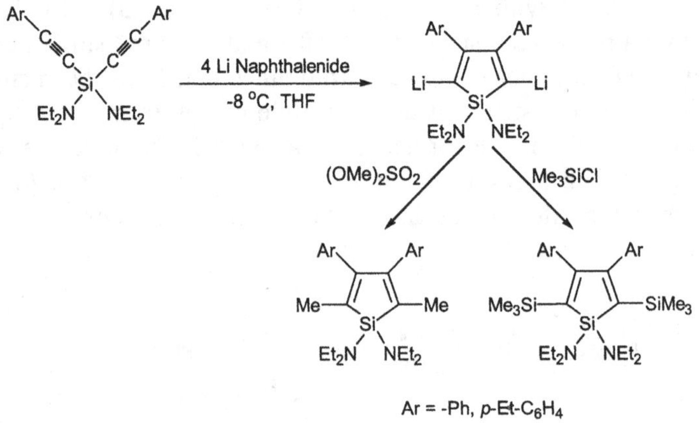

$$Ar = \text{-Ph, } p\text{-Et-C}_6H_4$$

Fig. 7.37. Synthesis of aryl/alkyl-substituted siloles

The reaction sequence shown in Fig. 7.37 first leads to the formation of 1,1-diamino-2,5-dilithiosilanes. The reaction of this species with Me₃SiCl or dimethylsulfate affords siloles that contain both aryl and alkyl substituents [115]. The amino units on these siloles can be readily replaced by chlorines by a reaction with dry hydrogen chloride (see Eq. 7.26).

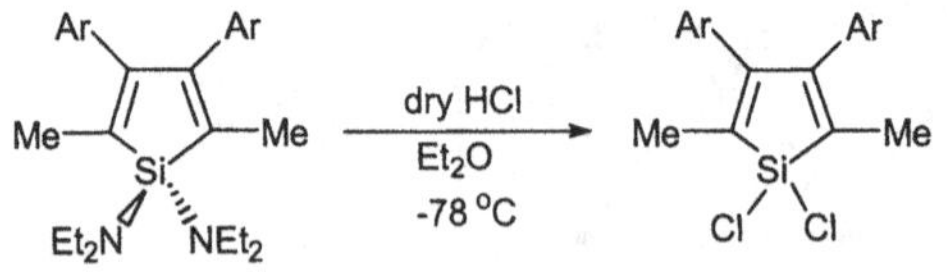

(7.26)

The dichlorosilole upon treatment with lithium metal in THF at -20 °C affords mainly the linear polysilole (M_w = 7200) along with small quantities of the cyclic hexamer (Fig. 7.38) [115]. The polysilole shown in Fig.

7.38 has a broad absorption in the UV-visible spectrum with a shoulder at 320 nm. This value is red-shifted by about 30-40 nm in comparison to the acyclic trimers and tetramers. The fluorescence spectrum of this polymer shows an emission at 460 nm. Interestingly, this poly(1,1-silole) decomposes upon treatment with lithium metal.

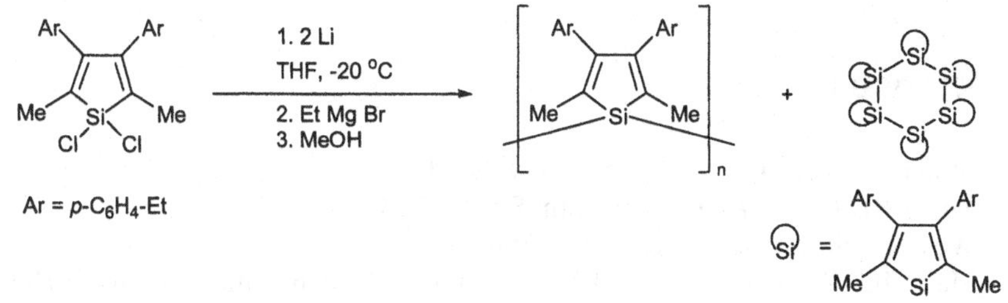

Fig. 7.38. Synthesis of poly(1,1-silole)s containing aryl and alkyl substituents

In contrast to poly(1,1-siloles) that are interconnected to each other by Si-Si bonds, poly(2,5-siloles) contain interconnected carbon chains. These are prepared by a multi-step synthetic methodology (Fig. 7.39) [116]. Poly(2,5-silole) shows a ^{29}Si NMR at δ 23.24 (major peak) and δ 21.57 and δ 24.12. Notice the downfield shift of this polymer which does not have Si-Si bonds in contrast to poly(1,1-siloles) where the polymer has Si-Si bonds.

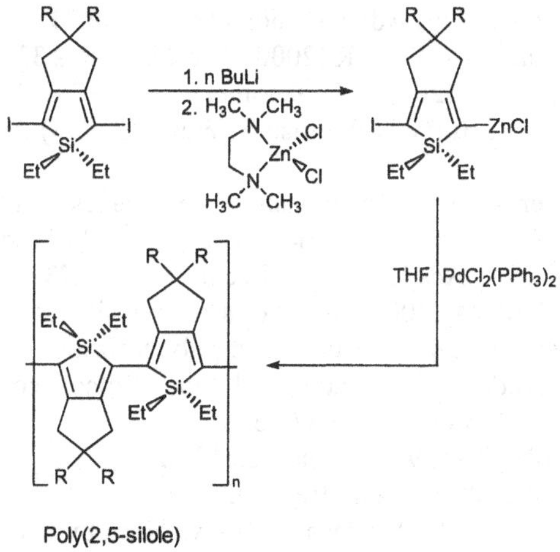

Poly(2,5-silole)

Fig. 7.39. Synthesis of poly(2,5-silole)

Poly(2,5-silole) has an optical absorption at 485 nm. This is considerably red-shifted in comparison to a dimer, di(2,5-silole). The latter has absorption at 340 nm. This indicates that considerable π-conjugation is present in this polymer. Another interesting aspect is that upon lowering the temperature to around 150 K the absorption is further red-shifted to 542 nm. It is speculated that this spectroscopic event is related to a change in the effective conjugation length of the polymer by modification in the conformation of the main chain.

References

1. Miller RD, Michl J (1989) Chem Rev 89:1359
2. Miller RD (1989) Angew Chem Int Ed 28:1733
3. West R (1986) J Organomet Chem 300:327
4. Mark JE, West R, Allcock HR (1992) Inorganic polymers. Prentice-Hall, Englewood Cliffs
5. Jones RG, Ando W, Chojnowski J (eds) (2000) Silicon containing polymers. Kluwer Academic, Dordrecht
6. Archer RD (2001) Inorganic and organometallic polymers. Wiley-VCH, Weinheim
7. Brough LF, Matsumura K, West R (1979) Angew Chem Int Ed 18:955
8. Burkhard CA (1949) J Am Chem Soc 71:963
9. Yajima S, Hayashi J,Omori M (1975) Chem Lett 931
10. Yajima S, Okamura K, Hayashi J, Omori M (1976) J Am Ceram Soc 59:324
11. Yajima S (1983) Ceram Bull 62:993
12. West R (1987) Angew Chem Int Ed 26:1201
13. Okazaki R, West R (1996) Adv Organomet Chem 39:231
14. Haaf M, Schmedake TA, West R (2000) Acc Chem Res 33:704
15. Gehrus B, Lappert MF (2001) J Organomet Chem 617/618:209
16. Koe JR, Powell DR, Buffy JJ, Hayase S, West R (1998) Angew Chem Int Ed 37:1441
17. Jones RG, Holder SJ (2000). Synthesis of polysilanes by the Wurtz reductive coupling reaction. In: Jones RG, Ando W, Chojnowski J (eds) Silicon containing polymers. Kluwer Academic, Dordrecht, pp 353-373
18. Sakurai H, Yoshida M (2000). Synthesis of polysilanes by new procedures: Part 1 Ring-opening polymerizations and polymerization of masked disilenes. In: Jones RG, Ando W, Chojnowski J (eds) Silicon containing polymers. Kluwer Academic, Dordrecht, pp 375-399
19. Imori T, Don Tilley T (1994) Polyhedron 13:2231
20. Don Tilley T (1993) Acc. Chem. Res 26:22
21. Gray GM, Corey J (2000). Synthesis of polysilanes by new procedures:Part 1: Catalytic dehydropolymerization of hydrosilanes. In: Jones RG, Ando W, Chojnowski J (eds) Silicon containing polymers. Kluwer Academic, Dordrecht, pp 401-418
22. Ishifune M, Kashimura S, Kogai Y, Fukuhara Y, Kato T, Bu HB, Yamashita N, Murai Y, Murase H, Nishida R (2000) J Organomet Chem 611:26

23. Miller RD, Farmer BL, Fleming W, Sooriyakumaran R, Rabolt J (1987) J Am Chem Soc 109:2509
24. Rabolt JF, Hofer D, Miller RD, Fickes GN (1986) Macromolecules 19:611
25. Chunwachirasiri W, West R, Winokur MJ (2000) Macromolecules 33:9720
26. Miller RD, Sooriyakumaran R (1987) J Polym Sci Polym Lett 25:321
27. Trefonas III P, Djurovich PI, Zhang XH, West R, Miller RD, Hoffer D (1983) J Polym Sci Polym Lett 21:819
28. Fujino M, Hisaki T, Fujiki M, Matsumoto N (1992) Macromolecules 25:1079
29. Seki S, Kunimi Y, Nishida K, Aramaki K, Tagawa S (2000) J Organomet Chem 611:64
30. West R, Lawrence DD, Djurovich PI, Stearley KL, Srinivasan KSV, Yu H (1981) J Am Chem Soc 103:7352
31. Zhang XH, West R (1984) J Polym Sci Polym Chem 22:159
32. Zhang XH, West R (1985) J Polym Sci Polym Lett 23:479
33. Zhang XH, West R (1984) J Polym Sci Polym Chem 22:225
34. Pannell KH, Rozell JM, Zeigler JM (1988) Macromolecules 22:276
35. Jones RG, Wong KC, Holder SJ (1998) Organometallics 17:59
36. Saxena A, Okoshi K, Fujiki M, Naito M, Guo G, Hagihara T, Ishikawa M (2004) Macromolecules 37:367
37. Benfield RE, Cragg RH, Jones RG, Swain AC (1991) Nature 353:340
38. Miller RD, Jenkner PK (1994) Macromolecules 27:5921
39. Miller RD, Thompson D, Sooriyakumaran R, Fickes GN (1991) J Polym Sci Polym Chem 29:813
40. Cragg RH, Jones RG, Swain AC, Webb SJ (1990) Chem Commun 1147
41. Kim HK, Matyjaszewski K (1988) J Am Chem Soc 110: 3321
42. Price GJ (1992) Chem Commun 1209
43. Lacave-Goffin B, Hevesi L, Devaux J (1996) Chem Commun 765
44. Jones RG, Benfield RE, Evans PJ, Swain AC (1995) Chem Commun 1465
45. Lacave-Goffin B, Hevesi L, Devaux J (1995) Chem Commun 769
46. Cleij TJ, Tsang SKY, Jenneskens, LW (1997) Chem Commun 329
47. Cypryk M, Gupta Y, Matyjaszewski K (1991) J Am Chem Soc 113:1046
48. Suzuki M, Kotani J, Gyobu S, Kaneko T, Saegusa T (1994) Macromolecules 27:2360
49. Matsumoto H, Arai T, Watanabe H, Nagai Y (1984) Chem Commun 724
50. Sakamoto K, Obata K, Hirata H, Nakajima M, Sakurai H (1989) J Am Chem Soc 111:7646
51. Sakamoto K, Yoshida M, Sakurai H (1990) Macromolecules 23:4494
52. Sanji T, Kawabata K, Sakurai H (2000) J Organomet Chem 611:32
53. Sanji T, Kitayama F, Sakurai H (1999) Macromolecules 32:5718
54. Sanji T, Takase K, Sakurai H (2001) J Am Chem Soc 123:12690
55. Sanji T, Nakatsu Y, Kitayama F, Sakurai H (1999) Chem Commun 2201
56. Aitken C, Harrod JF, Samuel E (1986) J Am Chem Soc 108:4059
57. Mu Y, Aitken C, Cote B, Harrod JF, Samuel E (1991) Can J Chem 69:264
58. Grimmond BJ, Rath NP, Corey JY (2000) Organometallics 19:2975
59. Grimmond BJ, Corey JY (1999) Organometallics 18:2223
60. Grimmond BJ, Corey JY (2000) Organometallics 19:3776

61. Went MJ, Sakurai H, Sanji T (2000). Modification and functionalization of polysilanes. In: Jones RG, Ando W, Chojnowski J (eds) Silicon containing polymers. Kluwer Academic, Dordrecht, pp 419-437
62. Hsio YL, Waymouth RM (1994) J Am Chem Soc 116:9779
63. Banovetz JP, Hsiao YL, Waymouth RM (1993) J Am Chem Soc 115:2540
64. Herzog U, West R (1999) Macromolecules 32:2210
65. Ban H, Sukegawa K, Tagawa S (1987) Macromolecules 20:1775
66. Matsui Y, Nishida K, Seki S, Yoshida Y, Tagawa S, Yamada K, Imahori H, Sakata Y (2002) Organometallics 21:5144
67. Furukawa K, Ebata K, Nakashima H, Kashimura Y, Torimitsu K (2003) Macromolecules 36:9
68. Michl J (1990) Acc Chem Res 23:128
69. Michl J, West R (2000). Electronic structure and spectroscopy of polysilanes. In: Jones RG, Ando W, Chojnowski J (eds) Silicon containing polymers. Kluwer Academic, Dordrecht, pp 499-529
70. Sun YP, Hamada Y, Huang LM, Maxka J, Hsiao JS, West R, Michl J (1992) J Am Chem Soc 114:6301
71. Obata K, Kira M (1999) Organometallics 18:2216
72. Schilling FC, Bovey FA, Zeigler JM (1986) Macromolecules 19:2309
73. Menescal R, West R (1990) Macromolecules 23:4492
74. Fujiki M, Koe JR (2000). Optically active silicon-containing polymers. In: Jones RG, Ando W, Chojnowski J (eds) Silicon containing polymers. Kluwer Academic, Dordrecht, pp 643-665
75. Michl J, West R (2000). In: Jones RG, Ando W, Chojnowski J (eds) Silicon containing polymers. Kluwer Academic, Dordrecht, pp 499-529
76. Harrah LA, Zeigler JM (1987) Macromolecules 20:601
77. Trefonas III P, West R, Miller RD, Hofer D (1983) J Polym Sci Polym Lett 21:823
78. Miller RD, Sooriyakumaran R (1988) Macromolecules 21:3120
79. Yuan CH, West R (1998) Macromolecules 31:1087
80. Sakamoto K, Yoshida, M, Sakurai H (1994) Macromolecules 27:881
81. Oka K, Fujiue N, Nakanishi S, Takata T, West R, Dohmaru T (2000) J Organomet Chem 611:45
82. Bukalov SS, Leites LA, West R (2001) Macromolecules 34:6003
83. Sanji T, Sakamoto K, Sakurai H, Ono K (1999) Macromolecules 32:3788
84. Nakanishi K, Berova N, Woody RW (1994) Circular dichroism: Principles and Applications. VCH, New York
85. Fujiki M (1994) J Am Chem Soc 116:11976
86. Nakashima H, Fujiki M, Koe JR (1999) Macromolecules 32:7707
87. Koe JR, Fujiki M, Nakashima H (1999) J Am Chem Soc 121:9734
88. Fujiki M (2000) J Am Chem Soc 122:3336
89. Fujiki M, Koe JR, Motonaga M, Nakashima H, Terao K, Teramoto A (2001) J Am Chem Soc 123:6253
90. Koe JR, Fujiki M, Motonaga M, Nakashima H (2001) Macromolecules 34:1082

91. Nakashima H, Koe JR, Torimitsu K, Fujiki M (2001) J Am Chem Soc 123:4847
92. Teramoto A, Terao, K, Terao Y, Nakamura N, Sato T, Fujiki M (2001) J Am Chem Soc 123:12303
93. Czubarow P, Sugimoto T, Seyferth D (1998) Macromolecules 31:229
94. Reichmanis E, Novembre AE, Nalamasu O, Dabbagh G (2000) Microlithographic applications of organosilicon polymers. In: Jones RG, Ando W, Chojnowski J (eds) Silicon containing polymers. Kluwer Academic, Dordrecht, pp 743-761
95. Matsumoto N, Suzuki H, Miyazaki H (2000) Electronic and optical properties in device applications of polysilanes. In: Jones RG, Ando W, Chojnowski J (eds) Silicon containing polymers. Kluwer Academic, Dordrecht, pp 531-552
96. Sekiguchi A, Sakurai H (1995) Adv Organomet Chem 37:1
97. Bianconi PA, Weidman TW (1988) J Am Chem Soc 110:2342
98. Bianconi PA, Schilling FC, Weidman TW (1989) Macromolecules 22:1697
99. Kobayashi T, Shimura H, Mitani S, Mashimo S, Amano A, Takano T, Abe M, Watanabe H, Kijima M, Shirakawa H, Yamaguchi H (2000) Angew Chem Int Ed 39:3110
100. Interrante LV, Shen Q (2000). In: Jones RG, Ando W, Chojnowski J (eds) Silicon containing polymers. Kluwer Academic, Dordrecht, pp 247-321
101. Interrante LV, Liu Q, Rushkin I, Shen Q (1996) J Organomet Chem 521:1
102. Wu HJ, Interrante LV (1992) Macromolecules 25:1840
103. Lienhard M, Rushkin I, Verdecia G, Wiegaud C, Apple T, Interrante LV (1997) J Am Chem Soc 117:12020
104. Boury B, Corriu RJP, Leclerq D, Mutin PH, Planeix JM, Vioux A (1991) Organometallics 10:1457
105. Matsumoto K, Matsuoka H (2003) Macromolecules 36:1474
106. Liao CX, Weber WP (1992) Polym Bull 28:281
107. Uchimaru Y, Tanaka Y, Tanaka M (1995) Chem Lett 164
108. Ohshita J, Kanaya D, Ishikawa M (1994) J Organomet Chem 486:55
109. Ohshita J, Takata K, Kai H, Kunai A, Komaguchi K, Shiotani M, Adachi A, Sakamaki K, Okita K, Harima Y, Kunugi Y, Yamashita K, Ishikawa M (2000) Organometallics 19:4492
110. Li H, West R (1998) Macromolecules 31:2866
111. Tamao K , Yamaguchi S (2000) J Organomet Chem 611:5
112. Joo, WC, Hong JH, Choi SB, Son HE, Kim CH (1990) J Organomet Chem 391:27
113. Sohn H, Huddleston RR, Powell DR, West R (1999) J Am Chem Soc 121:2935
114. Sanji T, Sakai T, Kabuto C, Sakurai H (1998) J Am Chem Soc 120:4552
115. Yamaguchi S, Jin RZ, Tamao K (1999) J Am Chem Soc 121:2937
116. Yamaguchi S, Jin RZ, Itami Y, Goto T, Tamao K (1999) J Am Chem Soc 121:10420

8 Organometallic Polymers

8.1 Introduction

Organometallic compounds have come to occupy a position of prominence in the last 50 years. Although the first organometallic compound is traditionally regarded as Zeise's salt, $K[PtCl_3CH_2=CH_2]$, the major impetus to this field is undoubtedly the discovery of ferrocene in the 1950's. This was followed by the landmark finding of Ziegler that organoaluminum compounds can function as catalysts towards the polymerization of ethylene. Later work showed that organoaluminum compounds in conjunction with titanium halides can be used for the preparation of high density polyethylene and isotactic polypropylene. As we had seen in Chap. 2 various developments in Ziegler-Natta catalysis have occurred since the original discovery. Historically the importance of Ziegler-Natta catalysts has been to move organometallic compounds from being laboratory curiosities to become industrially important compounds. Many other amazing developments have occurred in the field of organometallic chemistry and many textbooks deal with this subject [1-2]. In this chapter we will have a brief survey of polymers that incorporate organometallic fragments in their back-bone. Major motivations for attempts to prepare polymers that contain metals in the backbone are as follows:

1. The difficulty of assembling metal-containing polymers has been and continues to be a synthetic challenge. The process of attempting to prepare these polymers has led to a number of novel preparative routes.
2. The expectation that some members of this polymer family may have unusual electrical, magnetic or optical properties has also been a major motivation for studying these polymer systems [3-6].

Although many types of polymers are now known, in keeping with the objectives of this book, we will examine families of organometallic polymers that are now reasonably well developed and which already show promising properties. We will not be looking at coordination polymers or polymers where coordination metal complexes are linked to each other. This subject is dealt in much more detail elsewhere [7, 8]. In this chapter we will examine the following three types of polymers:

1. The first of these are *polygermanes* and *polystannanes*. Although germanium is not a metal we have included polygermanes in this discussion for the sake of continuity. In polygermanes and polystannanes the polymers contain catenated germanium and tin atoms, respectively [7]. Analogous to polysilanes that we have seen in Chap. 7 polygermanes and polystannanes have unusual electronic properties that result from a σ-delocalized backbone.

2. The second class of polymers that we will consider in this chapter is those that contain ferrocene in the backbone. Although *ferrocene- containing polymers* were among the first types of organometallic polymers that have been investigated, this area received a major boost by the discovery of the ring-opening polymerization of strained ferrocenophanes [3, 7].

3. The third class of polymers that we will investigate in this chapter is *polyynes* which are prepared by the involvement of acetylene functional groups in the polymerization reaction. These polymers are also called *rigid-rod organometallic polymers*. In these types of polymers various late transition metal ions alternate with a rigid organic unit.

8.2 Polygermanes and Polystannanes

Polygermanes and polystannanes are heavier Group 14 congeners of polysilanes (Fig. 8.1).

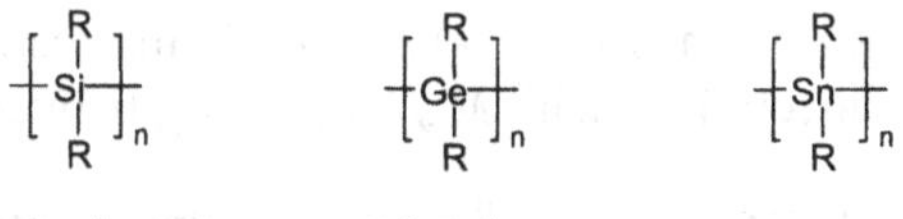

$$\left[\begin{array}{c} R \\ | \\ Si \\ | \\ R \end{array}\right]_n \qquad \left[\begin{array}{c} R \\ | \\ Ge \\ | \\ R \end{array}\right]_n \qquad \left[\begin{array}{c} R \\ | \\ Sn \\ | \\ R \end{array}\right]_n$$

Fig. 8.1. Polysilanes, polygermanes and polystannanes

As discussed in the last chapter, polysilanes have several important properties that emanate from σ-delocalization. Consequently, these polymers exhibit interesting behavior such as absorption in the UV-visible region, thermochromism, photoconductivity and semiconducting properties (particularly on doping) [9, 10]. These results have prompted investigations on the heavier Group 14 congeners of polysilanes viz., polygermanes and polystannanes. Polygermanes have been prepared mainly by the Wurtz- type coupling reaction of diorganogermaniumdichlorides, while polystannanes have been prepared principally by the catalytic dehydrogenation of diorganotindihydrides.

8.2.1 Synthesis of Polygermanes

Polygermanes have been prepared by the Wurtz-type coupling reaction of diorganogermaniumdichlorides, R_2GeCl_2 [11-13]. The latter are most readily synthesized by an equilibration between R_4Ge and $GeCl_4$ [12] (see Eq. 8.1).

$$GeCl_4 \ + \ RMgX \ \longrightarrow \ R_2Ge \tag{8.1}$$

$$R_4Ge \ + \ GeCl_4 \ \xrightarrow[\Delta]{AlCl_3} \ R_2GeCl_2$$

$$R = alkyl$$

The Wurtz-type coupling reaction of R_2GeCl_2 with sodium in high-boiling solvents such as toluene at reflux temperatures affords polygermanes (see Eq. 8.2).

$$R_2GeCl_2 \ \xrightarrow[\substack{toluene, \Delta \\ -NaCl}]{Na} \ [R_2Ge]_n \tag{8.2}$$

$$R = n\text{-Hex}, n\text{-Pen}, n\text{-Oct}$$

The crude polymers have polymodal molecular weight distribution and the higher-molecular-weight species can be separated from the lower-molecular-weight compounds by a reprecipitation process. In general, the yields of the high molecular weight polymers are low and range from 5-15%. The molecular weights (of the high-molecular-weight species) are quite high and range from 300,000-1,000,000. Interestingly, these polymers have a fairly narrow PDI with M_w/M_n ranging between 1.5-2.5 [12].

$$n\text{-Hex}_2GeCl_2 \ + \ n\text{-Hex}_2SiCl_2 \ \xrightarrow[-NaCl]{Na} \ \left[(n\text{Hex}_2Ge)_x \ (n\text{Hex}_2Si)_y\right] \tag{8.3}$$

$$n\text{-}R_2GeCl_2 \ + \ R'R''SiCl_2 \ \xrightarrow[-NaCl]{Na} \ \left[(n\text{Bu}_2Ge)_x \ (RR'Si)_y\right]$$

$$R = Ph, n\text{Bu}$$
$$R' = n\text{Hex}; R'' = Me$$
$$R' = c\text{Hex}; R'' = Me$$

High-molecular-weight germanium-silicon copolymers have been prepared by the sodium assisted cocondensation of R_2GeCl_2 and R_2SiCl_2 [11]
(see Eq. 8.3).

Polygermanes can also be synthesized by a dealkylation (demethanation) reaction [14-15]. Berry and coworkers have shown that
poly(dimethylgermane) and poly(arylmethylgermane)s can be synthesized
by an ambient-temperature polymerization process which is catalyzed by
an organometallic ruthenium complex. For example, the synthesis of
$H[Me_2Ge]_nH$ ($M_w = 2 \times 10^4 - 2 \times 10^5$) is carried out by the elimination of
methane from $HGeMe_3$. This reaction is catalyzed by the ruthenium complex $Ru(PMe_3)_4Me_2$. The polygermane prepared by this route is believed to
have a branched structure and possesses $(GeMe_2)_nGeMe_3$ pendant groups
[14] (see Eq. 8.4).

$$HGeMe_3 \xrightarrow[-CH_4]{Ru(PMe_3)_4Me_2} H{\left[\begin{matrix} Me \\ | \\ Ge \\ | \\ Me \end{matrix}\right]}_n Me \qquad (8.4)$$

In order to synthesize poly(arylmethylgermane)s the suitable starting
materials would be of the type $ArGeMe_2H$. These compounds are prepared
by a dearylation reaction of Ar_2GeMe_2 followed by reduction with lithium
aluminum hydride [15] (see Eq. 8.5).

$$Me_2GeCl_2 \xrightarrow[Et_2O]{2\ ArMgBr} Me_2GeAr_2 \qquad (8.5)$$

$$\downarrow CF_3SO_3H$$

$$Me_2GeArH \xleftarrow{LiAlH_4} Me_2GeAr(SO_3CF_3)$$

Demethanation of $ArGeMe_2H$ also proceeds in a facile manner to afford
high molecular weight polymers $H[MeArGe]_nH$ ($M_w = 5 \times 10^3 - 1 \times 10^4$;
$M_w/M_n = 1.3$) [15] (see Eq. 8.6).

$$Me_2GeArH \xrightarrow[-CH_4]{Ru(PMe_3)_4Me_2} H{\left[\begin{matrix} Me \\ | \\ Ge \\ | \\ Ar \end{matrix}\right]}_n Me \qquad (8.6)$$

$Ar = C_6H_5,\ p\text{-}CH_3\text{-}C_6H_4,\ p\text{-}CF_3\text{-}C_6H_4,\ p\text{-}OCH_3\text{-}C_6H_4$ etc

Polygermanes are soluble in many organic solvents. These polymers are
extremely sensitive to light, particularly in solution and in the presence of
oxygen.

The reduction of $RGeCl_3$ with sodium/potassium alloy under ultrasonic
conditions affords network polygermynes [$M_w = 3160$ (n-butylgermynes);

$M_w = 7040$ (phenylgermynes)] (see Eq. 8.7). Polygermynes have been suggested to possess network structures analogous to that of polysilynes.

$$RGeCl_3 \xrightarrow[\text{ultrasound}]{\text{Na/K}} \left[RGe\right]_n \qquad (8.7)$$

$$R = n\text{Bu, Ph}$$

8.2.2 Synthesis of Polystannanes

Although the initial efforts of the synthesis of polystannanes by the Wurtz-coupling methodology did not appear to be successful, Molloy and co-workers have shown that a carefully controlled reaction involving $n\text{Bu}_2\text{SnCl}_2$ with sodium dispersion in the presence of 15-crown-5-ether in toluene at 60 °C affords the polymer $[n\text{Bu}_2\text{Sn}]_n$ in optimum yields after 4 h of reaction [17] (see Eq. 8.8).

$$n\text{Bu}_2\text{SnCl}_2 \xrightarrow[\substack{\text{15-crown-ether} \\ \text{70 °C, -NaCl}}]{\text{Na}} \left[n\text{Bu}_2\text{Sn}\right]_n \qquad (8.8)$$

If the reaction is carried out for an extended period of time, cyclic oligomers such as $[n\text{Bu}_2\text{Sn}]_5$ and $[n\text{Bu}_2\text{Sn}]_6$ are formed. The polymeric products are characterized by high molecular weights ($\sim 10^6$). ^{119}Sn NMR is a useful diagnostic tool for distinguishing the polymers. Thus, while the chemical shift of the polymer $[n\text{Bu}_2\text{Sn}]_n$ is -178.9 ppm, the cyclic oligomers resonate further upfield ($\delta = -203.0$ and -204.2) [17].

In spite of the above successful example, the Wurtz-coupling strategy has not been successful for preparation of other types of polystannanes. Catalytic dehydrogenation has been shown to be more successful [18-20]. The monomers suitable for catalytic dehydrogenation reaction are of the type $R_2\text{SnH}_2$. These monomers are prepared by a two-step procedure. Thus, comproportionation reaction between $R_4\text{Sn}$ and SnCl_4 leads to the formation of $R_2\text{SnCl}_2$. These are then reduced by LiAlH_4 to afford the corresponding dihydrides, $R_2\text{SnH}_2$, as air- and temperature-sensitive substances. A number of poly(dialkylstannane)s are readily prepared by the catalytic action of organometallic zirconocene catalysts such as Cp_2ZrMe_2. Typically the polymerization is initiated by the addition of neat monomer to the catalyst. This leads to the evolution of hydrogen and the formation of polymeric materials along with cyclic products. The main cyclic products are five-membered compounds $[n\text{R}_2\text{Sn}]_5$ although small amounts of six-membered $[\text{R}_2\text{Sn}]_6$ are also formed. Molecular weights of the poly(dialkylstannane)s are fairly high as, for example, in $\text{H}[n\text{Bu}_2\text{Sn}]_n\text{H}$

$(M_w/M_n = 13,900/4600)$ or in $H[nOct_2Sn]H$ $(M_w/M_n = 92,600/21,700)$ (see Eq. 8.9) [18].

$$R_2SnH_2 \xrightarrow{\text{Metallocene catalyst}} H[SnR_2]_nH + cyc\text{-}[SnR_2]_m \qquad (8.9)$$

$$R = n\text{Bu}; \ n\text{Hex}; \ n\text{Oct} \qquad m = 5, 6$$

Poly(diarylstannane)s are also synthesized in a similar manner as that of poly(dialkylstannane)s. These monomers are prepared by a two-step procedure. Thus, comproportionation reaction between R_4Sn and $SnCl_4$ leads to the formation of R_2SnCl_2. These are then reduced by $LiAlH_4$ to afford the corresponding dihydrides, R_2SnH_2, as air- and temperature-sensitive substances (see Eq. 8.10).

$$4\text{ArMgBr} + SnCl_4 \longrightarrow Ar_4Sn \qquad (8.10)$$

$$Ar_4Sn + SnCl_4 \longrightarrow 2Ar_2SnCl_2$$

$$Ar_2SnCl_2 \xrightarrow{\text{LiAlH}_4} Ar_2SnH_2$$

$$Ar = p\text{-}t\text{Bu-C}_6\text{H}_4; \ p\text{-}n\text{Hex-C}_6\text{H}_4$$
$$o\text{-Et-C}_6\text{H}_4; \ p\text{-}n\text{BuO-C}_6\text{H}_4;$$
$$p\text{-}\{(\text{Me}_3\text{Si})_2\text{N}\}\text{-C}_6\text{H}_4$$

Dehydrogenation of Ar_2SnH_2 can also be accomplished by the use of metallocene catalysts (see Eq. 8.11). High-molecular-weight polymers along with cyclic products (mainly hexameric products) are formed. Separation of the polymeric products from cyclic rings is effected by fractionation.

$$(8.11)$$

The molecular weights of the poly(diarylstannanes) are quite high; for example, $H[(p\text{-}t\text{Bu-C}_6\text{H}_4)_2Sn]_nH$ has a $M_w/M_n = 56,000/16,700$.

The ^{119}Sn-NMR chemical shifts of the poly(alkyl-) and poly(arylstannanes), surprisingly, are quite similar; the cyclic products have up field chemical shifts (Table 8.1). In contrast to the fairly linear polystannanes prepared by the catalytic action of zirconocene catalysts, branched polymers are obtained in the reaction of $n\text{Bu}_2SnH_2$ with the rhodium(I) catalyst $HRh(CO)(PPh_3)_3$. This has been explained based on a dehydropolymerization-rearrangement process [21].

Table 8.1. ^{119}Sn-NMR chemical shifts of poly- and cyclostannanes

S.No	Polymer	$\delta^{119}Sn$
1	$H-[(p\text{-}t\text{Bu-}C_6H_4)_2Sn]_nH$	-197.0
2	$H-[(p\text{-}n\text{Hex-}C_6H_4)_2Sn]_nH$	-196.0
3	$H-[(p\text{-}n\text{BuO-}C_6H_4)_2Sn]_nH$	-183.5
4	$H[n\text{Bu}_2Sn]_nH$	-189.6
5	$H[n\text{Hex}_2Sn]_nH$	-190.9
6	$H[n\text{Oct}_2Sn]_nH$	-190.7
7	$Cyc\text{-}[n\text{Bu}_2Sn]_5$	-200.9
8	$Cyc\text{-}[n\text{Bu}_2Sn]_6$	-202.1
9	$Cyc\text{-}[(p\text{-}t\text{Bu-}C_6H_4)_2Sn]_6$	-221.0

Other methods of polystannane synthesis include electrochemical methods [22]. Thus, $[n\text{Bu}_2Sn]_n$ ($M_w = 10,900$; $M_w/M_n = 2.6$) and $[n\text{Oct}_2Sn]_n$ ($M_w = 5900$; $M_w/M_n = 1.7$) were prepared by the electrochemical polymerization of dibutyldichlorostannane and dioctyldichlorostannane, respectively, by using a platinum cathode and a silver anode. Tetrabutylammonium perchlorate is used as the supporting electrolyte and dimethoxy ethane is used as the solvent.

Polystannanes are extremely sensitive to light and moisture, particularly in solution. However, it has been shown that polystannanes are fairly stable from a thermal point of view [18, 20]. Decomposition to tetragonal tin is reported to occur on bulk pyrolysis of polystannanes in a flowing nitrogen atmosphere. On the other hand, heating polystannanes in an oxygen atmosphere affords a high ceramic residue, identified as SnO_2 (cassiterite) [18, 20].

8.2.3 Electronic Properties of Polygermanes and Polystannanes

The electronic spectra of polygermanes and polystannanes show a σ-delocalization analogous to that found in polysilanes. The HOMO-LUMO gap in these polymers is lower than in polysilanes. Thus, the absorptions in polygermanes are about 20-30 nm red-shifted in comparison to the corresponding polysilanes [11-12]. Analogous polystannanes show even further bathochromic shifts, about 70 nm in comparison to the corresponding polysilanes [18-20]. Thus, for example, in a series of dibutyl polymers a continuous red shift is observed: $[n\text{Bu}_2Si]_n$ (314 nm); $[n\text{Bu}_2Ge]_n$ (333 nm); $[n\text{Bu}_2Sn]_n$ (384 nm). Similar to polysilanes the extinction coefficients in polygermanes and polystannanes are also quite high. Also, following the similar trend as in polysilanes, poly(alkyl/arylstannane)s have further red-shifted absorptions in comparison to poly(alkylstannane)s. Some of the optical absorption data are summarized in Table 8.2.

Table 8.2. Optical absorption data for polystannanes, polygermanes and some polysilanes

S.No	Polymer	λ_{max}
1	H-[(p-tBu-C$_6$H$_4$)$_2$Sn]$_n$H	432
2	H-[(p-nHex-C$_6$H$_4$)$_2$Sn]$_n$H	436
3	H-[(p-nBuO-C$_6$H$_4$)$_2$Sn]$_n$H	448
4	H[(o-Et-p-nBuO-C$_6$H$_3$)$_2$Sn]$_n$H	506
5	H[nBu$_2$Sn]$_n$H	384
6	H[nHex$_2$Sn]$_n$H	384
7	H[nOct$_2$Sn]$_n$H	388
	[Ph(nHex)Ge]$_n$	355
8	[nBu$_2$Ge]$_n$	333
9	[nBu$_2$Si]$_n$	314
10	[nHex$_2$Ge]$_n$	340
11	[nPen$_2$Ge]$_n$	339
12	[nOct$_2$Ge]$_n$	342
13	[PhMeGe]$_n$	332
14	[(p-MeO-C$_6$H$_4$)(Me)Ge]$_n$	338
13	[(nHex$_2$Si)$_x$(nHex$_2$Ge)$_y$]	322
14	[(Ph$_2$Ge)$_x$ (n-Hex(Me)Si)$_y$]	354, 305
15	[(n-Hex)$_2$Si]$_n$	306

Poly(dihexylgermane) and poly(dioctylgermane) are strongly thermochromic. For example, a thin film of [nHex$_2$Ge]$_n$ absorbs at -11 °C at 370 nm, while at 22 °C the absorption shifts to 337 nm [12]. These changes are reversible as a function of temperature. On the other hand, poly(diarylstannane)s do not seem to be thermochromic, although poly(dialkylstannane)s are thermochromic [18-20].

8.3 Ferrocene-Containing Polymers

Ferrocene is both thermally and electrochemically very robust. These properties have motivated attempts to incorporate the ferrocene unit as part of a polymer framework. This section deals with various types of polymers containing ferrocene.

8.3.1 Synthetic Strategies

Ferrocene is an 18-valence electron compound. Consequently, intermolecular linkages of the iron centers of ferrocene cannot be carried out. This
will increase the number of ligands around iron and hence the number of
valence electrons will exceed 18 (in contrast, bent-metallocenes such as
Cp_2TiCl_2, Cp_2ZrCl_2 and Cp_2HfCl_2 can be considered as difunctional
monomers and can be condensed with a variety of other difunctional reagents such as diols, dicarboxylic acids etc. to afford medium-molecular-
weight polymers). Thus, in order to prepare polymers containing ferrocenes, the cyclopentadienyl units have to be utilized as the motifs to be
linked. Alternatively, the ferrocene groups can be present as side-chains or
pendants of a main-chain polymer (Fig. 8.2).

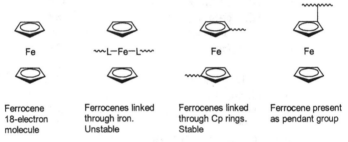

| Ferrocene 18-electron molecule | Ferrocenes linked through iron. Unstable | Ferrocenes linked through Cp rings. Stable | Ferrocene present as pendant group |

Fig. 8.2. Ferrocene and ways of incorporating it in a polymer

Using the above strategies many types of ferrocene polymers have been
successfully prepared. Thus, vinylferrocene can be polymerized by a variety of ways which include radical, cationic or anionic initiators to afford
poly(vinylferrocene) (see Eq. 8.12).

$$\text{(8.12)}$$

Vinylferrocene → Polymerization → Poly(vinylferrocene)

Poly(vinylferrocene) is soluble in solvents such as THF and shows an
electronic absorption (λ_{max} = 440 nm). This value is similar to that found
for ferrocene. The electrochemical studies on poly(vinylferrocene) reveal
that the pendant ferrocene units do not interact with each other as revealed
by the presence of a single oxidation potential [7-8]. While virgin
poly(vinylferrocene) is an insulator, its conductivity increases to about 10^{-8}
– 10^{-6} Scm^{-1} (from 10^{-12} Scm^{-1}) upon doping with dopants such as I_2.

Other ferrocene pendant-containing polymers include polymers such as poly(ferrocenyl methyl acrylate) and poly(ferrocenyl methyl methacrylate) [23]. These polymers have been prepared from the free-radical polymerization of the corresponding ferrocenyl acrylate monomers (Fig. 8.3) [23].

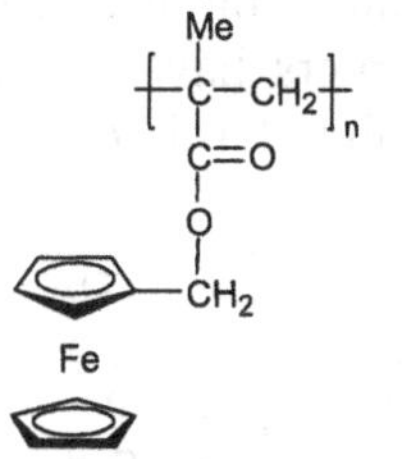

Poly(ferrocenyl methyl methacrylate)

Fig. 8.3. Acrylate polymers containing ferrocene pendant groups

Condensation polymerization methods have been used for the incorporation of ferrocenes into polymeric backbones. The synthetic strategy involved in this methodology is the preparation of suitable difunctional ferrocenyl monomers. A large number of functional groups can be readily introduced on the cyclopentadienyl unit without affecting the overall integrity of the ferrocene molecule. Representative examples of the application of this methodology are given in the following account.

One of the most successful early methodologies of preparing condensation polymers of ferrocene involves the condensation reaction of the difunctional reagents 1,1'-dilithioferrocene with 1,1'-diiodoferrocene. This reaction occurs with the elimination of lithium iodide (see Eq. 8.13) [24]. This synthetic protocol affords poly(1,1'-ferrocenylene)s with molecular weights that approach about M_n= 4000. The difficulty with this kind of a synthetic method is the need to have extremely high-purity starting materials for achieving high-molecular-weight polymers. This requirement is all the more difficult to achieve for highly reactive and sensitive monomers such as 1,1'-dilithioferrocene.

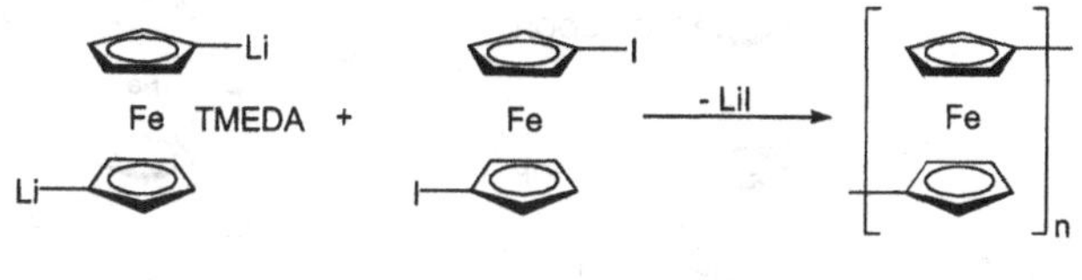

(8.13)

Poly(ferrocenylene)

Other approaches to prepare poly(ferrocenylene)s have been to use dehalogenation reactions of 1,1'-dibromoferrocene or 1,1'-diiodoferrocene using magnesium. Virgin poly(ferrocenylene)s are also insulators [25-26].

However, upon oxidation with TCNQ the electrical conductivity increases to 10^{-2} Scm^{-1}.

Other approaches of preparing condensation polymers of ferrocene involve the use of conventional organic functional groups. The resulting polymers are essentially organic polymers. In an example that typifies this approach Rausch and coworkers have prepared a number of ferrocene-based difunctional reagents (Fig. 8.4) [27].

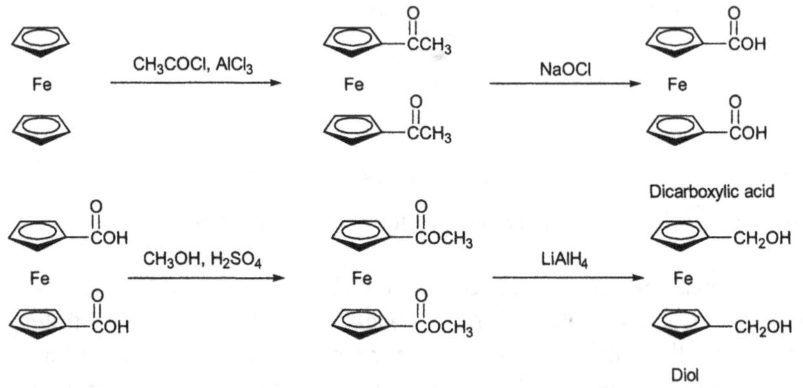

Fig. 8.4. Preparation of difunctional reagents based on ferrocene

The spacer methylene units between the functional group and the ferrocene motif can be increased by utilizing the synthetic strategy as shown in Fig. 8.5.

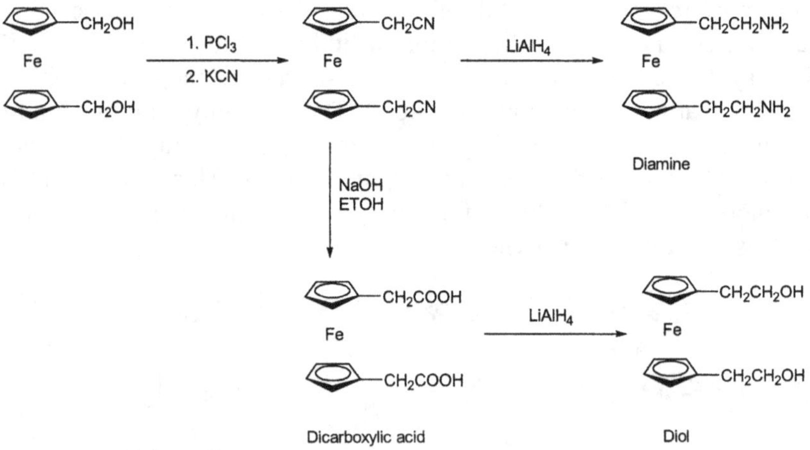

Fig. 8.5. Synthesis of spacer-separated difunctional reagents of ferrocene

The difunctional reagents containing spacer groups have been found to be more effective for the preparation of polymers. Utilizing these monomers ferrocene-containing polyamides and polyureas have been prepared. The molecular weights (M_n) of these polymers have been estimated from intrinsic viscosity measurements to be between 10,000 to 18,000. An example of a polyamide prepared by an interfacial condensation reaction between 1,1'-bis(β-aminoethyl)ferrocene, $Fc(CH_2CH_2NH_2)_2$, and diacidchloride is shown in Fig. 8.6.

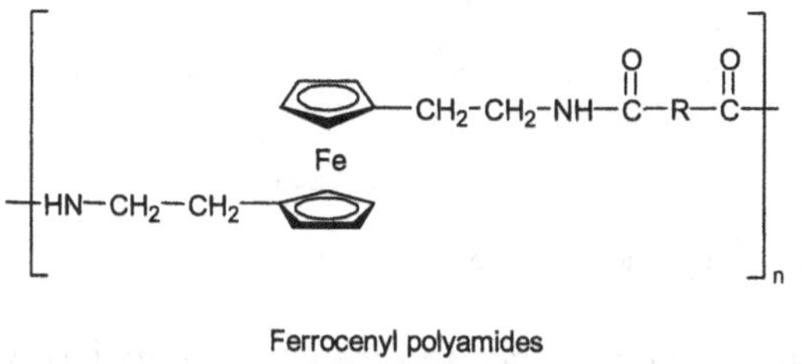

Ferrocenyl polyamides

Fig. 8.6. Polyamides containing ferrocene units

A different strategy from the above consists in the incorporation of silylamino groups on ferrocene. Condensation of such difunctional reagents with disilanols leads to the formation of polymers where the ferrocene groups are separated by organosilicon spacers [28] (Fig. 8.7).

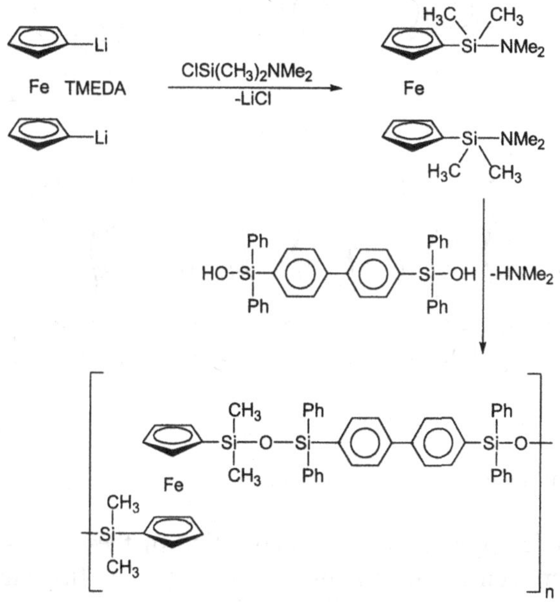

Fig. 8.7. Polymers containing ferrocene and organosiloxane units

Ferrocenes that are separated by other type of spacer units such as CH=CH groups, [poly(ferrocenylenevinylene)s] (M_n's of 3,000-10,000; M_w/M_n = 2.2-2.8) have been prepared by a Zn/TiCl$_4$-catalyzed condensation reaction of 1,1'-ferrocenyl dialdehydes (Fig. 8.8) [3]. Analogous to poly(vinylferrocene)s, these polymers are also insulators; however, they can be converted to semiconductors upon doping [29].

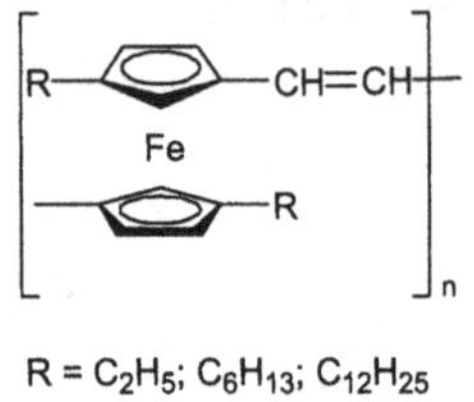

R = C$_2$H$_5$; C$_6$H$_{13}$; C$_{12}$H$_{25}$

Fig. 8.8. Poly(ferrocenylenevinylene)s

A novel design of *face-to-face* ferrocene polymers has been reported by Rosenblum and coworkers (Fig. 8.9) [30].

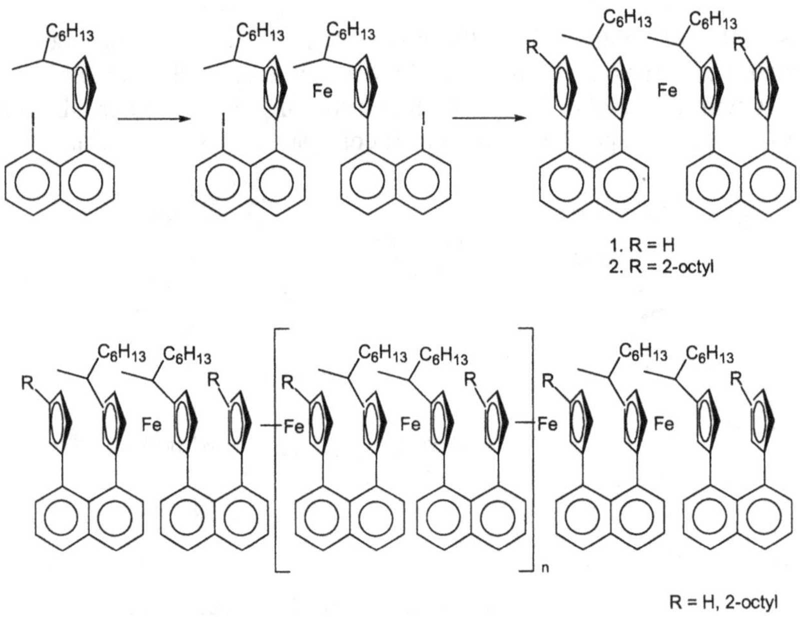

1. R = H
2. R = 2-octyl

R = H, 2-octyl

Fig. 8.9. Face-to-face ferrocenyl polymers

The synthetic approach for the preparation of these face-to-face ferrocenyl polymers consists of a multistep strategy. The first step consists of constructing a cyclopentadienyl ligand on a naphthalene support. Two such ligands were used for the preparation of a ferrocene derivative. This is followed by construction of the second cyclopentadienyl ring on each of the naphthalene units. Such a compound can be viewed as a rigid difunctional monomer (the terminal cyclopentadiene units are the functional groups). Reaction of this monomer with sodium hexamethylsilyl amide followed by treatment with $FeCl_2$ affords a purple-colored polymer. Molecular weight estimation by end-group analysis reveals the polymer to have a M_n of about 18,000. Such polymers are expected to have new magnetic and optical properties [30].

The other approach for the preparation of ferrocene containing polymers is by the ring-opening polymerization of suitable strained ferrocene monomers. This method provides a pathway for the preparation of the majority of ferrocene-containing polymers and will be discussed in the next section. However, by a suitable design unstrained ferrocene monomers also can be converted into polymers. Thus, Rauchfuss and coworkers have found that [3] trithiaferrocenes undergo a desulfurization reaction with $P(nBu_3)$ to afford high-molecular-weight poly(ferrocenylene persulfide)s (see Eq. 8.14) [31-34].

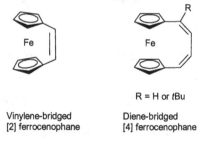 (8.14)

Trisulfide-bridged
[3]ferrocenophane

The presence of at least one alkyl group on the cyclopentadienyl motif is required to induce solubility in these poly(ferrocenylene persulfide)s. These polymers show two reversible oxidations. This type of electrochemical behavior is fairly common in ferrocene polymers where the ferrocene units are separated by short spacers as will be pointed out in the next section.

Vinylene-bridged
[2] ferrocenophane

R = H or tBu

Diene-bridged
[4] ferrocenophane

Fig. 8.10. Bridged ferrocenophanes which can be polymerized by ring-opening metathesis reaction

Vinylene-bridged   [2]ferrocenophanes   [35]   or   diene-bridged [4]ferrocenophanes [36] have been utilized as monomers in a ring-opening metathesis polymerization by the use of molybdenum catalysts (Fig. 8.10).

8.3.2 Strained Ferrocenophanes

Ferrocenophanes are a group of compounds where the two cyclopentadi-enyl rings of the ferrocene molecule are linked to each other by means of a bridging atom or a group. The number of bridging atoms is indicated as a prefix (see Eq. 8.14 and Fig. 8.10). Several types of ferrocenophanes are now known. These can be usually prepared by the reaction of 1,1'-dilithioferrocene with a halide of a main-group element-containing com-pound. For example, the reaction of 1,1'-dilithioferrocene with silicon tet-rachloride or dimethyldichlorosilane affords the corresponding silicon-bridged [1]ferrocenophanes (Fig. 8.11) [3, 7].

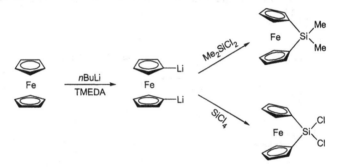

Fig. 8.11. Preparation of silicon-bridged [1]ferrocenophanes

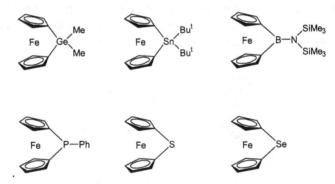

Fig. 8.12. [1]Ferrocenophanes containing various types of bridging elements

Using a similar methodology many other main-group element-containing ferrocenophanes have been prepared [7]. Representative examples of such compounds are shown in Fig. 8.12. While the germanium [37-38], tin [39], boron [40] or phosphorus-bridged ferrocenophanes are prepared by the reaction of 1,1'-dilithioferrocene and the corresponding halides [41-42], the synthesis of the sulfur-and-selenium bridged ferrocenophanes are accomplished in a reaction involving 1,1'-dilithioferrocene with $S(O_2SPh)_2$ or $Se(S_2CNEt_2)_2$ [43].

A characteristic feature of all the single-atom bridged ferrocenophanes is the ring-tilting of the cyclopentadienyl rings. Thus, in ferrocene the planes of the two cyclopentadienyl rings are parallel with respect to each other. However, in these bridged ferrocenophanes the cyclopentadienyl rings are tilted towards each other. The extent of tilting can be estimated by a tilt angle which gives an idea of the ring strain that is present in such compounds. For example, in the silicon-bridged ferrocenophane $[Fe(\eta^5C_5H_4)_2SiMe_2]$ the tilt angle is 20.8(5)° and the ring-strain as experimentally determined by a DSC (Differential Scanning Calorimetry) experiment is 80 kJ mol^{-1} [44-45]. This type of strain which is characteristic of most single-atom bridged ferrocenophanes can be compared to that present in organic rings: cyclobutane, 110 kJ mol^{-1}; 1,1,3,3-tetramethyldisilacyclobutane, 84 kJ mol^{-1}; cyclopentane, 31 kJ mol^{-1}. Thus, single-atom bridged ferrocenophanes do not have as much strain as cyclobutane rings but have more ring-strain than cyclopentane. In contrast to the single-atom bridged ferrocenophanes, double-atom bridged derivatives have less strain. Thus, the disilicon-bridged compound $[Fe(\eta^5-C_5H_4)_2(Me_2SiSiMe_2)$ has no ring strain and the planes of the cyclopentadienyl rings of this ferrocenophane are parallel (Fig. 8.13).

Ring strain of 80 kJmol^{-1} No Ring strain

Fig. 8.13. Estimate of ring strain present in single-atom and double-atom bridged ferrocenophanes

As will be shown in the next section, strained ferrocenophanes can be used as monomers for ring-opening polymerization to afford ferrocene-containing polymers.

8.3.3 Poly(ferrocenylsilane)s

The relief of ring-strain present in silicon-bridged ferrocenophanes can be utilized as the driving force for its ring-opening polymerization (ROP). Manners and coworkers have shown that it is possible to induce the ROP of silicon-bridged ferrocenophanes by at least three different synthetic approaches [3, 7, 44-46]. These are: (a) thermal ROP (b) anionic ROP and (c) transition-metal-complex catalyzed ROP.

8.3.3.1 Thermal Ring-opening Polymerization of Silicon-bridged Ferrocenophanes

Many silicon-bridged [1]ferrocenophanes can be polymerized by simply heating them. The temperatures of polymerization vary over a wide range and depend on the substituents on silicon as well as on the cyclopentadienyl rings. For example, $[Fe(\eta^5C_5H_4)_2SiMe_2]$ can be polymerized at 130 °C to afford a high-molecular-weight (M_w= 520,000; M_n= 340,000) amber colored polymer [47]. In contrast, the corresponding diphenyl derivative $[Fe(\eta^5C_5H_4)_2SiPh_2]$ can be polymerized only at 230 °C. The double-atom bridged ferrocenophane $[Fe(\eta^5C_5H_4)_2(Me_2SiSiMe_2)]$, on the other hand, cannot be polymerized (Fig. 8.14) [47].

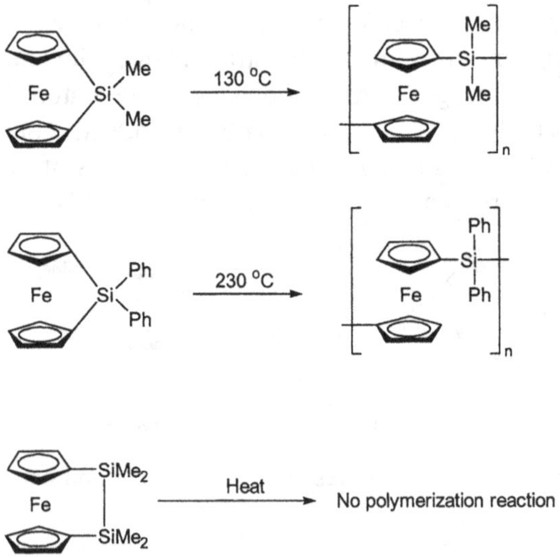

Fig. 8.14. Polymerization of silicon-bridged [1]ferrocenophanes

Using the thermal ROP methodology a large number of poly(ferrocenylsilane)s have been prepared. Representative examples are shown in Fig. 8.15. The range of substituents on silicon that can be present in poly(ferrocenylsilane)s is quite large and include alkyl, aryl, chloro, hy-drido, alkoxy as well as amino groups [3, 7, 44-46].

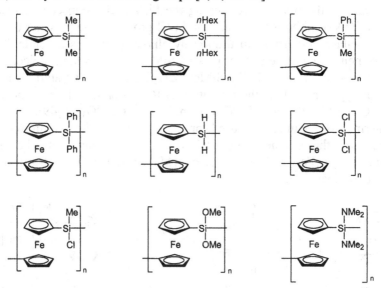

Fig. 8.15. Various types of poly(ferrocenylsilane)s

Poly(ferrocenylsilane)s that contain reactive chlorine groups can be used for further elaboration (Fig.8.16).

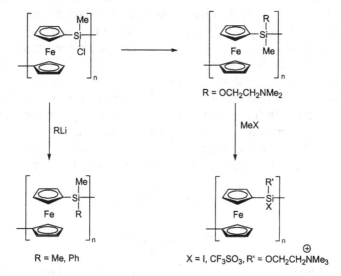

Fig. 8.16. Substitution of Si-Cl bonds in poly(ferrocenylsilane)s

Macromolecular substitution reactions on chlorine-containing poly(ferrocenylsilane)s can generate new types of polymers (Fig. 8.16). For example, the polymer containing the Si(Me)Cl group can be reacted with aryl/alkyl lithiums to replace the Si-Cl groups in the polymer by Si-R [48]. Similarly, chlorine substitution by alkoxy groups is possible. Thus, replacement of chlorine in $[Fe(\eta^5C_5H_4)_2SiMeCl]_n$ by an amino alcohol affords the polymer $[Fe(\eta^5C_5H_4)_2SiMe(OCH_2CH_2NMe_2)]_n$. Quaternization of the amino group leads to the formation of the water-soluble ionic polymer $[Fe(\eta^5C_5H_4)_2SiMe(OCH_2CH_2NMe_3)]^+_n\,n[X]^-$ (X= CF_3SO_3, I) [49].

Thermal ROP can also be used for copolymerizing the cycloferrocenylsilane with other strained rings such as cyclotetrasilanes. Thus, heating a 1:1 mixture of $[Fe(\eta^5C_5H_4)_2SiMe_2]$ and $[MePhSi]_4$ at 150 °C affords a random copolymer (Fig. 8.17) [50].

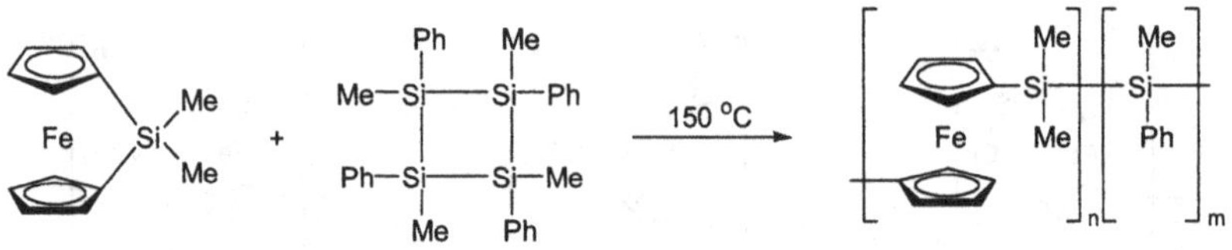

Fig. 8.17. Copolymerization of silicon-bridged [1]ferrocenylsilane with cyclotetrasilane

The mechanism of thermal ring-opening polymerization has been investigated. It is a chain-growth process. Thermal ROP of $[Fe(\eta^5C_5H_4)(\eta^5C_5Me_4)SiMe_2]$ has shown that a heterolytic cleavage of the Si-Cp (Cp = $\eta^5C_5H_4$) or Si-Cp'(Cp' = ($\eta^5C_5Me_4$) bonds are involved in the initiation reaction [44]. This leads to a positive charge on silicon and a negative charge on the cyclopentadienyl ring. Propagation of the polymerization is then continued by the possible attack of the silicon-centered cation on another molecule. It is likely that such a mechanism of polymerization is applicable for other types of silicon-bridged ferrocenophanes as well. The polymerization is readily quenched with many types of protic reagents including water. Consistent with this mechanism, in some instances dimeric products have been isolated and characterized (Fig. 8.18) [51].

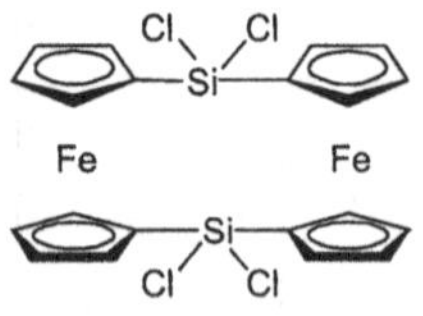

Fig. 8.18. Dimeric products formed in the polymerization of $[Fe(\eta^5C_5H_4)_2SiCl_2]$

The mechanism of the ring-opening polymerization of $[Fe(\eta^5C_5H_4)_2SiMe_2]$ is shown in Fig. 8.19.

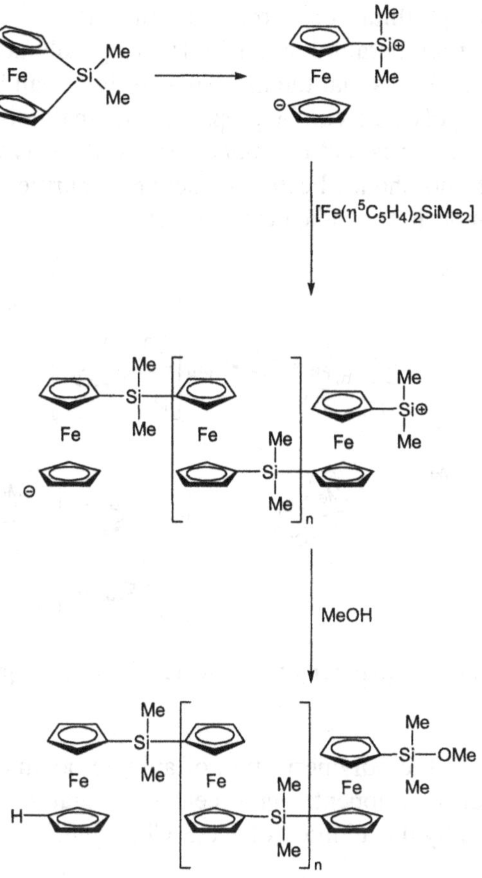

Fig. 8.19. Mechanism of the thermal ROP of $[Fe(\eta^5C_5H_4)_2SiMe_2]$

The use of spirocyclic ferrocenylsilanes allows the formation of cross-linked polymers. The extent of crosslinking can be controlled by the amount of the spirocyclic silanes that is used (Fig. 8.20) [52].

Fig. 8.20. Spirocyclic ferrocenophanes that can be used for crosslinking poly(ferrocenylsilane)s

8.3.3.2 Anionic Ring-opening Polymerization

Many ferrocenylsilanes can be polymerized by anionic initiators such as *n*-butyllithium, phenyllithium or ferrocenyllithium (Fig. 8.21). The reaction occurs at ambient temperature and affords living polymers. The utility of anionic polymerization is that the molecular weights can be controlled and also that block copolymers can be prepared. The main disadvantage of the anionic polymerization is that the monomer and the solvent should be rigorously purified and should be free of acidic impurities including water. Even traces of impurities can be detrimental [53].

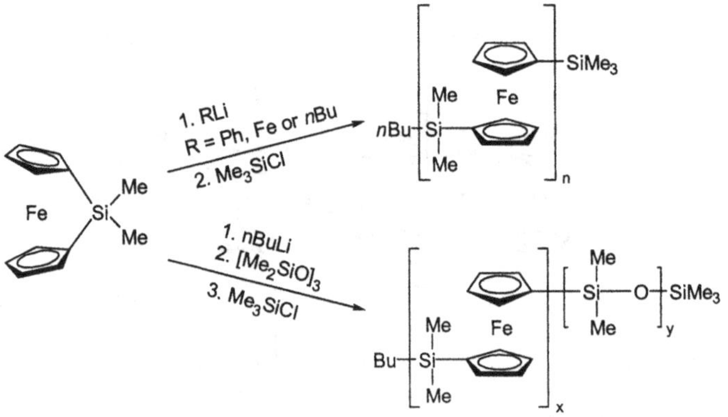

Fig. 8.21. Anionic polymerization of [Fe(η^5C$_5$H$_4$)$_2$SiMe$_2$] and preparation of block copolymers

The mechanism of the polymerization of anionic polymerization involves the generation of a cyclopentadienyl centered negative ion. The reaction can be terminated by the addition of Me$_3$SiCl (Fig. 8.22).

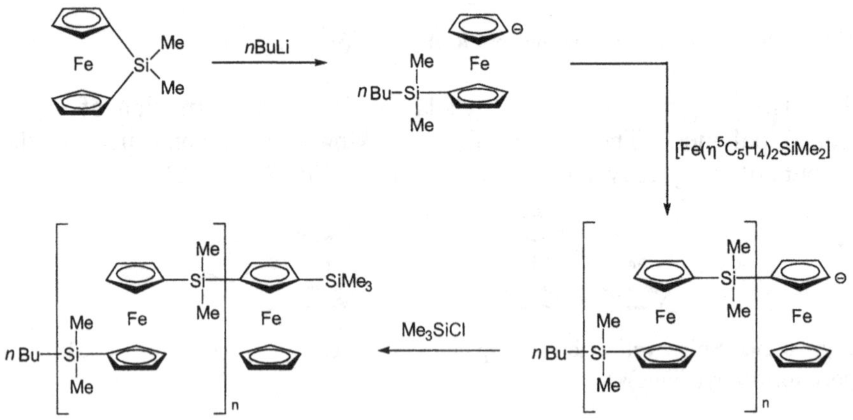

Fig. 8.22. Mechanism of anionic polymerization of [Fe(η^5C$_5$H$_4$)$_2$SiMe$_2$]

8.3.3.3 Transition Metal Catalyzed Ring-opening Polymerization

A number of transition metal salts or transition metal complexes have been very widely used for ambient temperature polymerization of a number of ferrocenylsilanes [54]. The successful catalysts that have been employed include Rh(I), Pd(II), Pt(II), Pd(0), Pt(0) complexes. To some extent the stringent requirements of the anionic polymerization are not required in the polymerization effected by transition metal catalysts. Molecular weight control is generally achieved by the addition of hydrosilane derivatives such as Et_3SiH. This polymerization methodology is quite general and has been applied to a large number of ferrocenophanes containing diverse substituents on the silicon center. For example, $[Fe(\eta^5C_5H_4)_2SiMe_2]$ can be readily polymerized by a variety of Pt(II) or Pt(0) complexes to afford a high-molecular-weight polymer (Fig. 8.23).

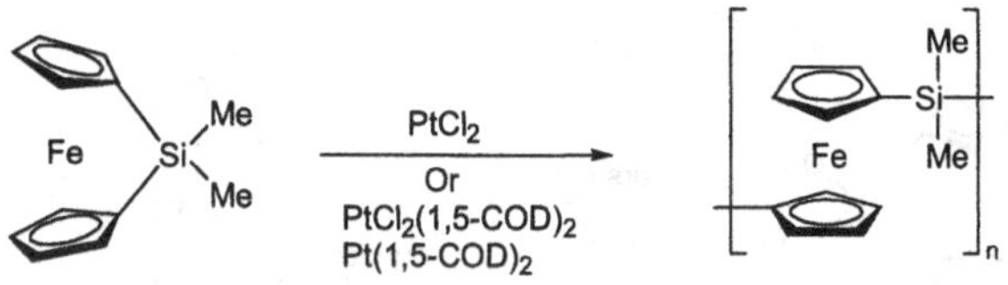

Fig. 8.23. Metal-catalyzed polymerization of $[Fe(\eta^5C_5H_4)_2SiMe_2]$

Similarly, a variety of ferrocenophanes containing alkoxy or aryloxy groups have been polymerized by the use of $PtCl_2$ (Fig. 8.24) [55].

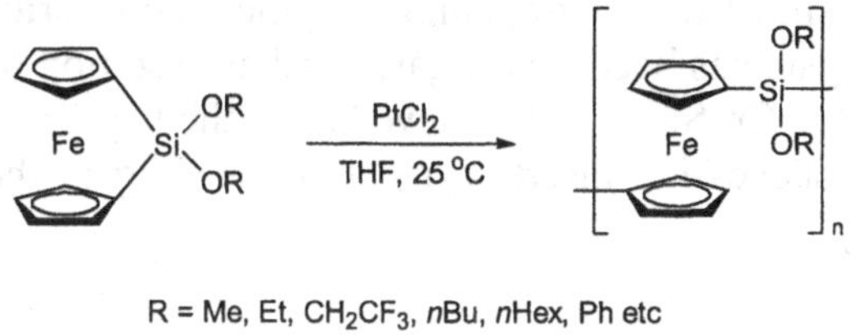

R = Me, Et, CH_2CF_3, nBu, nHex, Ph etc

Fig. 8.24. $PtCl_2$-catalyzed ROP of $[Fe(\eta^5C_5H_4)_2Si(OR)_2]$

Transition metal complex catalyzed ring-opening polymerization is fairly general and many other types of ferrocenylsilanes have also been polymerized by this methodology. For example, ferrocenophanes containing acetylide substituents as well as etheroxy substituents have been polymerized by the use of Karstedt's catalyst (platinum-divinyltetramethyldisiloxane complex) (Fig. 8.25) [56].

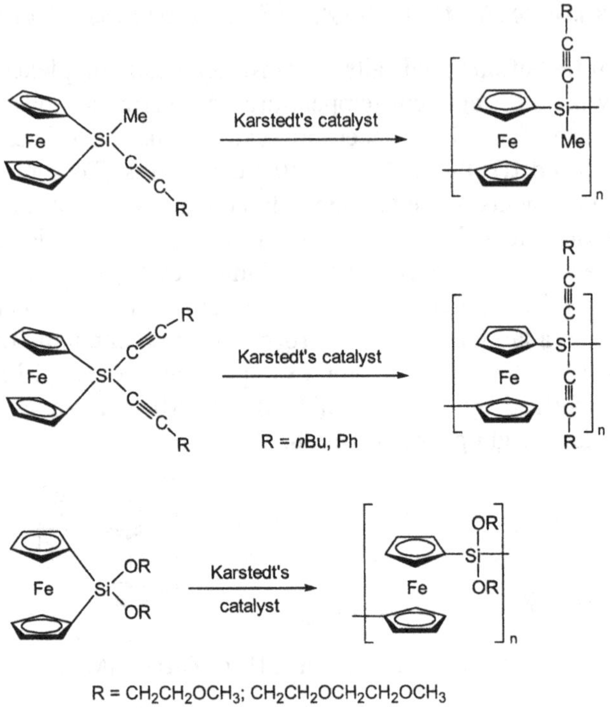

R = CH$_2$CH$_2$OCH$_3$; CH$_2$CH$_2$OCH$_2$CH$_2$OCH$_3$

Fig. 8.25. ROP of ferrocenylsilanes catalyzed by Karstedt's catalyst

The mechanism of the ring-opening polymerization effected by transition metal catalysts has been investigated and an insertion of the transition metal between the Cp-Si bond is believed to be the key initiation step. This is followed by successive insertion reactions leading to the formation of the polymer [7].

8.3.4 Other Ferrocenyl Polymers Prepared from Strained Ferrocenophanes

Many other types of strained ferrocenophanes have been successfully polymerized [37-43, 57-58]. These include polymers that contain boron, carbon, germanium, tin, phosphorus, sulfur as well as selenium. Some of the representative examples of the polymers of this type are shown in Fig. 8.26.

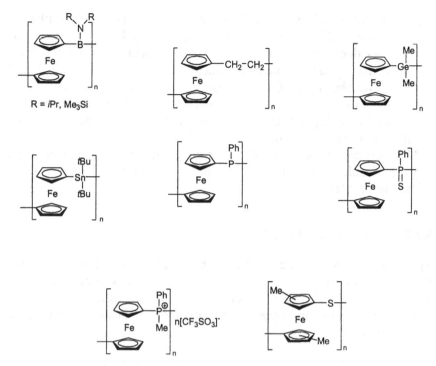

Fig. 8.26. Polyferrocenes containing alternate ferrocene and main-group elements

Interestingly, while the disilicon-bridged derivative $[Fe(\eta^5\text{-}C_5H_4)_2(Me_2SiSiMe_2)]$ has been shown to resist polymerization the corresponding carbon compounds can be polymerized (Fig. 8.27) [57]. Since carbon atoms are smaller in size than silicon atoms, the dicarbon-bridged derivative is sterically more strained than the corresponding compound having a disilicon bridge. However, ROP of the dicarbon-bridged ferrocenophane occurs at 300 °C.

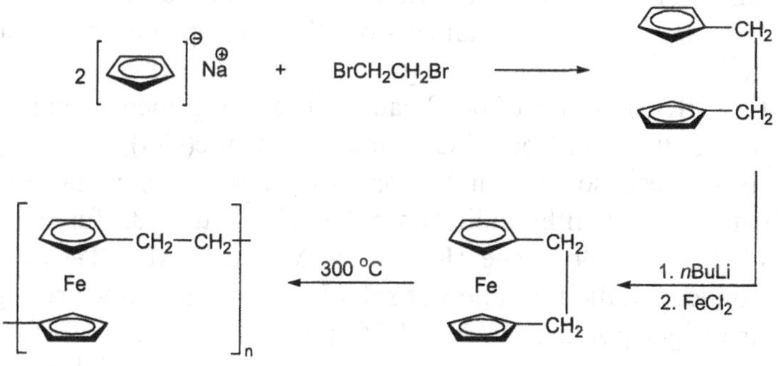

Fig. 8.27. Synthesis of ethylene-bridged ferrocenophane and its ROP

Polymerization of ferrocenylphosphines such as [Fe(η^5C$_5$H$_4$)$_2$PPh] is brought about readily by anionic initiators [3,41-42]. However, the resulting polymer does not elute on the GPC column and has to be converted into the corresponding sulfide by treatment with sulfur for determination of molecular weights by GPC. Poly(ferrocenylphosphine)s can function as ligands towards low-valent transition metal ions. A further interesting application of the poly(ferrocenylphosphine)s is that they can be quaternized readily to afford ionic polymers [58].

Although poly(ferrocenyl sulfide)s containing simple cyclopentadienyl rings are insoluble the choice of methylcyclopentadienyl ligands overcomes this problem [43].

8.3.5 Other Related Polymers

Hyperbranched poly(ferrocenylsilyne)s have been prepared by the reaction of 1,1'-dilithioferrocene with alkyltrichlorosilanes or vinyltrichlorosilane (Fig. 8.28) [59]. Although polymers with short side-chains were found to be insoluble in common organic solvents, those with longer side-chains were freely soluble.

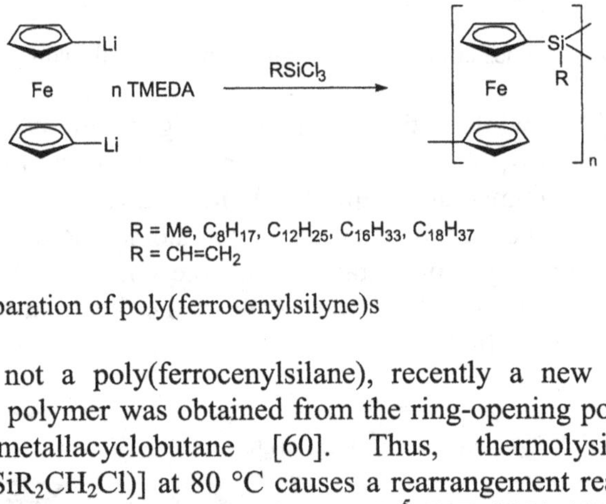

R = Me, C$_8$H$_{17}$, C$_{12}$H$_{25}$, C$_{16}$H$_{33}$, C$_{18}$H$_{37}$
R = CH=CH$_2$

Fig. 8.28. Preparation of poly(ferrocenylsilyne)s

Although not a poly(ferrocenylsilane), recently a new class of organometallic polymer was obtained from the ring-opening polymerization of a silametallacyclobutane [60]. Thus, thermolysis of [η^5-CpFe(CO)$_2$(SiR$_2$CH$_2$Cl)] at 80 °C causes a rearrangement reaction (Fe-Si to Fe-C) and the corresponding isomer [η^5-CpFe(CO)$_2$(CH$_2$SiR$_2$Cl)] is formed. Both these isomers can be converted to the silametallacyclobutane upon treatment with lithiumdiisopropylamide. Tetrahydrofuran solutions of the silametallacyclobutane (R = R' = Me; R = Me, R'= nBu) upon evaporation lead to the formation of soluble, air-stable, film-forming, high-molecular-weight polymers (M_w = 1.75 x 10^5; M_n = 7.2 x 10^4 for the dimethyl polymer) (Fig. 8.29) [60]. The silametallacyclobutanes with slightly sterically encumbered substituents (R = R'= nBu; R = Me, R' = Ph)

do not undergo such spontaneous polymerization. However, these monomers also should be viable compounds for metal-catalyzed polymerization
reaction.

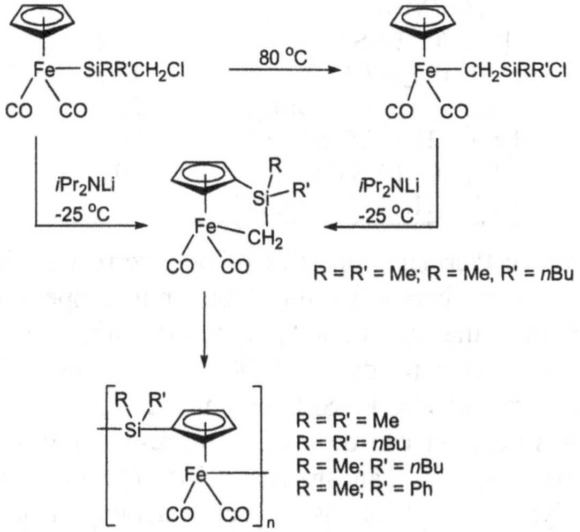

Fig. 8.29. Polymerization of ferrocenylsilacyclobutane

8.3.6 Properties and Applications of Poly(ferrocenylsilane)s and Related Polymers

Poly(ferrocenylsilane)s are high-molecular-weight polymers. The molecular weights of polymers prepared by the thermal ROP range from 10^5 to
10^6 with polydispersity indices of 1.5-2.5 [3, 7, 44-46]. The molecular
weights of the polymers prepared by the anionic polymerization method
can be controlled by varying the amount of initiator. Further, polymers
with narrow PDI's (less than 1.3) can be obtained by the anionic polymerization [3].

The ^{29}Si-NMR chemical shifts of various types of
poly(ferrocenylsilane)s are summarized in Table 8.3. The presence of oxygen substituents moves the chemical shifts upfield in comparison to alkyl
substituents. But it can be seen that most chemical shifts span a fairly narrow range.

Table 8.3. ^{29}Si NMR and T_g data for some selected poly(ferrocenylsilane)s

S.No	Polymer	δ^{29}Si	$T_g(T_m)$ °C
1	$[Fe(\eta^5C_5H_4)_2SiMe_2]_n$	-6.4	33 (122)
2	$[Fe(\eta^5C_5H_4)_2Si(nHex)_2]_n$	-2.3	-26
3	$[Fe(\eta^5C_5H_4)_2SiPh_2]_n$	-12.9	-

Table 8.3. (contd.)

4	$[Fe(\eta^5C_5H_4)_2SiPhMe]_n$	-10.9	54
5	$[Fe(\eta^5C_5H_4)_2SiCl_2]_n$	-	27
6	$[Fe(\eta^5C_5H_4)_2SiMeCl]_n$	-	59
7	$[Fe(\eta^5C_5H_4)_2SiOMe_2]_n$	-17.2	19
8	$[Fe(\eta^5C_5H_4)_2Si(OCH_2CF_3)_2]_n$	-13.8	16
9	$[Fe(\eta^5C_5H_4)_2Si(OnBu)_2]_n$	-20.9	-43
10	$[Fe(\eta^5C_5H_4)_2Si(OnHex)_2]_n$	-20.5	-51
11	$[Fe(\eta^5C_5H_4)_2Si(OR)_2]_n$	-20.8	-53
	$R = CH_2CH_2OCH_2CH_2OCH_3$		

The study of the thermal properties of poly(ferrocenylsilane)s reveals that these polymers are thermally quite stable up to temperatures as high as 400 °C. For example, the best studied polymer $[Fe(\eta^5C_5H_4)_2SiMe_2]_n$ retains as much as 35-40% char residue at 1000 °C [3]. This residue has been identified as a ferromagnetic Fe-Si-C ceramic. Elegant experiments have revealed that this polymer can be used as a pre-ceramic polymer for obtaining super paramagnetic nanostructures [3]. The glass-transition temperatures of poly(ferrocenylsilane)s vary considerably. Thus, low T_g's can be obtained by the use of long-chain alkyl group substituents (Table 8.3, entry 2) or long-chain alkoxy substituents (Table 8.3, entries 9-11). On the other hand, placing phenyl substituents increases the T_g (Table 8.3, entry 4). Because of the variation of thermal properties vis-à-vis the substituents that are present on silicon the polymer properties of poly(ferrocenylsilane)s depend on the substituents. Thus, while $[Fe(\eta^5C_5H_4)_2SiMe_2]_n$ is a thermoplastic, polymers such as $[Fe(\eta^5C_5H_4)_2Si(OnHex)_2]_n$ are elastomers [3, 7].

Poly(ferrocenylsilane)s have been investigated extensively in terms of their electrochemical behavior. The general observation of these investigations is that under cyclic voltammetric conditions these polymers show *two* reversible oxidation (Fe(II)/Fe(III)) peaks separated by a $\Delta E_{1/2}$ that varies from about 0.16 to 0.29 V. The observation of two oxidation peaks has been explained as two successive events where the first oxidation involves a set of alternate ferrocene units followed by the second oxidation of the other alternate set. The second oxidation occurs at a higher potential than the first one [3, 7].

Unlike polysilanes, poly(ferrocenylsilane)s do not have electron delocalization in their backbone. The electronic absorption of these polymers is typical of those known for unstrained ferrocene derivatives. The monomer ferrocenophanes, on the other hand, show bathochromic absorption (470-480 nm) in comparison to bis-trimethylsilylferrocene (448 nm) [44-46].

Virgin poly(ferrocenylsilane)s are insulators (10^{-13}-10^{-14} Scm^{-1}). However, upon doping the polymers with I_2 the conductivity increases to 10^{-7} Scm^{-1}, indicative of a hopping mechanism [3]. Some polymers such as $[Fe(\eta^5\text{-}C_5H_4)_2SinBu_2]_n$ have been reported to show conductivity of the order of 10^{-4} Scm^{-1} [7].

The structure of crystalline poly(ferrocenylsilane)s have attracted attention. Studies on model compounds have also been carried out to augment the understanding of the polymeric derivatives. These studies reveal that polymers such as $[Fe(\eta^5\text{-}C_5H_4)_2SiMe_2]_n$ have *trans*-planar structures [61].

The thermal properties of some of the ferrocenyl polymers containing other heteroelements such as Ge, Sn, and P are summarized in Table 8.4. Some of these polymers have very high T_g's (Table 8.4, entries 3-7). It is noted that in all these examples either bulky alkyl or aryl substituents such as phenyl or mesityl are present on the main-group element.

Table 8.4. T_g Data for some poly(ferrocenylgermane)s, poly(ferrocenylstannane)s and poly(ferrocenylphosphine)s

S.No	Polymer	$T_g(T_m)$ °C
1	$[Fe(\eta^5C_5H_4)_2GeMe_2]_n$	28 (125)
2	$[Fe(\eta^5C_5H_4)_2Ge(nBu)_2]_n$	-7 (74)
3	$[Fe(\eta^5C_5H_4)_2GePh_2]_n$	114
4	$[Fe(\eta^5C_5H_4)_2Sn(tBu)_2]_n$	124
5	$[Fe(\eta^5C_5H_4)_2SnMes_2]_n$	208
6	$[Fe(\eta^5C_5H_4)_2PPh]_n$	126
7	$[Fe(\eta^5C_5H_4)_2P(S)Ph]_n$	206

Poly(ferrocenylsilane)s are being investigated as organometallic polyelectrolytes [63]. These can be prepared by stepwise synthetic procedures (Figs. 8.30 and 8.31).

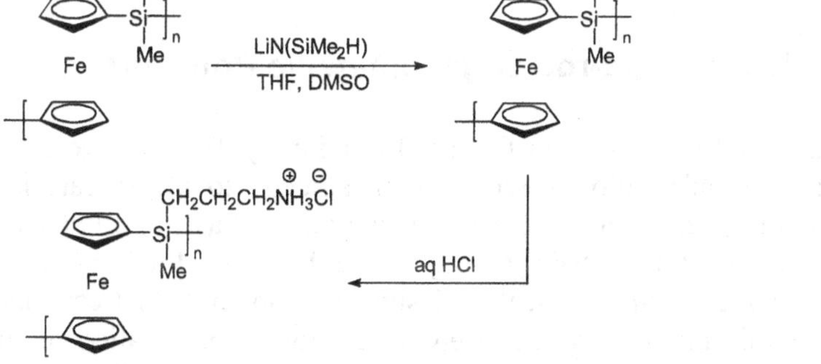

Fig. 8.30. Preparation of polyelectrolytes based on poly(ferrocenylsilane)s

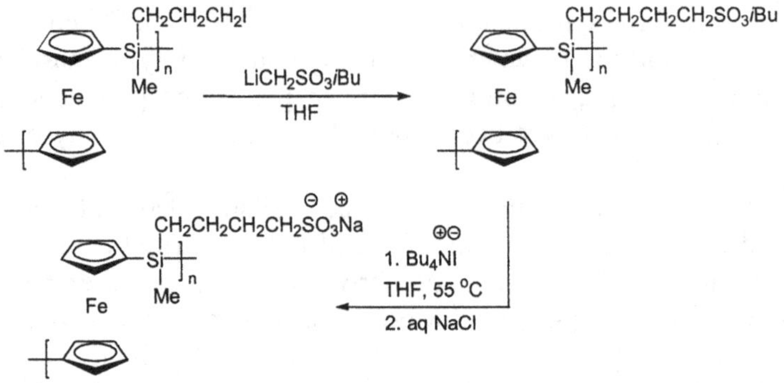

Fig. 8.31. Poly(ferrocenylsilane) polyelectrolytes containing sulfonate anions

Polyferrocenes have a wide range of refractive indices which can be tuned by changing the main-group element/and or the substituent on the main-group element. Such a tunability is important for applications in photonic devices. The refractive indices of some selected polyferrocenes are summarized in Table 8.5. [64].

Table 8.5. Refractive indices of various polyferrocenes at $\lambda = 589$ nm

S.No	Polymer	n
1	$[Fe(\eta^5C_5H_4)_2Si(CH_3)(C_2H_5CF_3)]_n$	1.60
2	$[Fe(\eta^5C_5H_4)_2Si(CH_3)(C_2H_5)]_n$	1.66
3	$[Fe(\eta^5C_5H_4)_2Si(CH_3)_2]_n$	1.68
4	$[Fe(\eta^5C_5H_4)_2Si(CH_3)(C_6H_5)]_n$	1.68
5	$[Fe(\eta^5C_5H_4)_2Ge(CH_3)_2]_n$	1.69
6	$[Fe(\eta^5C_5H_4)_2Sn(tBu)_2)]_n$	1.64
7	$[Fe(\eta^5C_5H_4)_2Sn(2,4,6-Me_3-C_6H_2)_2]_n$	1.66
8	$[Fe(\eta^5C_5H_4)_2P(C_6H_5)]_n$	1.74
9	$[Fe(\eta^5C_5H_4)_2P(S)(C_6H_5)]_n$	1.72

8.4 Polyyne Rigid-rod Organometallic Polymers

Hagihara and coworkers in the 1970's and early 1980's have reported a successful condensation polymerization strategy to incorporate late second- and third-row transition metal ions (mainly Pt(II) and Pd(II)) as part of a polymeric linear chain [65-67]. Since these metal ions prefer square-planar geometric structures, they designed compounds that contained two reactive chlorine groups in a *trans* orientation. Condensation of such difunctional monomers with *trans*-diacetylides afforded linear polymers which are called *polyynes* (Fig. 8.32).

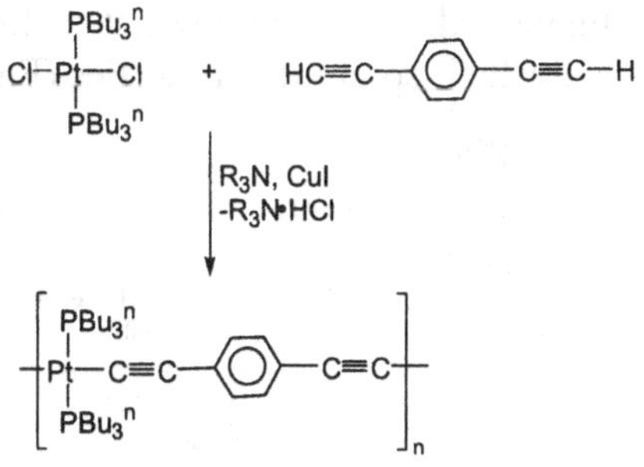

Fig. 8.32. Polyynes prepared from the reaction of *trans*-Pt(nBu$_3$P)$_2$Cl$_2$ with aromatic diynes

The condensation reaction between the metal salt and the diyne is catalyzed by Cu(I) salts and is usually carried out using amines such as diethylamine as solvents under reflux conditions. As in all condensation reactions the important criterion that allows high-molecular-weight formation is the purity of the monomers. Soluble polymers with high-molecular-weights (~10^5) were isolated using this protocol.

A variation of the above procedure consists of preparing the transition metal acetylides and using them for condensation. Thus, first the *trans*-bis(tri-*n*-butylphosphine)bis(1,4-butadiynyl)platinum can be prepared by the reaction of *trans*-Pt(PnBu$_3$)$_2$Cl$_2$ with 1,4-butadiyne (Fig. 8.33).

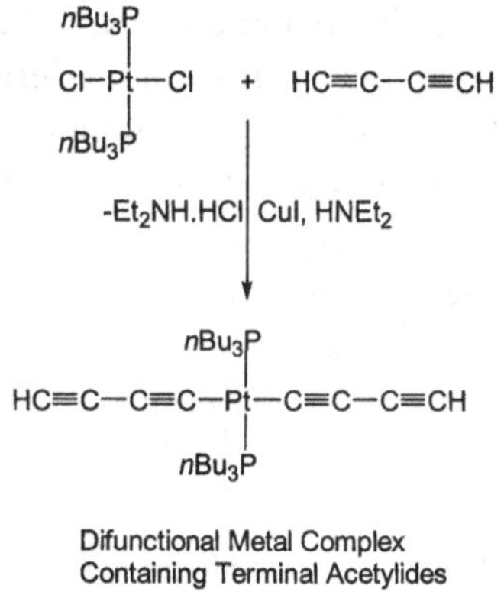

Difunctional Metal Complex
Containing Terminal Acetylides

Fig. 8.33. Synthesis of a difunctional Pt(II) complex containing two -C≡C-C≡C-H units arranged in a *trans* orientation

The terminal -C≡C-C≡C-H containing platinum monomer can be condensed with *trans*-Pt(PnBu$_3$)Cl$_2$ to afford the linear rod-like polymer where the platinum metal alternates with the butadiyne unit. The presence of the tri-*n*-butylphosphine ligands allows some amount of lipophilicity to these polymers and accounts for their solubility properties (Fig. 8.34).

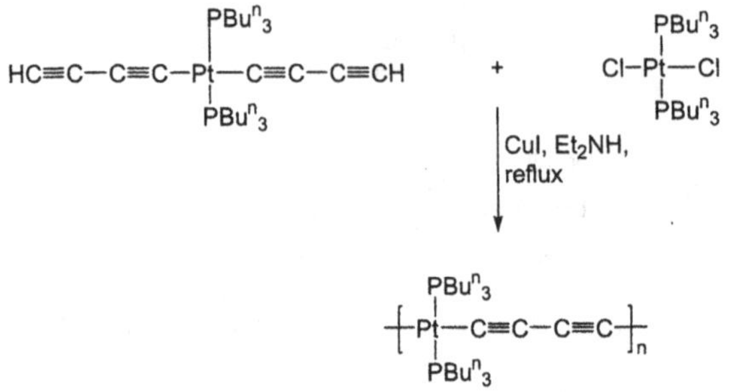

$$HC\equiv C-C\equiv C-\underset{\underset{PBu^n_3}{|}}{\overset{\overset{PBu^n_3}{|}}{Pt}}-C\equiv C-C\equiv CH \quad + \quad Cl-\underset{\underset{PBu^n_3}{|}}{\overset{\overset{PBu^n_3}{|}}{Pt}}-Cl$$

CuI, Et$_2$NH, reflux

$$\left[\!\!\!\left[\underset{\underset{PBu^n_3}{|}}{\overset{\overset{PBu^n_3}{|}}{Pt}}-C\equiv C-C\equiv C\right]\!\!\!\right]_n$$

Fig. 8.34. Condensation of [*trans*-Pt(*n*Bu$_3$P)$_2$(C≡C-C≡C-H)$_2$] with *trans*-Pt(*n*Bu$_3$P)$_2$Cl$_2$

Solution studies of these polymers suggest that these retain their rod-like structures in solution. For example, the intrinsic viscosities of randomly coiled polymers depend on the type of solvent used. On the other hand, the intrinsic viscosities of stiff rigid-rod polymers remain nearly invariant upon change of solvents. Based on such criteria and others it was possible to establish the rigid-rod nature of these polymers in solution. However, more recent studies suggest that these polymers are *worm-like* indicating a less stiff character [68].

Several variations of the above approach have been tried. For example, condensation of a butadiyne-linked diplatinum derivative with aryl diacetylenes leads to the formation of polymers containing the transition metal and two types of acetylide motifs in the chain structure (Fig. 8.35) [67].

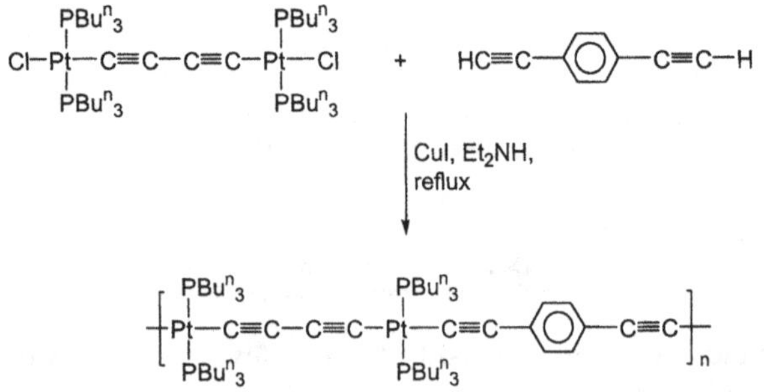

$$Cl-\underset{\underset{PBu^n_3}{|}}{\overset{\overset{PBu^n_3}{|}}{Pt}}-C\equiv C-C\equiv C-\underset{\underset{PBu^n_3}{|}}{\overset{\overset{PBu^n_3}{|}}{Pt}}-Cl \quad + \quad HC\equiv C-\!\!\!\bigcirc\!\!\!-C\equiv C-H$$

CuI, Et$_2$NH, reflux

$$\left[\!\!\!\left[\underset{\underset{PBu^n_3}{|}}{\overset{\overset{PBu^n_3}{|}}{Pt}}-C\equiv C-C\equiv C-\underset{\underset{PBu^n_3}{|}}{\overset{\overset{PBu^n_3}{|}}{Pt}}-C\equiv C-\!\!\!\bigcirc\!\!\!-C\equiv C\right]\!\!\!\right]_n$$

Fig. 8.35. Condensation of [ClPt(*n*Bu$_3$P)$_2$PtC≡C-C≡-CPt(*n*Bu$_3$P)$_2$Cl] with an aromatic diyne

Alternative variations of the procedure shown in Fig. 8.35 include the preparation of Pt(II) monomers where the terminal C≡CH units are separated from the metal by means of an aromatic spacer group. Condensation of such monomers with other square planar Pt(II) complexes containing *trans* chloride ligands also leads to the formation of polymers (Fig. 8.36) [67].

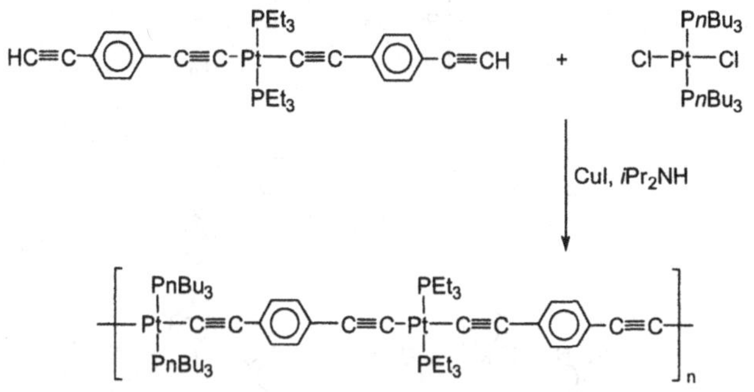

Fig. 8.36. Rigid-rod polymers containing aromatic spacer groups

Use of similar synthetic methodologies as above leads to the generation of rigid-rod polyynes containing extended spacer units (Fig 8.37) [67].

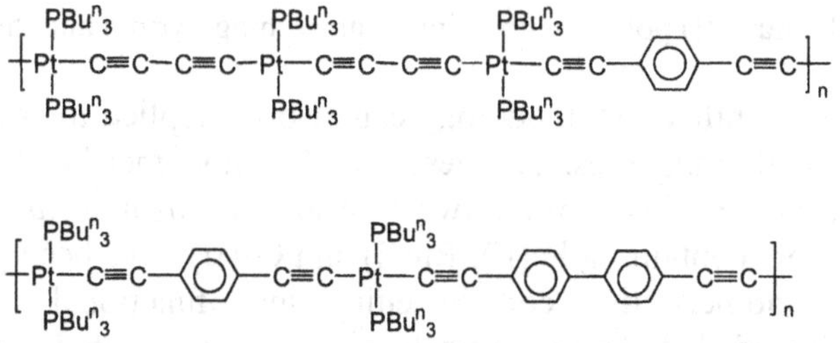

Fig. 8.37. Rigid-rod platinum (II) polyynes containing long spacer units

More recently, platinum complexes containing pyridyl type of ligands have been used to afford polyynes (Fig 8.38) [69].

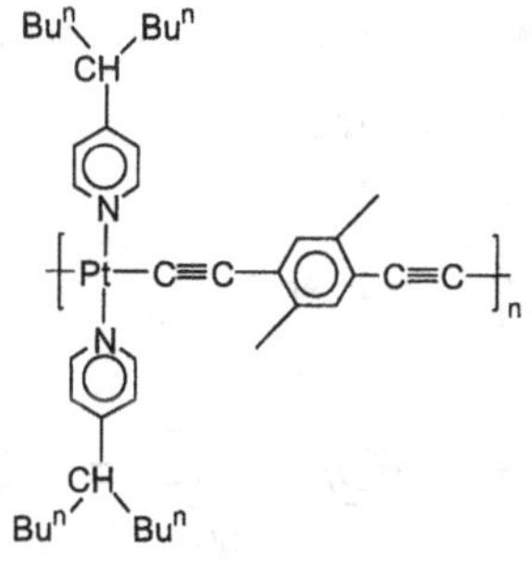

Fig. 8 38. Platinum(II) polyynes containing *p*-alkyl-pyridyl ligands

Several other types of polyynes containing pyridyl groups have also been synthesized (Fig 8.39) [70-71]. All of these polymers have fairly high molecular weights (above $M_n = 60,000$; PDI's of 1.3 to 1.8).

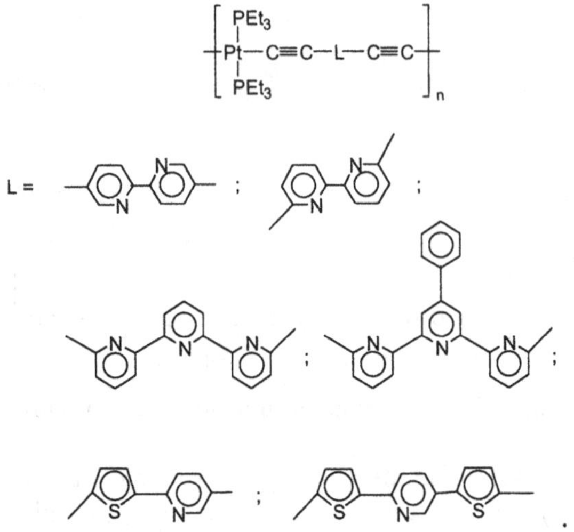

Fig. 8.39. Platinum(II) polyynes containing intervening pyridyl and thienyl groups

Hagihara's synthetic methodology can also be applied to six-coordinate transition metal complexes. The design of the monomer has to be carried out to ensure the presence of the two chloride ligands in a *trans* geometry. Thus, polymers containing Fe(II), Ru(II) and Os(II) have been prepared by choosing monomers that contain either monofunctional or chelating phosphine ligands [3]. Another variation that was warranted in this synthesis is to convert the terminal C≡CH unit of the diacetylide ligands to C≡CSnMe₃. This allowed the removal of the Me₃SnCl as the by-product of the condensation reaction (Figs. 8.40 and 8.41).

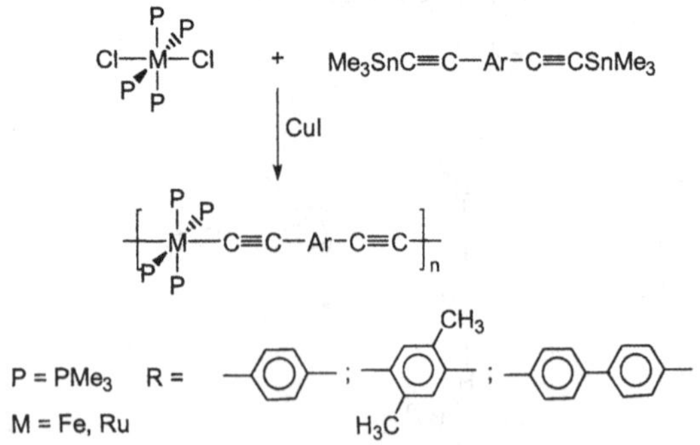

Fig. 8.40. Polyynes containing six-coordinate Fe(II) and Ru(II) centers

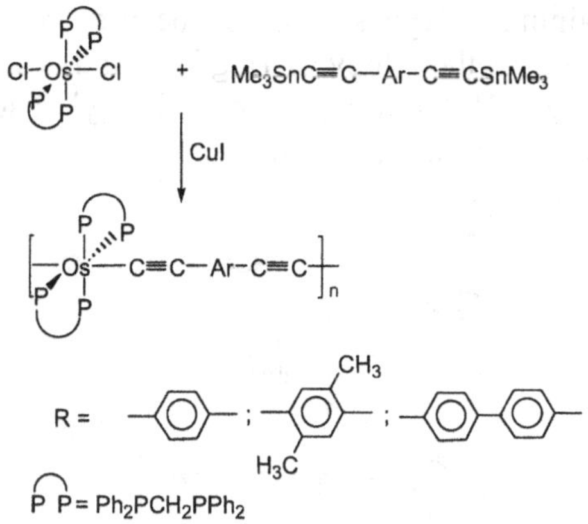

Fig. 8.41. Polyynes containing six-coordinate Os(II)

The above preparative technique has been extended to polymers containing rhodium also (Fig. 8.42) [3].

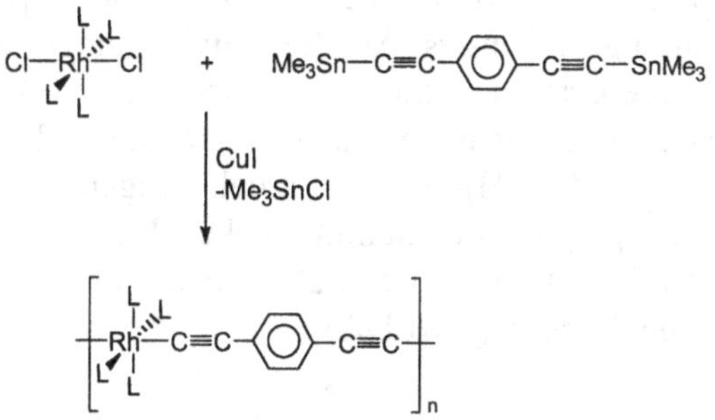

Fig. 8.42. Rh(III) containing polyynes

Interestingly, the condensation methodology allows the preparation of polymers containing two different types of metal centers. Thus, polymers containing rhodium and palladium have been prepared by the synthetic methodology shown in Fig. 8.43 [3].

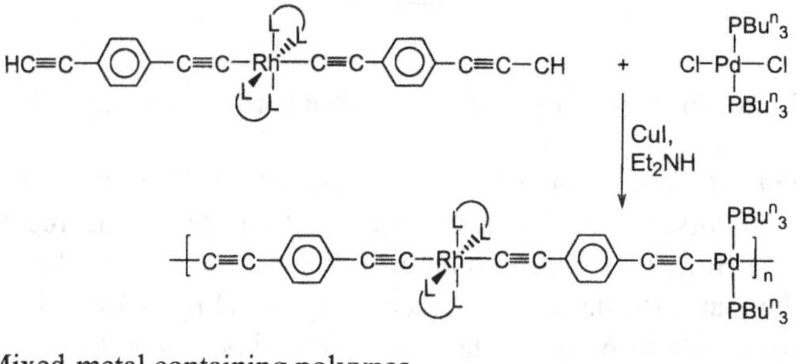

Fig. 8.43. Mixed-metal containing polyynes

Ferrocene-containing polyynes can also be prepared by adopting the original condensation methodology of Hagihara and coworkers. Thus, the condensation of the 1,1'-diiodoferrocene with aryl diacetylenes affords polyynes containing ferrocenes (Fig.8.44) [3].

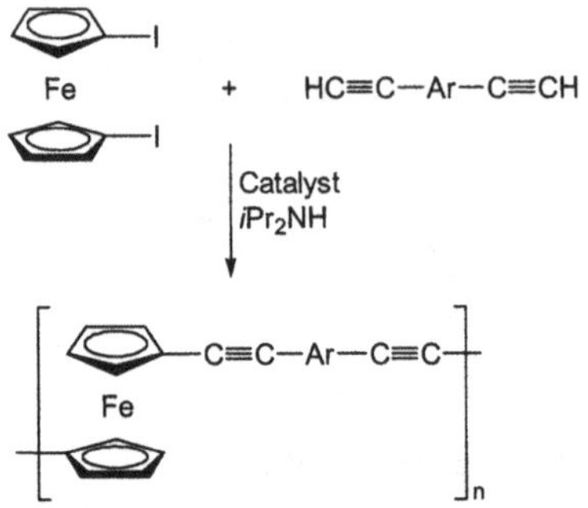

Fig. 8.44. Polyynes containing ferrocene units

Nickel-containing rigid-rod polymers could not be prepared by the use of chloride-containing monomers. An alternative strategy was successfully employed for the preparation of these polymers. This consists of an oxidative coupling reaction or a ligand exchange reaction [3]. Thus, oxidation of *trans*-Ni(PnBu$_3$)$_2$(C≡C-C≡CH)$_2$ leads to a dehydrogenative coupling reaction to afford linear polymers containing nickel (Fig. 8.45) [3]. This reaction is catalyzed by a Cu(I) halide and is carried out in the presence of tetramethylethylene diamine (TMEDA).

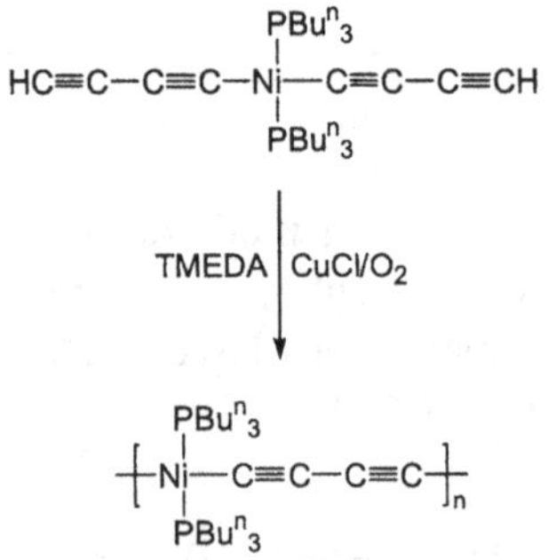

Fig. 8.45. Oxidative coupling reaction to afford Ni(II)-containing polyynes

Another approach for the preparation of Ni(II)-containing rigid-rod polymers involves a ligand exchange reaction. Thus, the reaction of the *trans*-Ni(PnBu$_3$)$_2$(C≡CH)$_2$ with an aryl diacetylene leads to the elimination of acetylene and formation of a linear polymer (Fig. 8.46) [3]. Other types of polyynes include Au(I)-containing rigid-rod polymers [4].

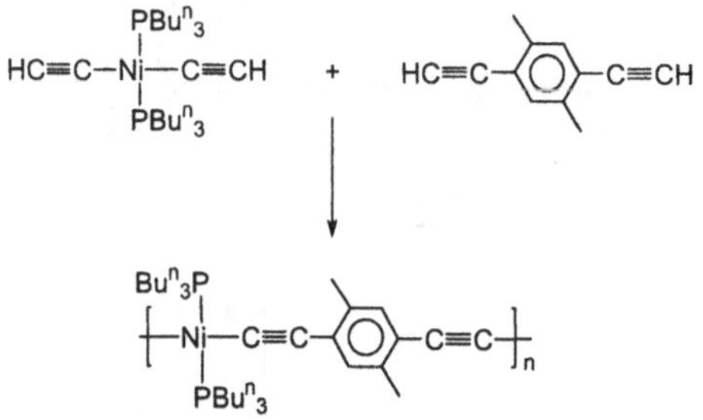

Fig. 8.46. Ligand exchange reaction to afford Ni(II)-containing polyynes

More recently, a new class of polymers has been synthesized involving cyclization reactions. This method of polymerization called *metallacyclization* has led to the formation of polymers containing metal units linked to aromatic units. Thus, the reaction of [{η^5C$_5$H$_4$(C$_6$H$_{13}$)}- Ru(COD)Br] with 4,4′-diethynylbiphenyl leads to an air-sensitive, thermally unstable polymeric product by a metallacyclization reaction (Fig. 8.47) [72]. This ruthenium polymer undergoes a reversible reduction. It has also been shown that such reduced ruthenium centers interact with each other in a ferromagnetic manner.

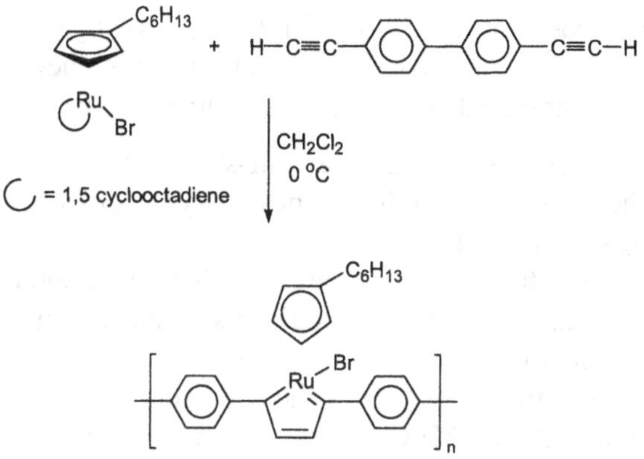

Fig. 8.47. Metallacyclization polymerization

Bunz and coworkers have developed synthetic routes for the preparation of polymers containing the [(η^5C$_5$H$_5$)(η^4C$_4$H$_4$)Co] moiety. Thus, Heck-coupling of 1,3-diethynyl-cyclobutadiene(cyclopentadienyl)cobalt with diiodobenzene derivatives affords polymers with moderate molecular weights (Fig. 8.48). Such polymers can also be prepared by an alternative route involving the acyclic diyne metathesis reaction. These organometallic polymers have been shown to possess novel aggregation behavior in the solution as well as solid state [73].

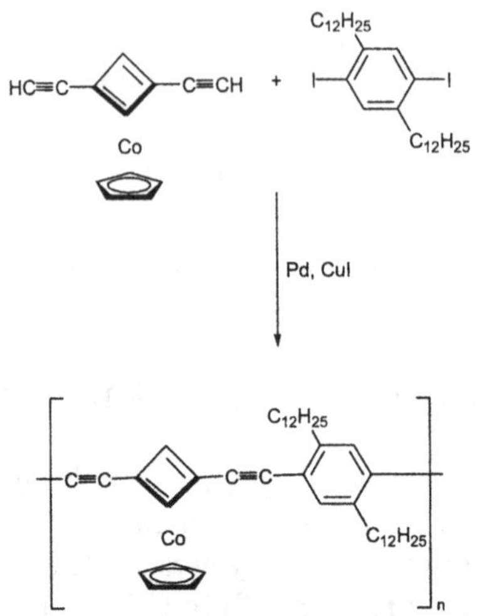

Fig. 8.48. Polymers containing $[(\eta^5C_5H_5)(\eta^4C_4H_4)Co]$ unit

8.4.1 Properties and Applications of Rigid-Rod Polyynes

Although rigid-rod polyynes are still being investigated some of their properties in general and those of the platinum polyynes in particular are beginning to be unraveled. These can be summarized as follows:

1. Many of these polymers are air-stable solids.
2. The molecular weights of these polymers can be fairly high with M_n's in order of 10^5.
3. Many of these polymers are soluble in common organic solvents particularly when long-chain alkyl substituents are present as side-chains.
4. The electronic spectra of the rigid-rod polymers are quite interesting. Strong MLCT (metal-ligand charge transfer) transitions are seen in their optical spectra.
5. As a function of chain-length the λ_{max} undergoes a bathochromic shift. Evidence of π-conjugation across the backbone chain is found in the bathochromic shifts seen in polyynes containing aromatic spacer groups. Thus, for example, polymers containing thienyl-pyridine linker groups, have λ_{max} values at 430-450 nm [71]. In the most well-studied of the platinum-containing polyynes the absorption seems to reach an optimum value by the time a pentameric oligomer is formed [7].

6. Many platinum-containing polyynes are also photoluminescent and some of them show strong triplet emission [7].
7. Many metal polyynes have third-order nonlinear optical properties and these polymers are of interest from the point of view of new types of optical devices. Other types of potential applications of these polymers include light-emitting diodes, lasers, photocells, field-effect transistors, low-dimensional conductors etc. [70].

References

1. Elschenbroich C, Salzer A (1992) Organometallics: A concise introduction. 2nd ed. VCH, Weinheim
2. Crabtree RH (2001) The organometallic chemistry of transition metals. 3rd ed. Wiley, New York
3. Nguyen P, Gomez-Elipe P, Manners I (1999) Chem Rev 99:1515
4. Puddephatt RJ (1998) Chem Commun 1055
5. Manners I (2001) Science 294:1664
6. Manners I (1996) Angew Chem Int Ed 35:1602
7. Manners I (2003) Synthetic metal containing polymers. Wiley-VCH, Weinheim
8. Archer RD (2001) Inorganic and organometallic polymers. Wiley-VCH, New York
9. Jones RG, Ando W, Chojnowski J (eds) (2000) Silicon containing polymers. Kluwer Academic, Dordrecht
10. Miller RD, Michl J (1989) Chem Rev 89:1359
11. Trefonas P, West R (1985) J Polym Sci Polym Chem 23:2099
12. Miller RD, Sooriyakumaran R (1987) J Polym Sci Polym Chem 25:111
13. Mochida K, Chiba H (1994) J Organomet Chem 473:45
14. Reichl JA, Popoff CM, Gallgher LA, Remsen EE, Berry DH (1996) J Am Chem Soc 118:9430
15. Katz SM, Reichl JA, Berry DH (1998) J Am Chem Soc 120:9844
16. Szymanski WJ, Visscher GT, Bianconi PA (1993) Macromolecules 23:869
17. Devylder N, Hill M, Molloy KC, Price GJ (1996) Chem Commun 711
18. Imori T, Lu V, Cai H, Don Tilley T (1995) J Am Chem Soc 117:9931
19. Lu V, Don Tilley T (1996) Macromolecules 29:5763
20. Lu V, Don Tilley T (2000) Macromolecules 33:2403
21. Babcock JR, Sita LR (1996) J Am Chem Soc 118:12481
22. Okano M, Matsumoto N, Arakawa M, Tsuruta T, Hamano H (1998) Chem Commun 1799
23. Pittman CU, Lai JC, van der Pool DP, Good M, Prado R (1970) Macromolecules 3:746
24. Neuse EW, Bednarik L (1979) Macromolecules 12:187

25. Sanechika K, Yamamoto T, Yamamoto A (1981) Polym J 13:255
26. Yamamoto T, Sanechika K, Yamamoto A, Katado M, Motoyama I, Sano H (1983) Inorg Chim Acta 73:75
27. Gonsalves K, Zhan-Ru L, Rausch MD (1984) J Am Chem Soc 106:3862
28. Patterson WJ, McManus SP (1974) J Polym Sci Polym Chem 12:837
29. Itoh T, Saitoh H, Iwatsuki S (1995) J Polym Sci Polym Chem 33:1589
30. Nugent HM, Rosenblum M (1993) J Am Chem Soc 115:3848
31. Brandt PF, Rauchfuss TB (1992) J Am Chem Soc 114:1926
32. Galloway CP, Rauchfuss TB (1993) Angew Chem Int Ed 32:1319
33. Compton DL, Rauchfuss TB (1994) Organometallics 13:4367
34. Compton DL, Brandt PF, Rauchfuss TB, Rosenbaum DF, Zukoski CF (1995) Chem Mater 7:2342
35. Buretea MA, Don Tilley T (1997) Organometallics 16:1507
36. Heo RW, Somoza FB, Lee TR (1998) J Am Chem Soc 120:1621
37. Peckham TJ, Massey JA, Edwards M, Manners I, Foucher DA (1996) Macromolecules 29:2396
38. Kapoor RN, Crawford GM, Mahmoud J, Dementev VV, Nguyen MT, Diaz AF, Pannell KH (1996) Organometallics 15: 2848
39. Jäkle F, Rulkens R, Zech G, Foucher DA, Lough AJ, Manners I (1998) Chem Eur J 4:2117
40. Berenbaum A, Braunschweig H, Dirk R, Englert U, Green JC, Jäkle F, Lough AJ, Manners I (2000) J Am Chem Soc 122:5765
41. Honeyman CH, Foucher DA, Dahmen FY, Rulkens R, Lough AJ, Manners I (1995) Organometallics 14:5503
42. Peckham TJ, Massey JA, Honeyman CH, Manner I (1999) Macromolecules 32:2830
43. Rulkens R, Gates DP, Balaishis D, Pudelski JK, McIntosh DF, Lough AJ, Manners I (1997) J Am Chem Soc 119:10976
44. Manners I (1996) Polyhedron 15:4311
45. Manners I (1995) Adv Organomet Chem 37:131
46. Manners I (1999) Chem Commun 857
47. Foucher DA, Tang BZ, Manners I (1992) J Am Chem Soc 114:6246
48. Zechel DL, Hultzsch KC, Rulkens R, Balaishis D, Ni Y, Pudelski JK, Lough AJ, Manners I (1996) Organometallics 15:1972
49. Nicole Power-Billard K, Manners I (2000) Macromolecules 33:26
50. Rulkens R, Resendes R, Verma A, Manners I (1997) Macromolecules 30:8165
51. Calleja G, Carre F, Cerveau G (2001) Organometallics 20:4211
52. MacLachlan MJ, Lough AJ, Manners I (1996) Macromolecules 29:8562
53. Ni Y, Rulkens R, Manners I (1996) J Am Chem Soc 118:4102
54. Gomez-Elipe P, Resendes R, McDonald PM, Manners I (1998) J Am Chem Soc 120:8348
55. Nguyen P, Stojcevic G, Kulbaba K, MacLachlan MJ, Liu XH, Lough AJ, Manners I (1998) Macromolecules 31:5977
56. Berenbaum A, Lough AJ, Manners I (2002) Organometallics 21:4415
57. Nelson JM, Rengel H, Manners I (1993) J Am Chem Soc 115:7035

58. Peckham TJ, Lough AJ, Manners I (1999) Organometallics 18:1030
59. Sun Q, Xu K, Peng H, Zheng R, Häussler M, Tang BZ (2003) Macromolecules 36:2309
60. Sharma HK, Cervantes-Lee F, Pannell KH (2004) J Am Chem Soc 126:1326
61. Papkov VS, Gerasimov MV, Dubovik II, Sharma S, Dementiev VV, Pannell KH (2000) Macromolecules 33:7107
62. Chen Z, Foster MD, Zhou W, Fong H, Reneker DH, Resendes R, Manners I (2001) Macromolecules 34:6156
63. Hempenius MA, Brito FF, Vancso GJ (2003) Macromolecules 36:6683
64. Paquet C, Cyr PW, Kumacheva E, Manners I (2004) Chem Commun 234
65. Sonogashira K, Takahashi S, Hagihara N (1977) Macromolecules 10:879
66. Takahashi S, Kariya M, Yatake T, Sonogashira K, Hagihara N (1978) Macromolecules 11:1063
67. Takahashi S, Ohyama Y, Murata, Sonogashira K, Hagihara N (1980) J Polym Sci Polym Chem 18:349
68. Abe A, Kimura N, Tabata S (1991) Macromolecules 24:6238
69. Adams CJ, James SL, Raithby PR (1997) Chem Commun 2155
70. Khan MS, Al-Mandary MRA, Al-Suti MK, Hisehm AK, Raithby PR, Ahrens B, Mahon MF, Male L, Marbeglia EA, Tedesco E, Friend RH, Köhler A, Feeder N, Teat SJ (2002) Dalton Trans 1358
71. Khan MS, Al-Mandary MRA, Al-Suti MK, Feeder N, Nahar S, Köhler A, Friend RH, Wilson PJ, Raithby PR (2002) Dalton Trans 2441
72. Kurashina M, Murata M, Watanabe T, Nishihara H (2003) J Am Chem Soc 125:12420
73. Steffen W, Köhler B, Altmann M, Scherf U, Stitzer K, Loye HCZ, Bunz UHF (2001) Chem Eur J 7:117

Index

^{31}P-NMR chemical shifts of
   polyphosphazenes, 143
^{31}P-NMR spectra of persubstituted
   cyclophosphazenes, 106

acyclic diene metathesis
   polymerization, 179
alkoxycyclophosphazenes, 93
amorphous polymers, 78
anionic initiators, 316
anionic polymerization, 37, 265
applications of polysiloxanes, 244
arylcyclophosphazenes, 94
aryloxycyclophosphazenes, 92
atom transfer radical
   polymerization, 58

benzoyl peroxide, 32
bonding in cyclophosphazenes, 108
Borazine, 206
boron nitride, 11
branching, 34
Bromocyclophosphazenes, 89
butadiyne-linked diplatinum, 326

cationic polymerization, 40
chain transfer, 35
chlorocyclophosphazene monomers,
   167
chloro-cyclophosphazenes, 86
condensation polymerization, 63
condensation reactions, 4
crosslinked cyclophosphazene, 178
curing of polysiloxane, 234
cyclocarbophosphazene, 184
Cyclomatrix polymers, 179

cyclophosphazene, 82
cyclophosphazene monomer, 162
cyclophosphazene pendant groups,
   164
cyclophosphazenes as ligands, 105
cyclosiloxanes, 220

dehydrogenative polymerization,
   266
difunctional reagents, 96
direct process, 219
disilene monomers, 252

elastomer, 78
Electroluminescence, 281
electron delocalization, 273

face-to-face ferrocene polymers, 308
ferrocene, 304
ferrocene polymers, 304
ferrocenophanes, 310
fibers, 79
flame-retardant, 174
flexible polymers, 147
fluorocyclophosphazene monomers,
   166
fluorocyclophosphazenes, 89
Free-radical polymerization, 30

glass transition temperature, 76, 147
glass transition temperatures, 172
Group-transfer polymerization, 62

helical polysilanes, 279
helical structures, 273
heterogeneous polymeric ligand,
   178

high-temperature stability, 174
hydridophosphazenes, 90
hydrosilylation, 269

initiators, 31
iodocyclophosphazenes, 89
ionic initiators, 36

macromolecular substitution, 127
masked disilene, 262
metal carbenes, 57
metal-containing polymers, 296
metal-olefin complex, 46
metal-olefin complexes, 46
methylalumoxane, 52
mineral silicate structures, 211
molecular structures of
    cyclophosphazenes, 106
molecular weights, 144

natural polymers, 28
nonlinear optical properties, 152
non-metallocene catalysts, 54
nucleophilic substitution, 100
number average molecular weight,
    75

olefin metathesis, 56
olefin metathesis polymerization, 57
organic polymers, 2, 29
organometallic polymers, 296
organosilicon polymers, 215

phosphine-containing
    polyphosphazene, 136
poly(1,1'-ferrocenylene)s, 305
poly(1,1'-silole)s, 15
Poly(alkyl/aryloxothiazenes), 206
poly(alkyl/arylphosphazene)s, 135
poly(carbocyclophosphazene), 20
poly(carbophosphazene)s, 184
poly(chlorothionylphosphazene),
    194
poly(dichlorophosphazene), 113
poly(difluorophosphazene), 114

poly(dimethylphosphazene), 131
poly(dimethylsilane), 251
poly(dimethylsiloxane), 8, 241, 246
poly(ferrocenylenesilyne)s, 320
poly(ferrocenylphosphine)s, 320
poly(ferrocenylsilane)s, 19, 313
poly(ferrocenylvinylene)s, 15
poly(heterophosphazene)s, 183
poly(metallaphosphazene)s, 200
poly(organocarbophosphazene)s,
    188
poly(organophosphazene)s, 126
poly(oxothiazene)s, 12
poly(*p*-phenylene phosphaalkene),
    15
poly(*p*-phenylenephosphaalkene),
    205
poly(silaethylene), 18
poly(silylenemethylene)s, 284
poly(thionylphosphazene)s, 196
poly(thiophosphazene), 20, 190
poly(vinylferrocene), 304
polyborazylene, 206
polycarbosilane, 252
polycarbosilanes, 284
polydichlorophosphazene, 12
polydispersity index, 76
polyferrocenes, 324
polygermanes, 298
polymeric sulfur, 17
polymerization methods, 30
polymers containing
    cyclophosphazene rings, 172
polymers that contain
    cyclophosphazenes, 156
polyphosphazenes, 82, 112
Polyphosphazenes, 12
polyphosphazenes containing
    carboxy groups, 137
polyphosphinoboranes, 203
Polyphosphinoboranes, 11
polysilane electronic structure, 277
polysilane-organic polymer blocks,
    271
polysilanes, 9, 249
polysiloles, 289

polysiloxane, 14, 210, 224
polysiloxanes, 227
polysilsesquioxane, 8
polysilynes, 281
polystannane, 10
polystannanes, 300
polythiazyl, 17
preparation of polysilanes, 255
pseudohalogenocyclophosphazenes, 89

rigid-rod polymers, 16
rigid-rod polyynes, 327, 332
rigid-rod structure, 288
ring-opening polymerization, 4, 60, 317
rod-like structures, 326

self-extinguishing polymer, 170
self-extinguishing polymers, 167
silarylene-siloxane copolymers, 239
silica, 211
silicon-bridged [1]ferrocenophanes, 310
silicon-containing polymers, 288
silicones, 216
siloxane polymers, 234
Si-Si bond energy, 273

solid electrolytes, 151
solubility properties of polyphosphazenes, 145
step-growth polymerization, 63
supported catalysts, 50

thermal properties, 146
thermochromic behavior, 277
thermochromism, 278
thermoset resins, 72

Vinyl monomers, 2

weight average molecular weight, 76
Wurtz-coupling, 300
Wurtz-coupling reaction, 258

X-ray crystal structures, 107
X-ray data, 142

Zeise's salt, 296
Ziegler-Natta polymerization, 43

σ-bonded chain, 273
σ-σ* transitions, 273